MOLECULAR TARGETING
AND SIGNAL TRANSDUCTION

Cancer Treatment and Research

Steven T. Rosen, M.D., *Series Editor*

Winter, J.N. (ed.): *Blood Stem Cell Transplantation.* 1997. ISBN 0-7923-4260-7.
Muggia, F.M. (ed): *Concepts, Mechanisms, and New Targets for Chemotherapy.* 1995. ISBN 0-7923-3525-2.
Klastersky, J. (ed): *Infectious Complications of Cancer.* 1995. ISBN 0-7923-3598-8.
Kurzrock, R., Talpaz, M. (eds): *Cytokines: Interleukins and Their Receptors.* 1995. ISBN 0-7923-3636-4.
Sugarbaker, P. (ed): *Peritoneal Carcinomatosis: Drugs and Diseases.* 1995. ISBN 0-7923-3726-3.
Sugarbaker, P. (ed): *Peritoneal Carcinomatosis: Principles of Management.* 1995. ISBN 0-7923-3727-1.
Dickson, R.B., Lippman, M.E. (eds.): *Mammary Tumor Cell Cycle, Differentiation and Metastasis.* 1995. ISBN 0-7923-3905-3.
Freireich, E.J, Kantarjian, H. (eds): *Molecular Genetics and Therapy of Leukemia.* 1995. ISBN 0-7923-3912-6.
Cabanillas, F., Rodriguez, M.A. (eds): *Advances in Lymphoma Research.* 1996. ISBN 0-7923-3929-0.
Miller, A.B. (ed.): *Advances in Cancer Screening.* 1996. ISBN 0-7923-4019-1.
Hait , W.N. (ed.): *Drug Resistance.* 1996. ISBN 0-7923-4022-1.
Pienta, K.J. (ed.): *Diagnosis and Treatment of Genitourinary Malignancies.* 1996. ISBN 0-7923-4164-3.
Arnold, A.J. (ed.): *Endocrine Neoplasms.* 1997. ISBN 0-7923-4354-9.
Pollock, R.E. (ed.): *Surgical Oncology.* 1997. ISBN 0-7923-9900-5.
Verweij, J., Pinedo, H.M., Suit, H.D. (eds): *Soft Tissue Sarcomas: Present Achievements and Future Prospects.* 1997. ISBN 0-7923-9913-7.
Walterhouse, D.O., Cohn, S. L. (eds.): *Diagnostic and Therapeutic Advances in Pediatric Oncology.* 1997. ISBN 0-7923-9978-1.
Mittal, B.B., Purdy, J.A., Ang, K.K. (eds): *Radiation Therapy.* 1998. ISBN 0-7923-9981-1.
Foon, K.A., Muss, H.B. (eds): *Biological and Hormonal Therapies of Cancer.* 1998. ISBN 0-7923-9997-8.
Ozols, R.F. (ed.): *Gynecologic Oncology.* 1998. ISBN 0-7923-8070-3.
Noskin, G. A. (ed.): *Management of Infectious Complications in Cancer Patients.* 1998. ISBN 0-7923-8150-5
Bennett, C. L. (ed): *Cancer Policy.* 1998. ISBN 0-7923-8203-X
Benson, A. B. (ed): *Gastrointestinal Oncology.* 1998. ISBN 0-7923-8205-6
Tallman, M.S. , Gordon, L.I. (eds): *Diagnostic and Therapeutic Advances in Hematologic Malignancies.* 1998. ISBN 0-7923-8206-4
von Gunten, C.F. (ed): *Palliative Care and Rehabilitation of Cancer Patients.* 1999. ISBN 0-7923-8525-X
Burt, R.K., Brush, M.M. (eds): *Advances in Allogeneic Hematopoietic Stem Cell Transplantation.* 1999. ISBN 0-7923-7714-1
Angelos, P. (ed): *Ethical Issues in Cancer Patient Care* 2000. ISBN 0-7923-7726-5
Gradishar, W.J., Wood, W.C. (eds): *Advances in Breast Cancer Management.* 2000. ISBN 0-7923-7890-3
Sparano, Joseph A. (ed): *HIV & HTLV-I Associated Malignancies.* 2001. ISBN 0-7923-7220-4.
Ettinger, David S. (ed): *Thoracic Oncology.* 2001. ISBN 0-7923-7248-4.
Bergan, Raymond C. (ed): *Cancer Chemopre*vention. 2001. ISBN 0-7923-7259-X.
Raza, A., Mundle, S.D. (eds): *Myelodysplastic Syndromes & Secondary Acute Myelogenous Leukemia* 2001. ISBN: 0-7923-7396.
Talamonti, Mark S. (ed): *Liver Directed Therapy for Primary and Metastatic Liver Tumors.* 2001. ISBN 0-7923-7523-8.
Stack, M.S., Fishman, D.A. (eds): *Ovarian Cancer.* 2001. ISBN 0-7923-7530-0.
Bashey, A., Ball, E.D. (eds): *Non-Myeloablative Allogeneic Transplantation.* 2002. ISBN 0-7923-7646-3
Leong, Stanley P.L. (ed): *Atlas of Selective Sentinel Lymphadenectomy for Melanoma, Breast Cancer and Colon Cancer.* 2002. ISBN 1-4020-7013-6
Andersson , B., Murray D. (eds): *Clinically Relevant Resistance in Cancer Chemotherapy.* 2002. ISBN 1-4020-7200-7.
Beam, C. (ed.): *Biostatistical Applications in Cancer Research.* 2002. ISBN 1-4020-7226-0.
Brockstein, B., Masters, G. (eds): *Head and Neck Cancer.* 2003. ISBN 1-4020-7336-4.
Frank, D.A. (ed.): *Signal Transduction in Cancer.* 2003. ISBN 1-4020-7340-2.
Figlin, Robert A. (ed.): *Kidney Cancer.* 2003. ISBN 1-4020-7457-3.
Kirsch, Matthias; Black, Peter McL. (ed.): *Angiogenesis in Brain Tumors.* 2003. ISBN 1-4020-7704-1.
Kumar, Rakesh (ed.): *Molecular Targeting and Signal Transduction.* 2004. ISBN 1-4020-7822-6.

MOLECULAR TARGETING AND SIGNAL TRANSDUCTION

edited by

Rakesh Kumar, Ph.D.
The University of Texas
M. D. Anderson Cancer Center
Houston, TX
USA

KLUWER ACADEMIC PUBLISHERS
Boston / New York / Dordrecht / London

Distributors for North, Central and South America:
Kluwer Academic Publishers
101 Philip Drive
Assinippi Park
Norwell, Massachusetts 02061 USA
Telephone (781) 871-6600
Fax (781) 681-9045
E-Mail: kluwer@wkap.com

Distributors for all other countries:
Kluwer Academic Publishers Group
Post Office Box 322
3300 AH Dordrecht, THE NETHERLANDS
Telephone 31 786 576 000
Fax 31 786 576 254
E-Mail: services@wkap.nl

 Electronic Services <http://www.wkap.nl>

Library of Congress Cataloging-in-Publication Data

A C.I.P. Catalogue record for this book is available
from the Library of Congress.

Molecular Targeting and Signal Transduction
Rakesh Kumar
ISBN: 1-4020-7822-6
e-Book ISBN: 1-4020-7847-1

CONTENTS

PREFACE

Epidemiological studies suggest that cancer is the most common cause of mortality in the western world. Our limited understanding of cellular signal-transduction-networks, the molecular pathways that control the expression and function of critical regulatory gene products in the physiology of normal and cancerous cells, has been a barrier to progress in improving the overall cure-rate of human cancers. Many current clinical interventions focus on the use of cytotoxic and cytostatic agents as cancer therapies, but these treatments are not necessarily specific to a given tumor type. Although such broad therapeutic approaches have been very fruitful in the management of human cancers, it is now clear that further significant gain will only come from a more complete understanding of key signaling pathways feeding into phenotypic alterations characteristic of cancer cells. Delineation of the physiologic roles of the specific regulatory signaling components, with known association with metastatic phenotypes, is a highly promising area which will likely provide the next generation of targeted strategies in the future of molecular cancer medicine. These signaling components are likely to be used in diagnosis, prognosis, and as novel targets for therapeutic development.

Cancer cells harbor characteristic changes that are directly linked to enhanced cell division, survival, and movement. A finely coordinated balance between cell survival and programmed cell death is critical for the normal development and maintenance of tissue homeostasis. Alteration of this balance is one of the underlying mechanisms behind diseases of cell proliferation, including cancers. Cancer mortality usually stems from the propensity for tumor cells to metastasize, rather than from the primary tumor itself, which may remain small and undetected. The process of cancer metastasis requires, among other steps, alterations in target gene products, increased angiogenesis, increased directional motility, dysfunction in the expression and functions of cell adhesion components, enhanced cell survival, resistance to therapeutic apoptosis, and increased energy requirement. Since each of these phenotypic changes are dependent on a key signaling nodule which is generally hyperactivated in cancer, new molecular therapeutic strategies are aimed to develop small inhibitory molecules to target these critical signal transduction integrators. These approaches are likely to offer an added arsenal to effectively fight cancers, and improve the efficacy of currently used anti-cancer agents when combined.

In this book, we bring together major principles of cancer cell biology: survival, apoptosis, adhesion, and cell cycle deregulation. Research leaders from prominent cancer centers in the United States and around the world summarize up-to-date accounts of major discoveries that highlight the significance of signal transduction in cancer cell physiology and in the treatment of human cancers. This book is directed at clinicians and scientists working in the areas of experimental and molecular therapeutics, molecular medicine, translation cancer research, and bio-medical sciences in general.

Rakesh Kumar, Ph.D.
December 2003

BIOLOGY OF THE EPIDERMAL GROWTH FACTOR RECEPTOR FAMILY

CHRISTOPHER J. BARNES AND RAKESH KUMAR

1. INTRODUCTION

The epidermal growth factor receptor was the first receptor tyrosine kinase to be discovered and remains the most investigated. Most of the mechanistic principles of receptor tyrosine kinases first established with the EGFR family as a model. EGFR is a single pass transmembrane receptor with two extracellular, cysteine-rich regions involved in ligand binding, and intervening region important for receptor dimerization, an intracellular tyrosine kinase domain, and a number of intracellular sites for autophosphorylation, phosphorylation by other kinases, and docking of intracellular signaling components. Three additional EGFR family members have been identified, human epidermal growth factor receptor (erbB or HER) 2, 3 and 4. A range of growth factors serves as ligands for these receptors, but none have been identified for the HER2 receptor. HER receptors exist both as monomers and dimers, either homo- or heterodimers. Ligand binding to HER1, HER3 or HER4 induces rapid receptor dimerization, with a marked preference for HER2 as a dimer partner [1]. Moreover, HER2-containing heterodimers generate intracellular signals that are significantly more potent than signals emanating from other HER combinations.

Activating mutations, gene amplification and overexpression of HER family kinases have been implicated as integral contributors to a variety of cancers, including breast, colon, pancreas, lung, and squamous cancers and glioblastomas. The molecular mechanisms by which HER kinases transform cells involves direct effects on components of the cell-cycle regulatory apparatus, in particular cyclin D1 and p27Kip1 [2, 3]. Attenuation of HER family signaling is a developing strategy for the management of human malignancies and is the subject of a number of ongoing clinical trials.

There are numerous excellent recent reviews of HER family biology, signaling and therapeutic targeting [4-7]. This chapter will provide an overview of several key developments in the rapidly evolving area of HER family biology and the integral role of these receptors in malignant transformation and as targets of cancer therapy.

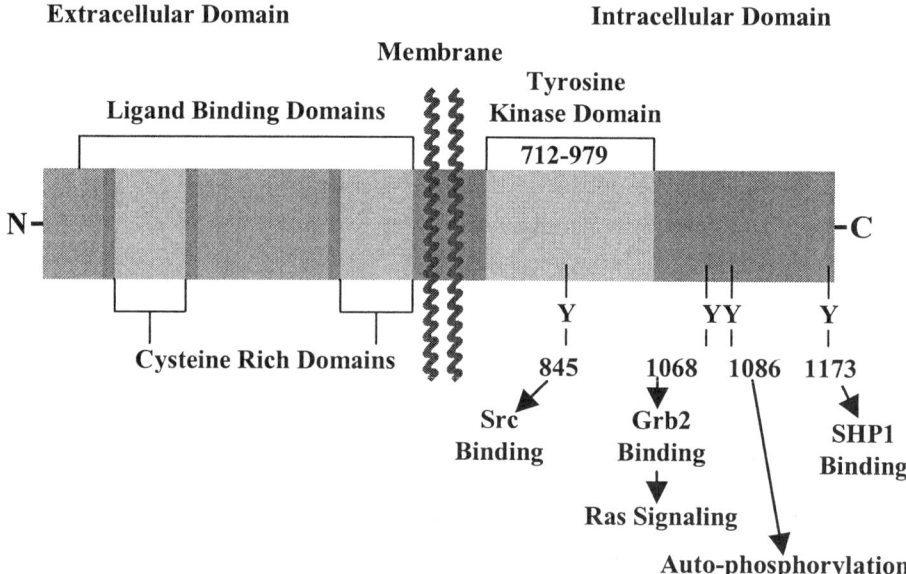

Figure 1. Epidermal Growth Factor Receptor Domain Structure and Selected Phosphorylation sites. EGFR is a single pass transmembrane receptor with two extracellular, cysteine-rich regions involved in ligand binding, and intervening region important for receptor dimerization, an intracellular tyrosine kinase domain, and a number of intracellular sites for autophosphorylation, phosphorylation by other kinases, and docking of intracellular signaling components.

2. THE STRUCTURE OF EGFR AND HER3 EXTRACELLULAR DOMAINS

As with other receptor tyrosine kinases, the HER family receptors are composed of an extracellular ligand binding domain, a single transmembrane domain, and a cytoplasmic tyrosine kinase domain flanked by multiple regulatory sequences (Figures 1 and 2). Several models have been proposed as to the domains involved in ligand binding, the stoichiometry of ligand and receptor in various receptor dimers, and the mechanism underlying the preferred HER2 heterodimerization among HER family members. Recent reports of the crystal structures of HER3 [8] and EGFR [9,10] shed new light on these long standing questions and are the subject of an excellent review [11]. In brief, dimerization of EGFR requires the binding of two molecules of monomeric ligand (i.e., EGF) to two molecules of EGFR in a 2:2 EGFR:EGFR complex formed from stable 1:1 EGF:EGFR intermediates [9].

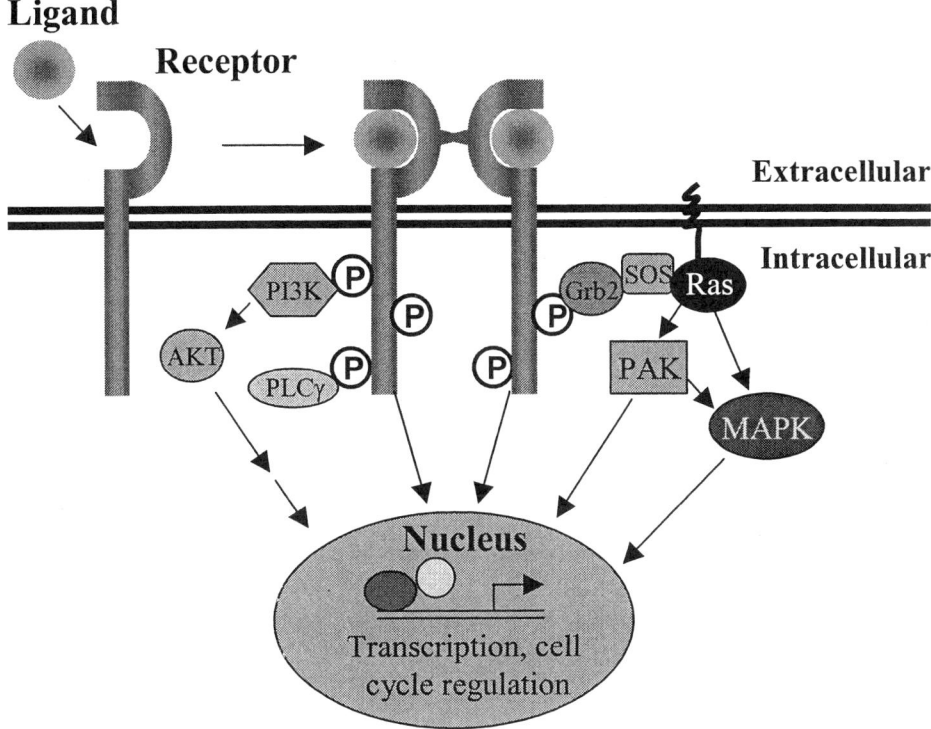

Figure 2. Receptor Tyrosine Kinase Signaling. Using the epidermal growth factor receptor as an example, ligand binding initiates receptor homo-(shown) or heterodimerization and subsequent intracellular tyrosine kinase activation, autophosphorylation, and numerous protein-protein interactions mediating signaling cascades that dictate cell function. PI3K, phosphoinositol-3 kinase; PLCγ, phospholipase 3 gamma; SOS, son of sevenless; PAK, p21-activated kinase; MAPK, mitogen-activated protein kinase.

The two receptors adopt a 'back-to-back' conformation, with the ligand binding pockets facing outward and interaction mediated by a critical dimerization loop which becomes exposed following intramolecular conformational changes upon ligand binding. This dimerization loop sequence is conserved in the other HER family members. However, HER2, which has no known ligand, may be in a constitutively activated conformation, thus facilitating heterodimerization of HER2 with other family members upon ligand binding and subsequent conformational change. These new protein structures facilitate a greater understanding of the mechanisms of ligand-induced receptor activation and dimerization and may thus lead to novel therapeutic strategies in the treatment of deregulated HER family signaling.

3. NUCLEAR LOCALIZATION AND FUNCTION

In addition to ligand recognition, binding, and initiation of intracellular signaling cascades, receptor tyrosine kinases such as the EGFR family have newly emerging roles in the regulation of nuclear functions through their direct translocation to the cell nucleus. Although nuclear localization of EGFR had been noted in previous publications, Lin et al. [12] demonstrated the presence of a strong transactivation domain in the carboxy terminus of EGFR, which mediated nuclear transactivation functions. Although an EGFR mutant which lacked the transmembrane domain could also translocate to the nucleus, this movement required both wild type EGFR and active ligand signaling [13]. Both intracellular and extracellular domains of EGFR appear to move to the nucleus in a ligand-bound form, although the mechanism this translocation of the full length protein remains to be determined.

HER4, the most recently identified HER family member, also translocates to the nucleus and has transactivation functions, but through a separate mechanism. In addition to matrix metalloprotease cleavage of the HER4 extracellular domain upon receptor activation, it was recently shown that the intracellular domain is also cleaved, by gamma secretase, releasing a protein fragment that translocates to the nucleus [14]. This carboxyl-terminal HER4 fragment then associates with the WW-domain-containing transcriptional regulatory protein YAP (Yes-associated protein) and acts as a co-transcriptional activator [15]. Thus, in addition to initiating numerous cytoplasmic signaling cascades upon activation, HER family members may directly influence transcriptional activity and nuclear function via translocation to the nuclear compartment.

4. TRANSACTIVATION FUNCTIONS

The classical portrayal of signal transduction by HER family receptor tyrosine kinases consists of a multi-step process including ligand binding and receptor dimerization, trans-phosphorylation of cytoplasmic tyrosine residues, and sequential recruitment and activation of cytoplasmic effector proteins which signal transcriptional and regulatory changes [reviewed in 7]. However, there is increasing evidence that the HER family receptors propagate not only signals initiated by their own ligands but also those initiated by multiple other signaling pathways. These trans-regulatory interactions can be mediated through increased receptor ligand availability, direct phosphorylation of HERs by protein tyrosine kinases, and via novel heterodimerization partners and transactivation. The net result of these alternative activation strategies is enhanced HER signaling to multiple cell regulatory pathways.

The best-studied mechanism of HER family transactivation involves activation of G protein coupled receptors (GPRs). Multiple ligands such as lysophosphatidic acid (LPA) and thrombin activate specific GPRs. Activated GPRs can in turn initiate signaling cascades which influence HER family signaling. These cascades include activation of matrix metalloproteases (MMPs), which cleave EGF-like ligands and free them for receptor activation, and cytoplasmic kinases such as Src and Jak2, which directly phosphorylate and activate EGFR. Additional signaling molecules can also indirectly increase or decrease receptor phosphorylation through activation of additional kinases or phosphatases and thus influence HER-mediated cell signaling [7].

4.1. Cyclooxygenase and Prostaglandin E_2 Transactivation

Another recent report reveals prostaglandin (PG) E_2 as another mediator of EGFR transactivation [16]. PGs have long been implicated in tissue growth regulation and cancer progression, but the mechanisms underlying their observed trophic effects were unclear. New data indicate that PGE_2-induced EGFR transactivation involves Src kinase-activated MMPs releasing a tethered EGFR ligand, transforming growth factor alpha (TGFα from the cell membrane that subsequently initiates ligand-mediated receptor activation [16]. This novel activation pathway functions similarly to that described for GPRs.

    This new data may complete a positive feedback loop for HER family-mediated cellular growth regulation. HER2 overexpression upregulates the cyclooxygenase 2 (COX2) enzyme [17]. COX enzymes are rate limiting for prostaglandin production from the PGE_2 precursor arachidonic acid, and COX2 is overexpressed in a number of epithelial cancers. One scenario for a feedback loop involves increased PGE_2 via COX2 triggering release of TGFα, which binds to and activates EGFR. Subsequent heterodimerization of EGFR with HER2 further increases HER2-mediated mitogenic signaling and COX2 protein production. Of interest, the selective COX2 inhibitor celecoxib was recently reported to decrease mammary tumor incidence and PGE_2 levels in mouse mammary tumor virus/HER2 transgenic mice [18]. These data may suggest new molecular targeted cancer treatment regimens.

4.2. uPA/uPAR Transactivation

Cytokine receptors for growth hormone and interleukin 6 have been shown to interact with and activate EGFR and HER2, respectively, initiating HER-dependent signaling cascades [7]. New data indicate that EGFR also plays a central role in propagating cellular signals initiated by the extracellular serine protease urokinase plasminogen activator (uPA), its cell surface receptor (uPAR), and integrins. uPA/uPAR are overexpressed in most types of malignant cancers and their overexpression is a predictor of poor prognosis. There appears to be a causal link between uPAR overexpression and tumor cell invasion and metastasis [19]. Overexpression of uPAR in Hep3 human carcinoma cells stimulates the $\alpha_5\beta_1$ integrin complex to activate EGFR upon binding to fibronectin [20]. This activation may involve focal adhesion kinase. Of interest, dual inhibition of focal adhesion kinase and EGFR signaling was recently reported to cooperatively enhance apoptosis in breast cancer cells [21]. The fibronectin-stimulated EGFR activity was independent of EGFR overexpression or the release of EGF-like ligands [20], but was rather dependent upon overexpression of uPAR and a functional uPA/uPAR/integrin complex. Thus receptor overexpression may not account for all aberrant HER family receptor activation. Rather, screening for the expression and activity of HER family transactivation partners in developing malignancies may help direct targeted therapies. The emerging central role of HER family kinases as integrators of diverse signals stresses the importance of these receptors as therapeutic targets.

4.3. Direct STAT Activation

In classical EGFR signaling, signals begin at the cell surface with the receptor tyrosine kinase and cascade into the nucleus though intermediary proteins. There is now evidence that EGFR and HER2 may bypass established signaling cascades and directly phosphorylate and activate transcriptional regulators. One example lies in direct activation of signal transducer and activator of transcription (STAT) proteins. STATs are a family of latent signaling transcription factors that are activated by a variety of extracellular signals. These proteins have critical and highly diverse functions in regulating development, differentiation, cell cycle progression, and apoptosis. A recent report indicates that both STAT1 and STAT3 can bind to multiple tyrosine residues within the EGFR cytoplasmic domain and become activated [22]. HER2 has also been recently reported to activate STAT3 [23], with activation dependent the activity of cytoplasmic kinases Src and Jak2. Finally, recent evidence shows that STAT3 activation abrogates growth factor dependence and contributes to tumor growth in vivo [24], raising the possibility of another possible novel positive feedback mechanism for enhanced cell growth and survival involving HER family receptors and direct regulation of transcriptional modulators.

5. CANCER THERAPEUTIC TARGETING OF HER FAMILY RECEPTORS

HER family receptor tyrosine kinase function is aberrantly regulated in a majority of the different epithelial cancers and plays an integral role in tumor growth, progression and metastasis. Attenuation of HER family signaling is a developing strategy for the management of human malignancies. Novel agents that modulate signaling through HER family receptors have recently emerged as promising therapies for primary or adjuvant cancer treatment. These new agents are the subjects of several recent reviews [4,5,6]. Various therapeutic strategies have been developed to block HER family signaling, with the most frequent strategies involving humanized monoclonal antibodies directed against one or more of the receptors, and small molecule tyrosine kinase inhibitors that inactivate the signaling tyrosine kinase domain of HER proteins. Anti-receptor antibodies including C225 (Erbitux, against EGFR) and 4D5 (Trastuzumab or Herceptin, against HER2) bind to the receptor extracellular domain and induce internalization and degradation, thus effectively blocking receptor activation of subsequent cellular signaling cascades. Small molecule inhibitors interact with the extracellular domain and effectively block ligand binding, or act against the cytoplasmic tyrosine kinase domain and inhibit receptor tyrosine phosphorylation and cytoplasmic signaling. Other approaches include toxins conjugated to anti-receptor antibodies or receptor ligands, antisense therapies, and directed transcriptional repression to down regulate receptor or ligand expression. Most of these agents target EGFR or HER2, since these receptors are most often deregulated in human cancers.

Research on HER family inhibitors is rapidly evolving, with many new compounds in preclinical and clinical development. Recent reports describe novel EGFR (EMD 55900) [25] and HER2 (2C4) [26] monoclonal antibodies, and small molecule inhibitors directed against EGFR (OSI-774) [27], both EGFR and HER2 (PKI-166 [28] and GW572016 [29]) and pan HER family inhibitors (CI-1033) [30] are in various stages of testing. Of the HER family-directed therapies, ZD1839 (gefitinib, IRESSA), a substituted aniloquinazoline, has progressed the furthest in clinical development. ZD1839 is an orally administered, selective, reversible inhibitor of EGFR that competes for ATP binding and blocks receptor autophosphorylation and kinase activation. This EGFR tyrosine kinase inhibitor is effective against numerous tumor types in preclinical testing [31,32] and has shown consistent and clinically meaningful disease stabilization and a low frequency of regression across a variety of tumor types, with manageable side effects [33-35]. ZD1839 recently received FDA approval for use as a third line therapy in treating non-small cell lung cancer [36]. However, research is needed to elucidate how these novel agents interact with standard chemotherapies and other molecularly targeted agents to provide a mechanistic rationale for combinatorial therapies. As clinical experience with these and other new inhibitors increases, the ability to direct therapies towards individuals with specific HER family and genetic alterations may be possible.

5.1. HER2 and Estrogen Receptor

One example of rationally directed therapy has recently been demonstrated with breast cancer. HER2 has emerged as one of the most important oncogenes in breast cancer, with gene amplification occurring in 20-30% of early stage breast cancer and a significant correlation between receptor overexpression and reduced survival. HER2 overexpression is also correlated with a lack of response to endocrine therapy and chemotherapeutic agents. Indeed, overexpression of HER2 is inversely associated with nuclear localization of the estrogen receptor and HER2/ER positive primary breast carcinomas show a poor anti-proliferative response to endocrine therapy [37].

The recent discovery that HER2 overexpression upregulates expression of an alternatively spliced form of a transcriptional corepressor molecule associated with metastasis (metastasis associated protein 1 short form or MTA1s) provides a mechanism for this phenomenon [38]. MTA1s is overexpressed in breast tumors that are negative for nuclear estrogen receptor (ER) and is highly correlated with HER2 overexpression. Studies indicate that MTA1s sequesters ER in the cell cytoplasm, effectively blocking nuclear import of ER and rendering breast cancer cells insensitive to hormonal therapies. Treatment of cancer cells with the anti-HER2 antibody Trastuzumab should down regulate HER2 and thus MTA1s expression, and restore sensitivity to tamoxifen therapy and nuclear ER [39]. Indeed, additional reports that both sequential treatment with anti-HER2 therapy prior to endocrine therapy and combination of Trastuzumab other standard chemotherapies [41, 40] are successful strategies against HER2 overexpressing cancers further highlight the potential of this mechanistic-based therapeutic approach.

5.2. HER2 and Vascular Endothelial Growth

Another example of rationale combinatorial chemotherapy is demonstrated by dual targeting of growth factor signaling and angiogenesis. Overexpression of HER2 in human tumor cells is closely associated with increased angiogenesis and increased expression of vascular endothelial growth factor (VEGF) [43, 42]. The anti-HER2 antibody Trastuzumab downregulates HER2 and VEGF expression in breast cancer and attenuates the aggressive phenotype and angiogenesis associated with EHR2 overexpression [44, 43]. This strategy of dual inhibition has also proven effective with antibodies against EGFR and VEGF in pancreatic cancer [45].

5.3. Natural Inhibitors and Targeted Degradation

Several recent reports add to the complexity of HER family protein signaling, regulation and homeostasis, but also reveal new potential for therapeutic intervention. Decorin, an endogenous, small leucine-rich proteoglycan, binds to a region of the EGFR partially overlapping the EGF binding site and effectively blocks EGF-mediated receptor activation [46]. This antagonist to EGFR signaling may be a key negative regulator of tumor growth. Future investigations may lead to the development of mimicking peptides capable of suppressing EGFR function as novel therapeutic agents. Also, a recent report of down regulation of EGFR-mediated growth-promoting signals by treatment with 1,25-dihydroxyvitamin D-3 [47] opens up new possibilities for EGFR regulation.

In addition, it was recently reported that the histone deacetylase inhibitors trichostatin A and sodium butyrate rapidly and selectively inhibit transcription from an amplified HER2 gene but not from normal copy number genes [48]. Targeted repression of HER2 transcription in HER2 overexpressing cells has been tried previously using gene therapy with the adenovirus 5 E1A gene [49]. Selective repression by histone deacetylase inhibitors may provide a novel route for reversing elevated HER2 expression levels in cells with gene amplification. Targeted disruption of transcriptional complexes essential for HER2 expression using short, cell-permeable peptides has also been demonstrated [50].

Finally, new insights into protein turnover and targeted degradation could lead to novel therapies. A newly discovered protein regulator of the EGFR, termed EGF receptor related protein (ERRP), normally attenuates EGFR activation, but is progressively downregulated in pancreatic ductal carcinomas that overexpress EGFR [51]. Also, the Csk homologous kinase (CHK) acts as a negative growth regulator of human breast cancer through specific interaction of its SH2 domain with HER2, effectively blocking Src family kinase activity [52 and references within]. Re-expression of novel negative regulatory proteins or induced expression of high affinity inhibitory proteins may restore normal receptor homeostasis in a deregulated setting and serve as potential future therapies. Pharmacologic manipulation of ubiquitination and degradation via ubiquitin ligases such as CHIP [53] and NEDD4 [54] also provide new routes for stimulated downregulation of deregulated HER family members.

6. CONCLUSION

Recent research has strengthened the basis for an intimate role of HER family kinases in a variety of cancers. In addition to propagating cytoplasmic signaling initiated by HER family receptor ligands, HER family members can also propagate signals initiated by multiple other signaling pathways and may serve as central nodes in conveying extracellular signals. Attenuation of HER family signaling is a developing strategy for the management of human malignancies and is the subject of ongoing clinical trials and preclinical mechanistic investigations.

7. ACKNOWLEDGEMENTS

This is work was supported by grants from NIH, CA88923, CA65746, CA80066, and CA97007.

Christopher J. Barnes and Rakesh Kumar
The University of Texas M. D. Anderson Cancer Center, Molecular and Cellular Oncology, Houston, TX

8. REFERENCES

1. Graus-Porta D., Beerli R.R., Daly J.M., Hynes N.E. ErbB-2, the preferred heterodimerization partner of all ErbB receptors, is a mediator of lateral signaling. EMBO J 1997; 16:1647-55
2. Hulit J., Lee R.J., Russell R.G., Pestell R.G. ErbB-2-induced mammary tumor growth: the role of cyclin D1 and p27Kip1. Biochem Pharmacol 2002; 64:827-836
3. Balasenthil S., Sahin A.A., Barnes C.J., Wang R.-A., Pestell R.G., Vadlamudi R.K., Kumar R. P21-activated kinase-1 signaling mediates cyclin D1 expression in mammary epithelial and cancer cells. J Biol Chem 2003; 10.1074/jbc.M309937200
4. Bange J., Zwick E., and Ullrich A. Molecular targets for breast cancer therapy and prevention. Nature Med 2001; 7:548-552
5. deBono J.S., Rowinsky E.K. The ErbB receptor family: a therapeutic target for cancer. Trends Mol Med 2002; 8:S19-S26
6. Shawver L.K., Slamon D., Ullrich A. Smart drugs: tyrosine kinase inhibitors in cancer therapy. Cancer Cell 2002; 1:117-123
7. Yarden Y., Sliwkowski M.X. Untangling the ErbB signaling network. Nature Reviews Molecular Cell Biology 2001; 2:127-137
8. Cho H.S. Leahy D.J. Structure of the extracellular region of HER3 reveals an interdomain tether. Science 2002; 297:1330-1333
9. Garrett T.P.J., McKern N.M., Lou M.Z., Elleman T.C., Adams T.E., Lovrecz G.O., Zhu H.J., Walker F., Frenkel M.J., Hoyne P.A., Jorissen R.N., Nice E.C., Burgess A.W., Ward C.W. Crystal structure of a truncated epidermal growth factor receptor extracellular domain bound to transforming growth factor alpha. Cell. 2002; 110:763-773
10. Ogiso H., Ishitani R., Nureki O., Fukai S., Yamanaka M., Kim J.H., Saito K., Sakamoto A., Inoue M., Shirouzu M., Yokoyama S. Crystal structure of the complex of human epidermal growth factor and receptor extracellular domains. Cell 2002; 110:775-787
11. Schlessinger J. Ligand-induced, receptor-mediated dimerization and activation of EGF receptor Cell 2002; 110:669-672
12. Lin S.Y., Makino K., Xia W., Matin A., Wen Y., Kwong K.Y., Bourguignon L., Hung M.C. Nulcear localization of EGF receptor and its potential new role as a transcription factor. Nat Cell Biol 2001; 3:802-808
13. Marti U., Wells A. The nuclear accumulation of a variant epidermal growth factor receptor (EGFR) lacking the transmembrane domain requires coexpression of a full-length EGFR. Mol Cell Biol Res Comm 2000; 3:8-14
14. Ni C.Y., Murphy M.P., Golde T.E., Carpenter G. gamma-Secretase cleavage and nuclear localization of ErbB-4 receptor tyrosine kinase. Science 2001; 294:2179-81
15. Komuro A., Nagai M., Navin N.E., Sudol M. WW domain-containing protein YAP associates with ErbB4 and acts as a co-transcriptional activator for the carboxyl-terminal fragment of ErbB4 that translocates to the nucleus. J Biol Chem 2003; 278:33334-41

16. Pai R., Soreghan B., Szabo I.L., Pavelka M., Baatar D., Tarnawski A.S. Prostaglandin E2 transactivates EGF receptor: a novel mechanism for promoting colon cancer growth and gastrointestinal hypertrophy. Nature Med 2002; 8:289-293

17. Vadlamudi R. Mandal M. Adam L. Steinbach G. Mendelsohn J. Kumar R. Regulation of cyclooxygenase-2 pathway by HER2 receptor. Oncogene 1999; 18:305-14

18. Howe L.R., Subbaramaiah K., Patel J., Masferrer J.L., Deora A,. Hudis C., Thaler H.T., Muller W.J., Du B.H., Brown A.M.C., Dannenberg A.J. Celecoxib, a selective cyclooxygenase 2 inhibitor, protects against human epidermal growth factor receptor 2 (HER-2)/neu-induced breast cancer Cancer Res 2002; 62:5405-5407

19. Andreasen P.A., Egelund R., Petersen H.H. The plasminogen activation system in tumor growth, invasion, and metastasis. Cell Mol Life Sci 2000; 57:25-40

20. Liu D., Ghiso J.A.A., Estrada Y., Ossowski L. EGFR is a transducer of the urokinase receptor initiated signal that is required for in vivo growth of a human carcinoma Cancer Cell 2002; 1:445-457

21. Golubovskaya V., Beviglia L., Xu L.H., Earp H.S., Craven R., Cance W. Dual inhibition of focal adhesion kinase and epidermal growth factor receptor pathways cooperatively induces death receptor-mediated apoptosis in human breast cancer cells J Biol Chem 2002; 277:38978-38987

22. Xia L., Wang L.J., Chung A.S., Ivanov S.S., Ling M.Y., Dragoi A.M., Platt A., Gilmer T.M., Fu X.Y., Chin Y.E. Identification of both positive and negative domains within the epidermal growth factor receptor COOH-terminal region for signal transducer and activator of transcription (STAT) activation. J Biol Chem 2002; 277:30716-30723

23. Ren Z.Y., Schaefer T.S. ErbB-2 activates Stat3 alpha in a Src- and JAK2-dependent manner. J Biol Chem 2002; 277:38486-38493

24. Kijima T., Niwa H., Steinman R.A., Drenning S.D., Gooding W.E., Wentzel A.L., Xi S.C., Grandis J.R. STAT3 activation abrogates growth factor dependence and contributes to head and neck squamous cell carcinoma tumor growth in vivo. Cell Growth & Differ 2002; 13:355-362

25. Solbach C. Roller M. Ahr A. Loibl S. Nicoletti M. Stegmueller M. Kreysch HG. Knecht R. Kaufmann M. Anti-epidermal growth factor receptor-antibody therapy for treatment of breast cancer. Int J Cancer. 2002; 101:390-394

26. Agus D.B., Akita R.W., Fox W.D., Lewis G.D., Higgins B., Pisacane P.I., Lofgren J.A., Tindell C., Evans D.P., Maiese K., Scher H.I., Sliwkowski M.X. Targeting ligand-activated ErbB2 signaling inhibits breast and prostate tumor growth. Cancer Cell 2002 2:127-137

27. Ng S.S.W., Tsao M.S., Nicklee T., Hedley D.W. Effects of the epidermal growth factor receptor inhibitor OSI-774, Tarceva, on downstream signaling pathways and apoptosis in human pancreatic adenocarcinoma. Mol Cancer Ther 2002; 1:777-783

28. Mellinghoff I.K., Tran C., Sawyers C.L. Growth inhibitory effects of the dual ErbB1/ErbB2 tyrosine kinase inhibitor PKI-166 on human prostate cancer xenografts. Cancer Res 2002; 62:5254-5259

29. Xia W.L., Mullin R.J., Keith B.R., Liu L.H., Ma H., Rusnak D.W., Owens G., Alligood K.J., Spector N.L. Anti-tumor activity of GW572016: a dual tyrosine kinase inhibitor blocks EGF activation of EGFR/erbB2 and downstream Erk1/2 and AKT pathways. Oncogene 2002 21:6255-6263

30. Allen L.F., Lenehan P.F., Eiseman I.A., Elliot W.L., Fry D.W. Potential benefits of the irreversible pan-erbB inhibitor, CI-1033, in the treatment of breast cancer. Semin Oncol 2002; 29:11-21

31. Sirotnak F.M. Studies with ZD1839 in preclinical models. Semin Oncol 2003; 30(*Suppl 1*):12-20

32. Barnes C.J., Yarmand-Bagheri R., Manda, M., Yang Z., Clayman G.L., Kumar R. Suppression of Epidermal Growth Factor Receptor, MAPK and Pak1 Pathways and Invasiveness of Human Cutaneous Squamous Cancer Cells by the Tyrosine Kinase Inhibitor ZD1839 ('Iressa'). Mol Cancer Ther 2003; 2:345-351

12

33. Baselga J., Rischin D., Ranson M., Calvert H., Raymond E., Kieback D.G., Kaye S.B., Gianni L., Harris A., Bjork T., Averbuch S.D., Feyereislova A., Swaisland H., Rojo F., Albanell J. Phase I safety, pharmacokinetic, and pharmacodynamic trial of ZD1839, a selective oral epidermal growth factor receptor tyrosine kinase inhibitor, in patients with five selected solid tumor types. J Clin Oncol 2002; 20:4292-4302

34. Wakeling A.E., Guy S.P., Woodburn J.R., Ashton S.E., Curry B.J., Barker A.J., Gibson K.H. ZD1839 (Iressa): An orally active inhibitor of epidermal growth factor signaling with potential for cancer therapy. Cancer Res 2002; 62:5749-5754

35. Herbst R.S., Maddox A.M., Small E.J., Rothenberg L., Small E.L., Rubin E.H., Baselga J., Rojo F., Hong W.K., Swaisland H., Averbuch S.D., Ochs J., LoRusso P.M. Selective oral epidermal growth factor receptor tyrosine kinase inhibitor ZD1839 is generally well-tolerated and has activity in non-small-cell lung cancer and other solid tumors: Results of a phase I trial. Journal of Clin Oncol 2002; 20:3815-3825

36. Twombly R. Despite Concerns, FDA panel backs EGFR inhibitor. JNCI 2002; 94:1596-1597

37. Dowsett M., Haper-Wynne C., Boeddinghaus I., Salter J., M Hills M., Dixon M., Ebbs S., Gui G., Sacks N., Smith I. HER-2 amplification impedes the antiproliferative effects of hormone therapy in estrogen receptor positive primary breast cancer. Cancer Res 2001; 61:8452-8458

38. Kumar R., WangR.-A., Mazumdar A., Talukder A.H., Mandal M., Yang Z., Bagheri-Yarmand R., Sahin A., Hortobagyi G., Adam L., Barnes C.J., Vadlamudi R.K. A naturally occurring MTA1 variant sequesters oestrogen receptor-□ in the cytoplasm. Nature 2002; 418:654 – 657

39. Kumar R. Another tie that binds the MTA family to breast cancer. Cell 2003 113:142-143

40. Lee S., Yang W.T., Lan K.H., Sellappan S., Klos K., Hortobagyi G., Hung M.C., Yu D.H. Enhanced sensitization to Taxol-induced apoptosis by Herceptin pretreatment in ErbB2-overexpressing breast cancer cells. Cancer Res 2002 62:5703-5710

41. Slamon D.J., Leyland-Jones B., Shak S., Fuchs H., Paton V., Bajamonde A., Fleming T., Eirmann W., Wolter J., Pegram M. et al. Use of chemotherapy plus a monoclonal antibody against HER2 for metastatic breast cancer that overexpresses HER2. N Engl J Med 2001; 344:783-792

42. Bagheri-Yarmand R., Vadlamudi R.K., Wang R.A., Mendelsohn J., Kumar R. Vascular endothelial growth factor up-regulation via p21-activated kinase-1 signaling regulates heregulin-beta1-mediated angiogenesis. J Biol Chem 2000; 275:39451-39457

43. Kumar R., Yarmand-Bagheri R. The role of HER2 in angiogenesis. Semin Oncol 2001; 28:27-32

44. Pegram M.D., Reese D.M. Combined biological therapy of breast cancer using monoclonal antibodies directed against HER2/neu protein and vascular endothelial factor. Semin Oncol 2002; 29:29-37

45. CH Baker C.H., Solorzano C.C., Fidler I.J. Blockade of vascular endothelial growth factor receptor and epidermal growth factor receptor signaling for therapy of metastatic human pancreatic cancer. Cancer Res 2002; 62:1996-2003

46. Santra M., Reed C.C., Iozzo R.V. Decorin binds to a narrow region of the epidermal growth factor (EGF) receptor, partially overlapping but distinct from the EGF-binding epitope. J Biol Chem 2002; 277:35671-35681

47. Cordero J.B., Cozzolino M., Lu Y., Vidal M. Slatopolsky E., Stahl P.D., Barbieri M.A., Dusso A. 1,25-dihydroxyvitamin D down-regulates cell membrane growth- and nuclear growth-promoting signals by the epidermal growth factor receptor. J Biol Chem 2002; 277(41):38965-38971

48. Scott G.K., Marden C., Xu F., Kirk L., Benz C.C. Transcriptional repression of ErbB2 by histone deacetylase inhibitors detected by a genomically integrated ErbB2 promoter-reporting cell screen Mol Cancer Ther 2002; 1:385-392

49. Hung M.C., Hortobagyi G.N., Ueno N.T. Development of clinical trial of E1A gene therapy targeting HER-2/neu-overexpressing breast and ovarian cancer. Adv Exper Med Biol 2000; 465:171-80

50. Asada S., Choi Y., Yamada M., Wang S.C., Hung M.C., Qin J., Uesugi M. External control of Her2 expression and cancer cell growth by targeting a Ras-linked coactivator. PNAS USA 2002; 99:12747-12752
51. Feng J. Adsay NV. Kruger M. Ellis KL. Nagothu K. Majumdar APN. Sarkar FH. Expression of ERRP in normal and neoplastic pancreata and its relationship to clinicopathologic parameters in pancreatic adenocarcinoma. Pancreas 2002; 25:342-349
52. Kim S., Zagozdzon R., Meisler A., Baleja J.D., Fu Y.G., Avraham S., Avraham H. Csk homologous kinase (CHK) and ErbB-2 interactions are directly coupled with CHK negative growth regulatory function in breast cancer. J Biol Chem 2002; 277:36465-36470
53. Xu W.P., Marcu M., Yuan X.T., Mimnaugh E., Patterson C., Neckers L. Chaperone-dependent E3 ubiquitin ligase CHIP mediates a degradative pathway for c-ErbB2 Neu. PNAS USA 2002; 99:12847-12852
54. Katz M., Shtiegman K., Tal-Or P., Yakir L., Mosesson Y., Harari D., Machluf Y., Asao H., Jovin T., Sugamura K., Yarden Y. Ligand-independent degradation of epidermal growth factor receptor involves receptor ubiquitylation and hgs, an adaptor whose ubiquitin-interacting motif targets ubiquitylation by Nedd4. Traffic 2002; 3:740-751

INTEGRIN SIGNALING IN CANCER

HIRA LAL GOEL AND LUCIA R. LANGUINO

1. INTRODUCTION

This chapter focuses on a family of proteins designated "Integrins", that mediate cell - extracellular matrix (ECM) or cell-cell adhesion, and on the mechanisms of signal transduction activated by integrins (Figure 1). Integrins have emerged as modulators of a variety of cellular functions. They have been implicated in cell migration, survival, normal and aberrant cellular growth, differentiation, gene expression and modulation of intracellular signal transduction pathways. Thus, they can control the events that characterize, phenotypically, a tumor: lack of differentiation, abnormal growth and increased survival, local invasion and infiltration of surrounding normal tissues, and finally, metastatic spread.

Neoplastic cells have a markedly different surrounding matrix than normal cells; thus, changes in the integrin profile may be functionally relevant and contribute to cancer progression.

The structural and functional characteristics of integrins, their expression and the potential of integrins and of integrin – downstream pathways as therapeutic targets is discussed in this chapter.

2. THE INTEGRIN FAMILY OF ADHESION RECEPTORS

Adhesive contacts between cells and ECM components play a crucial role in organ development, abnormal tissue growth, and tumor progression. These interactions are mediated by integrins, the most widely distributed gene superfamily of adhesion receptors, expressed by all mammalian cells [1]. Integrins can also mediate cell-cell interactions, although the ability to mediate cell-cell contact is restricted to a few members of the family [$\alpha_L\beta_2$, $\alpha_M\beta_2$, $\alpha_X\beta_2$, $\alpha_D\beta_2$, $\alpha_4\beta_1$, and the $\alpha_4\beta_7$.

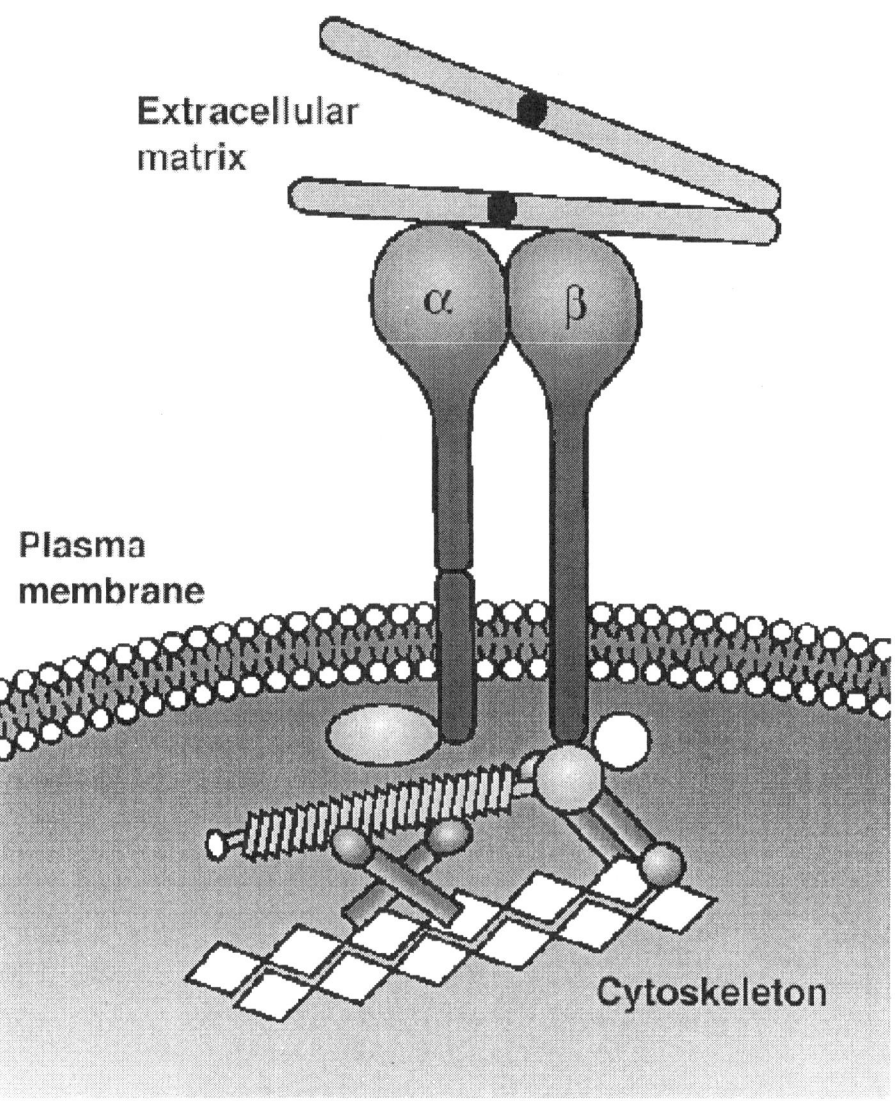

Figure 1. Integrins connect the extracellular matrix and the cytoskeleton. Integrin extracellular domains bind to specific sites in the extracellular matrix proteins, such as fibronectin. Their cytoplasmic domains bind to cytoskeletal proteins such as talin (long coiled coil) and α-actinin (parallel cylinders) and, through them, to other linkers such as vinculin (cross-shaped cylinders) or to actin, depicted as diamond-shaped microfilaments. Additional cytoplasmic proteins are also recruited; many of these function in signaling.

Table 1. The integrin family

Integrin	Ligand
$\beta_{1(A)}, \beta_{1B}, \beta_{1C}, \beta_{1C-2}, \beta_{1D}$	
α_1	Laminin, Collagen
α_2	Laminin, Collagen
α_{3A}, α_{3B}	Laminin, Collagen, Fibronectin, Entactin
α_4	Fibronectin, VCAM1, Osteopontin, Angiostatin
α_5	Fibronectin, L1, Tissue Transglutaminase
$\alpha_{6A}, \alpha_{6B},$ $\alpha_{6X1}, \alpha_{6X2}$ α_{7A}, α_{7B}	Laminin
α_8	Fibronectin, Tenascin, Nephronectin, Vitronectin, Osteopontin
α_9	Tenascin
α_{10}, α_{11}	Collagen
α_V	Fibronectin, Osteopontin, TGF β-LAP
β_{5A}, β_{5B}	
α_V	Vitronectin, Osteopontin, TGF β-LAP

Integrin	Ligand
β_2	
α_L	ICAM1, ICAM2, ICAM3, ICAM4
α_M	iC3b, Fibrinogen, Factor X, ICAM1, ICAM2, ICAM4
α_X	iC3b, Fibrinogen
α_D	ICAM3
β_6	
α_V	Vitronectin, Osteopontin, TGF β-LAP
β_7	
α_4, α_{IEL}	Fibronectin, Tenascin, Vitronectin, TGF β-LAP

Integrin	Ligand
$\beta_{3A}, \beta_{3B}, \beta_{3C}$	
$\alpha_{IIb}, \alpha_{IIbalt}$	Fibrinogen, Fibronectin, von Willebrand Factor, Vitronectin, Thrombospondin, Disintegrin, Osteopontin
α_V	Vitronectin, Fibrinoge, Fibronectin, Turnstatin, von Willebrand Factor, Thrombospondin, Disintegrin, L1, MMP2, Osteopontin, Endostatin
β_7	
α_4, α_{IEL}	Fibronecti VCAM MAdCAMl

Integrin	Ligand
$\beta_{4A}, \beta_{4B}, \beta_{4C}, \beta_{4D}$	
α_{6A}, α_{6B}	Laminin-5
β_8	
α_V	Vitronectin, Fibronectin, Laminin, TGF β-LAP

Table 1. The table shows the integrin family of adhesion receptors, comprised of 8 subfamilies; in each subfamily, integrins share the β subunit. Some of the integrin subunits are expressed as alternatively spliced forms, designated with A, B, C, C-2, D, X1, X2 and IIbalt. The ligands known to bind each integrin complex are indicated.

17

3. α and β Subunits

Based on immunochemical and molecular evidence, integrins are structurally organized into heterodimeric transmembrane complexes, variously assembled through the non-covalent association between an α and a β subunit [2]. So far, 18 α subunits, 8 β subunits, and 24 complexes have been identified and their expression and function characterized in various cell types. The integrin family is divided into subfamilies that share the β subunit [3]. Each β subunit associates with one to twelve α subunits and each α can associate with more than one β subunit. Functional specificity is determined by the specific associated subunits and by the cell type that expresses the heterodimeric complex (Table I). Integrins are expressed as constitutively active or inactive receptors for ECM ligands. Their functional state is cell type-dependent as well as ligand-dependent [4]. These different functional states might be crucial in modulating integrin-mediated functions *in vivo*.

4. Integrin Cytoplasmic Domains

Several experimental evidence obtained with recombinant deletion mutants and chimeric forms of integrin α and β cytoplasmic domains has demonstrated that cytoplasmic tails modulate receptor distribution, receptor surface expression, ligand binding affinity of the extracellular domain, cell adhesion, and cell spreading [2,5]. Therefore, structural differences in the primary sequences of the integrin intracellular domains are predicted to determine the specificity of a variety of integrin-mediated events. In support of this hypothesis, mutations and deletions in the integrin cytoplasmic domain have been found in the β_3 and β_4 integrin subgroups in, respectively, Glanzmann's thrombasthenia [6] and junctional epidermolysis Bullosa [7], thus pointing to the cytoplasmic domain as a key player in determining crucial cellular responses *in vivo*. Alternatively spliced forms of the α (α_3, α_6, α_7) and β (β_1, β_3, β_4) integrin cytoplasmic domains have been identified, thus adding further complexity to the regulatory pathways mediated by integrins. It is well established that the cytoplasmic domain of the β_1 subunit is required for integrins to modulate many cellular functions as well as to trigger signaling events which result in protein phosphorylation and interactions with intracellular proteins [5]. Four different β_1 isoforms containing alternatively spliced cytoplasmic domains have been identified (β_{1A}, β_{1B}, β_{1C}, and β_{1D}) and have been shown to differentially affect receptor localization, cell proliferation, cell adhesion and migration, interactions with intracellular proteins and, ultimately, phosphorylation and activation of signaling molecules [5]. The expression of integrin variants is tissue and cell-type specific [5]. A selective expression has been shown for the β_{1C} integrin subunit, an inhibitor of cell proliferation [5], in hematopoietic cells, platelets, activated endothelial cells, and epithelial cells of liver, kidney, lung and prostate. The β_{1B} isoform has been found to be restricted to skin and liver, while the β_{1D} subunit has been detected in striated muscle, where it replaces the common β_{1A} isoform. Similar to β_1 variants, a differential distribution of the variant forms β_{3A}, β_{3B}, α_{3A}, α_{3B}, α_{6A}, α_{6B}, α_{7A}, and α_{7B} in relationship to their wild type counterparts has also been described using protein and mRNA analysis [5]. The functional differences described for these variants suggest that modulation of splicing patterns of β_1 mRNA may provide an accessory mechanism to regulate signaling pathways initiated by integrins [8].

5. Integrins Modulate Cell Proliferation

By interacting with the ECM and, inside the cell, with the cytoskeleton, integrins transfer signals from the extracellular environment to intracellular compartments and control many cellular functions, such as proliferation, migration, differentiation, and gene expression [9-11]. These signals are initiated after integrin engagement with natural ligands or surrogate antibody ligands and include increases in cytosolic free $[Ca^{2+}]_i$, tyrosine phosphorylation, elevation of intracellular pH, and stimulated transcription and translation of immediate and early inflammatory genes [11]. Integrins can act synergistically with growth factors in modulating cellular functions [11]. The ability of integrins to modulate cell proliferation has been extensively characterized [9]. Several studies have shown that cell adhesion to the ECM is required for cell cycle progression and proliferation in different cell types [9]. Cell adhesion and spreading on fibronectin, vitronectin, and collagen activates mitogen-activated protein (MAP) kinase [12,13]. Ras-independent and -dependent pathways have been implicated in MAP kinase activation by integrins [8,14]. Cell adhesion mediated by integrins modulates the cell cycle, whereas detachment from the matrix induces cell cycle arrest [15]. Cyclin A expression and cyclin E-dependent kinase activity are also known to be induced by cell attachment to the matrix, adding to the evidence that complex pathways of growth control are mediated by integrins and their ligands [9].

Loss of cell anchorage to the ECM has been shown to upregulate the expression of the cyclin-dependent kinase inhibitors p27^{kip1} and p21$^{cip1/waf1}$, while at the same time decreasing the levels of cyclin A [9]. Changes in p27^{kip1} have also been observed in response to integrin expression [16]. Overall, these studies show that modulation of cell cycle regulators is mediated by adhesion- and spreading-dependent events as well as by integrin expression. Integrin ligation contributes to the abnormal proliferation of transformed cells [17] and, in the absence of their ligands, integrins block cell proliferation and down-regulate *c-fos* and *c-jun* early genes.

6. SIGNALING PATHWAYS ACTIVATED BY INTEGRINS

Integrin engagement and consequent cell adhesion and spreading are likely involved in cancer initiation and/or progression because of their ability not only to mediate interactions with ECM proteins, but also to regulate multiple intracellular signaling molecules that are necessary for cell motility, cell survival and proliferation [1, 3, 11, 18, 19]. The mechanisms of signaling that occur proximal to the membrane are poorly known; integrin clustering or association with members of the transmembrane 4 superfamily might be ways to trigger proliferation signals and, consequently, regulate tumor invasion and growth [20].

The best characterized pathways activated by integrins are the Focal Adhesion Kinase (FAK), the phosphatidylinositol 3-kinase (PI 3-kinase) and the Ras/MAP kinase pathways (Figure 2). This section focuses on gene products that have been shown to be involved in signaling events mediated by integrins whose expression levels and activity are altered in cancer. Here, we will refer to prostate cancer as a model system, although the discussion is relevant to all types of cancer.

7. FAK

FAK is a non-receptor protein tyrosine kinase that has been shown to co-localize with integrins at focal contact sites [21]. FAK becomes tyrosine phosphorylated in response to integrin engagement and other stimuli [21-23]. FAK inhibition induces apoptosis and overexpression of FAK prevents apoptosis induced either in absence of ECM survival signals or in response to other stimuli [23]. Several studies have suggested a role for FAK in controlling cell migration in response to integrin engagement or to growth factors [23]. Direct evidence on the role of FAK *in vivo* in regulating cell migration has been obtained with the generation of FAK null mice. Ablation of the FAK gene results in embryonic lethality at day 8.0-8.5 due to severe mesodermal defects [24]. Cells derived from these embryos show a decreased migration *in vitro* as compared to cells derived from wild type embryos [24, 25]. In addition, FAK overexpression in chinese hamster fibroblasts and perturbation of endogenous FAK signaling in different cell types using dominant negative forms of the molecule have confirmed its involvement in controlling cell motility [23, 26-28].

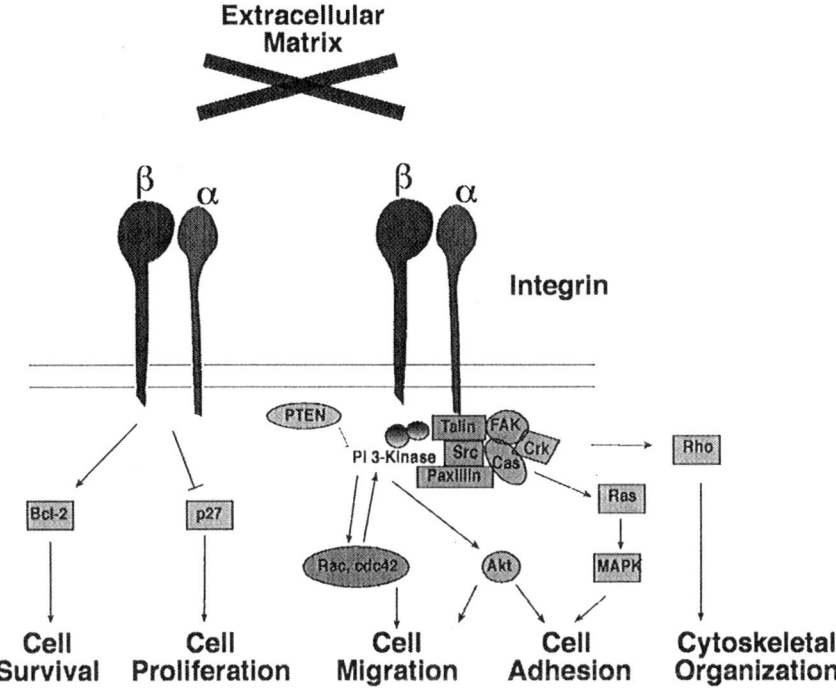

Figure 2. Integrin-signaling pathways deregulated in several types of cancer. Schematic drawing showing the signal transduction pathways activated by integrins that control cell migration, survival and proliferation.

In normal prostate, FAK expression is either low or absent but is significantly increased in high-grade adenocarcinomas and in invasive and metastatic prostate cancers compared to benign prostate and low grade adenocarcinoma [29, 30]. FAK association with Src, a cytoplasmic tyrosine kinase, is crucial for regulating cell migration *in vitro* [31]. Fibroblasts derived from Src, Fyn and Yes triple knock-out mice show impaired haptotactic migration in response to fibronectin and re-expression of Src increases their ability to migrate in response to fibronectin as compared to triple knock-out fibroblasts [32]. Recently, Slack *et al.* reported that inhibition of the FAK/Src signaling pathway significantly blocks migration of prostate carcinoma cells *in vitro*, demonstrating the crucial role exerted by these molecules in the regulation of prostate cell motility [33].

8. PI 3-kinase/AKT

In addition to stimulating FAK, integrins can also activate the PI 3-kinase pathway [11]. PI 3-kinases comprise a family of lipid kinases activated by a wide variety of extracellular stimuli. The lipid products of PI 3-kinases, specifically phosphatidylinositol([3,4)biphosphate [$PI(3,4)P_2$] and (3,4,5)triphosphate [$PI(3,4,5)P_3$], affect cell proliferation, survival, differentiation and migration by targeting specific signaling molecules such as the serine/threonine protein kinase B, also known as AKT [34-36]. Integrin-mediated adhesion to the ECM stimulates the production of $PI(3,4)P_2$ and $PI(3,4,5)P_3$ [37, 38], the association of the p85 PI 3-kinase subunit with FAK [39] and AKT activation [37,38]. AKT plays an important role in transducing survival signals in response to several growth factors and to integrin engagement [37, 40]. Recent studies have shown a significant increase in AKT kinase activity and phosphorylation associated with prostate cancer progression; specifically, the highest levels of phosphorylated AKT correlated with high Gleason grade, tumor stage III/IV and invasive cancer [41,42]. Several studies have reported that integrins control cancer cell motility through the PI 3-kinase pathway [43] which has been shown *in vitro* to be crucial for human prostate cancer cell migration [4]. Analysis of PI 3-kinase expression and activity on prostate cancer specimens is therefore needed to determine the clinicopathological significance of the PI 3-kinase pathway in prostate cancer initiation and/or progression.

9. PTEN

The tumor suppressor gene PTEN (or MMAC-1) encodes a dual specificity phosphatase but it has also the ability to dephosphorylate inositol phospholipids such as $PI(3,4,5)P_3$ and as a consequence to negatively regulate the PI 3-kinase/AKT pathway [44,45]. Interestingly, absence of PTEN expression was observed in 60% of the analyzed prostate tumors and correlated with high levels of AKT phosphorylation [41]. The PTEN gene is frequently deleted or mutated in a wide variety of human cancers and has been shown to be involved in regulation of cell migration on integrin substrates [45]. Tamura *et al.* have shown that FAK is one of the PTEN substrates [45]. PTEN inhibits cell migration and invasion by dephosphorylating FAK and the adapter protein Shc, thereby antagonizing integrin-triggered signaling [46]. The PTEN gene was located at 10q23.3 [47, 48]. Loss of heterozygosity (LOH) in the region 10q23.3 is present in 29-42% of clinically localized prostate cancers [49,50]. Similarly, LOH at 10q23 is present in more than 50% of metastatic prostate tumors [49-51]. PTEN is also frequently mutated or deleted in prostate cancer and prostate cancer cell lines [47, 49, 51, 52]. The PTEN protein is expressed in secretory epithelia in normal adult prostate and loss of PTEN protein expression in primary prostate cancers correlates with high Gleason grade and advanced pathological stage [53].

10. Ras/MAP kinase

Ras proteins belong to a large family of GTPases which function as signal transducers by cycling from an active GTP-bound form to an inactive GDP-bound form and activated ras stimulates numerous signaling cascades such as the MAP kinase pathway [54]. The Ras/MAP kinase pathway plays a pivotal role in modulating gene expression, cell cycle progression, survival and motility [55, 56]. Integrin clustering has been shown to stimulate Ras GTP-loading [14, 38, 57] and to activate specific effectors of the Ras/MAP kinase signaling cascade [58, 59] which results in increased cell proliferation, cell cycle progression and survival [60]. Some integrins exert a negative effect on the Ras/MAP kinase pathway which leads to cell cycle arrest and differentiation [61] and inhibition of cell proliferation [8]. There is evidence that cell motility is controlled by integrins via a signaling cascade involving Shc and MEK1 and the MAP kinases ERK1 and 2, respectively [62, 63]. Sustained activation of the Ras/MAP kinase pathway can also prevent apoptosis triggered by loss of cell-ECM contacts [64-66]. Recently it has been reported that migration and survival mechanisms promoted by integrin engagement are coordinately regulated through activation of pathways that involve ERK activity [67]. The activity and expression levels of MAP kinase are significantly higher in primary prostatic adenocarcinoma and in metastatic lesions than the levels detected in benign prostate [68-70]. Increased MAP kinase activation correlates with high Gleason score and tumor stage [68]. Since Ras mutations are uncommon in prostate cancer [71], chronic stimulation of the Ras/MAP kinase pathway is most likely achieved by alterations in the levels of upstream regulators such as integrins, growth factors and growth factor receptors during prostate cancer initiation and/or progression. Several studies suggest that integrin engagement activates members of the Rho-family of small GTPases [43, 72]. Specifically, Rho, Rac and Cdc42 have been shown to be required for cell motility [43, 72]. However, analysis of either Rho-family of small GTPases' gene alterations, protein expression or activity in prostate cancer tissues has not been performed.

11. Bcl-2

The Bcl-2 protein is a proto-oncogene that promotes cell survival [73] and is a member of a large family that consists of pro-apoptotic and pro-survival factors [74]. The Bcl-2 gene is activated by chromosomal translocation in the majority of non-Hodgkin's lymphomas and is also up-regulated in many solid tumors, indicating that it might contribute to resistance to apoptosis in response to chemotherapeutic agents and radiation therapy [74]. Adhesion to fibronectin through $\alpha_5\beta_1$ and $\alpha_v\beta_1$ and to vitronectin through $\alpha_v\beta_3$ integrins was shown to up-regulate Bcl-2 transcription and protein levels and resulted in protection from apoptosis induced by serum deprivation [75, 76].

Bcl-2 protein levels are low or absent in normal prostate and Bcl-2 expression is restricted to basal cells [77]. In prostate carcinoma Bcl-2 is up-regulated and its expression correlates with hormone-refractory disease [71, 77] and with poor survival [71]. Analysis of metastatic lesions obtained from prostate cancer patients after hormone treatment (hormone-refractory tumors) stained positive for Bcl-2 [78]. There is evidence for a role for Bcl-2 in promoting cancer cell motility and invasion. Overexpression of Bcl-2 increases migration and metastatic potential of breast cancer cells [79] and therefore its involvement in cancer cell invasion deserves to be investigated.

12. SYNERGISTIC ACTIVITY OF INTEGRINS AND GROWTH FACTOR RECEPTORS

The synergistic activity of integrins and growth factor receptors in stimulating intracellular signaling pathways is required for proper growth and differentiation (Figure 3) [80]. Although growth factors and ECM can independently stimulate several signaling pathways, the interactions between these signals is known to regulate several physiological functions including cell adhesion, migration, proliferation and differentiation [81]. Understanding of the basic mechanisms that mediate the interactions between integrins and growth factor receptors will help elucidating abnormal cellular functions in cancer.

Several reports have shown the association between integrins and growth factor receptors by co-immunoprecipitation; among others, $\alpha_v\beta_3$ with either platelet-derived growth factor receptor (PDGFR) or vascular endothelial growth factor receptor (VEGFR); $\alpha_6\beta_4$ with Erb-2 [80]. Src family kinases, PI 3-kinase, MEK/ERK as well as PKC are potential mediators of the cross-talk between integrins and growth factor receptors, since they can be activated synergistically by both integrins and growth factor receptors [82-84].

It has been shown that FAK may be a point of convergence in the actions of integrins and growth factors, like insulin, insulin-like growth factors (IGF),

PDGFs or hepatocyte growth factors [80, 85]. FAK, an important mediator of integrin signaling, is also known to be a substrate for the insulin receptor and IGF-I receptor (IGF-IR) [86, 87]. In adherent cells, IGF-I stimulation showed transient dephosphorylation of FAK with rapid disassembly of actin filaments involving PI 3-kinase signaling and phosphotyrosine phosphatase activities, suggesting a negative regulation of IGF on integrin signaling [86, 88].

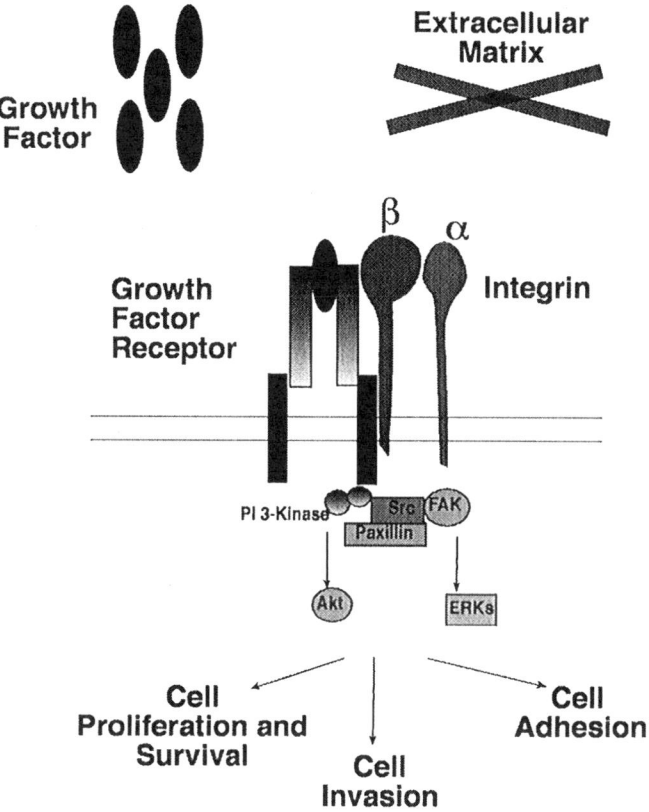

Figure 3. Synergistic activity of integrins and growth factor receptors. This figure illustrates the interaction between integrins and growth factor receptors and the signaling proteins involved in their cross-talk.

Several studies have shown that integrins and IGF-IR act synergistically. Tai et al. have shown that IGF-IR associates with β_1 transiently upon IGF stimulation and this increases AKT and MAPK activities [89]. Finally, blocking ligand occupancy of the $\alpha_V\beta_3$ integrin inhibited IGF-I stimulated biological actions including cell migration, DNA and protein synthesis as well as attenuates downstream signaling [90]. The studies by Clemmons group have shown that IGF-IR interacts with $\alpha_V\beta_3$ integrin [91]. Blocking of ligand occupancy of $\alpha_V\beta_3$ with echistatin decreases IGF stimulated tyrosine phosphorylation of its receptor and associated downstream signaling [92].

This effect is suggested to be mediated by increased recruitment of the phosphatase SHP-2 to IGF-IR [93]. In conclusion, growth factors (including PDGF and IGF) have a significant effect in the regulation of cell adhesion and migration. Similarly, cell stimulatory signals generated by various growth factors do also depend on proper cell-ECM interactions. Future studies will need to show how these synergistic signals can be differentially activated in cancer.

13. CONCLUSIONS

This review highlights the current knowledge of the alterations that occur in cancer and that involve integrins and integrin-activated pathways. Integrins are cell surface receptors for ECM proteins that mediate a variety of functions related to cell proliferation, differentiation, and survival. Although the specific functions of integrins and their ligands in cancer need to be further investigated, recent publications outlining their expression pave the way for future investigations describing integrin functional alterations and the signaling pathways involved in cancer progression. Similarly, changes in integrin affinity, avidity, or activation state are likely to control cell-ECM interaction; additional investigations on these topics will help understanding the role of integrins in cancer. Future research will focus on functional correlates, combining general knowledge of integrins and integrin signaling with an increasing appreciation for the role of the ECM in cancer progression. Integrins and their downstream signaling effectors appear to be promising targets for cancer therapy.

14. ACKNOWLEDGMENTS

This work was supported by grants from NIH, RO1 CA-89720 and from Army, PCRP DAMD17-98-1-8506.

Hira Lal Goel and Lucia R. Languino
Departments of Cancer Biology and Cell Biology, University of
Massachusetts Medical School, Boston, Massachusetts

15. REFERENCES

1. Hynes, R.O. 2002. Integrins: bidirectional, allosteric signaling machines. *Cell* 110:673-687.
2. Hemler, M.E., Weitzman, J.B., Pasqualini, R., Kawaguchi, S., Kassner, P.D., and Berdichevsky, F.B. 1995. Structure, Biochemical Properties, and Biological Functions of Integrin Cytoplasmic Domains. In *Integrins: The Biological Problems.* Y. Takada, editor. Boca Raton: CRC Press Inc. 1-35.

3. Ruoslahti, E. 1997. Integrins as signaling molecules and targets for tumor therapy. *Kidney International* 51:1413-1417.
4. Zheng, D.Q., Woodard, A.S., Tallini, G., and Languino, L.R. 2000. Substrate specificity of $\alpha_v\beta_3$ integrin-mediated cell migration and phosphatidylinositol 3-kinase/AKT pathway activation. *J. Biol. Chem.* 275:24565-24574.
5. Fornaro, M., and Languino, L.R. 1997. Alternatively spliced variants: a new view of the integrin cytoplasmic domain. *Matrix Biology* 16:185-193.
6. Williams, M.J., Hughes, P.E., O'Toole, T.E., and Ginsberg, M.H. 1994. The inner world of cell adhesion: integrin cytoplasmic domains. *Trends in Cell Biol.* 4:109-112.
7. Vidal, F., Aberdam, D., Miquel, C., Christiano, A.M., Pulkkinen, L., Uitto, J., Ortonne, J.-P., and Meneguzzi, G. 1995. Integrin $\beta 4$ mutations associated with junctional epidermolysis bullosa with pyloric atresia. *Nature Genetics* 10:229-234.
8. Fornaro, M., Steger, C.A., Bennett, A.M., Wu, J.J., and Languino, L.R. 2000. Differential role of β_{1C} and β_{1A} integrin cytoplasmic variants in modulating focal adhesion kinase, protein kinase B/AKT, and Ras/Mitogen-activated protein kinase pathways. *Mol. Biol. Cell* 11:2235-2249.
9. Bottazzi, M.E., and Assoian, R.K. 1997. The extracellular matrix and mitogenic growth factors control G1 phase cyclins and cyclin-dependent kinase inhibitors. *Trends in Cell Biology* 7:348-352.
10. Frisch, S.M., and Ruoslahti, E. 1997. Integrins and anoikis. *Current Op. Cell Biol.* 9:701-706.
11. Schwartz, M.A., Schaller, M.D., and Ginsberg, M.H. 1995. Integrins: emerging paradigms of signal transduction. *Annu. Rev. Cell Dev. Bio.* 11:549-599.
12. Chen, Q., Kinch, M.S., Lin, T.H., Burridge, K., and Juliano, R.L. 1994. Integrin-mediated cell adhesion activates mitogen-activated protein kinases. *J. Biol. Chem.* 269:26602-26605.
13. Zhu, X., and Assoian, R.K. 1995. Integrin-dependent activation of MAP kinase: a link to shape-dependent cell proliferation. *Mol. Biol. Cell* 6:273-282.
14. Clark, E., and Hynes, R. 1996. Ras activation is necesary for integrin-mediated activation of extracellular signal-regulated kinase 2 and cytosolic phospholipase A_2 but not for cytoskeletal organization. *J. Biol. Chem.* 271:14814-14818.
15. Fang, F., Orend, G., Watanabe, N., Hunter, T., and Ruoslahti, E. 1996. Dependence of cyclin E-CDK2 kinase activity on cell anchorage. *Science (Wash. D.C.)* 271:499-502.
16. Fornaro, M., Tallini, G., Zheng, D.Q., Flanagan, W.M., Manzotti, M., and Languino, L.R. 1999. p27kip1 acts as a downstream effector of and is coexpressed with the β_{1C} integrin in prostatic adenocarcinoma. *J. Clin. Invest.* 103:321-329.
17. Varner, J.A., Emerson, D.A., and Juliano, R.L. 1995. Integrin $\alpha_5\beta_1$ expression negatively regulates cell growth: reversal by attachment to fibronectin. *Mol. Biol. Cell* 6:725-740.
18. Manes, T., Zheng, D.Q., Tognin, S., Woodard, A.S., Marchisio, P.C., and Languino, L.R. 2003. alphavbeta3 integrin expression up-regulates cdc2, which modulates cell migration. *J. Cell Biol.* 161:817-826.
19. Fornaro, M., Plescia, J., Chheang, S., Tallini, G., Zhu, Y.-M., King, M., Altieri, D.C., and Languino, L.R. 2003. Fibronectin protects prostate cancer cells from tumor necrosis factor alpha-induced apoptosis via the AKT/Survivin pathway. *J. Biol. Chem.* 278:50402-50411.
20. Hemler, M., Mannion, B., and Berditchevski, F. 1996. Association of TM4SF proteins with integrins: relevance to cancer. *Biochim. Biophys. Acta* 1287:67-71.
21. Guan, J.L., Trevithick, J.E., and Hynes, R.O. 1991. Fibronectin/integrin interaction induces tyrosine phosphorylation of a 120 kDa protein. *Cell Regul.* 2:951.
22. Kornberg, L., Earp, H., Turner, C., Prockop, C., and Juliano, R. 1991. Signal transduction by integrins: increased protein tyrosine phosphorylation caused by clustering in β_1 integrins. *Proc. Natl. Acad. Sci. U S A* 88:8392-8396.
23. Schaller, M.D. 2001. Biochemical signals and biological responses elicited by the focal adhesion kinase. *Biochim. Biophys. Acta* 1540:1-21.

28

24. Ilic, D., Furuta, Y., Kanazawa, S., Takeda, N., Sobue, K., Nakatsuji, N., Nomura, S., Fujimoto, J., Okada, M., Yamamoto, T., et al. 1995. Reduced cell motility and enhanced focal adhesion contact formation in cells from FAK-deficient mice. *Nature* 377:539-543.

25. Ilic, D., Kanazawa, S., Furuta, Y., Yamamoto, T., and Aizawa, S. 1996. Impairment of mobility in endodermal cells by FAK deficiency. *Exp. Cell Res.* 222:298-303.

26. Zheng, D.Q., Woodard, A.S., Fornaro, M., Tallini, G., and Languino, L.R. 1999. Prostatic carcinoma cell migration via $\alpha_v\beta_3$ integrin is modulated by a focal adhesion kinase pathway. *Cancer Research* 59:1655-1664.

27. Cary, L.A., Chang, J.F., and Guan, J.-L. 1996. Stimulation of cell migration by overexpression of focal adhesion kinase and its association with Src and Fyn. *J. Cell Sci.* 109:1787-1794.

28. Gilmore, A., and Romer, H. 1996. Inhibition of focal adhesion kinase (FAK) signaling in focal adhesions decreases cell motility and proliferation. *Mol. Biol. Cell* 7:1209-1224.

29. Tremblay, L., Hauck, W., Aprikian, A.G., Begin, L.R., Chapdelaine, A., and Chevalier, S. 1996. Focal adhesion kinase pp[125FAK] expression, activation and association with paxillin and p[50CSK] in human metastatic prostate carcinoma. *Int. J. Cancer* 68:164-171.

30. Stanzione, R., Picascia, A., Chieffi, P., Imbimbo, C., Palmieri, A., Mirone, V., Staibano, S., Franco, R., De Rosa, G., Schlessinger, J., et al. 2001. Variations of proline-rich kinase Pyk2 expression correlate with prostate cancer progression. *Lab. Invest.* 81:51-59.

31. Sieg, D.J., Hauck, C.R., and Schlaepfer, D.D. 1999. Required role of focal adhesion kinase (FAK) for integrin-stimulated cell migration. *J. Cell Sci.* 112:2677-2691.

32. Klinghoffer, R.A., Sachsenmaier, C., Cooper, J.A., and Soriano, P. 1999. Src family kinases are required for integrin but not PDGFR signal transduction. *EMBO J.* 18:2459-2471.

33. Slack, J.K., Adams, R.B., Rovin, J.D., Bissonette, E.A., Stoker, C.E., and Parsons, J.T. 2001. Alterations in the focal adhesion kinase/Src signal transduction pathway correlate with increased migratory capacity of prostate carcinoma cells. *Oncogene* 20:1152-1163.

34. Rameh, L.E., and Cantley, L.C. 1999. The role of phosphoinositide 3-kinase lipid products in cell function. *J. Biol. Chem.* 274:8347-8350.

35. Jiang, B.H., Aoki, M., Zheng, J.Z., Li, J., and Vogt, P.K. 1999. Myogenic signaling of phosphatidylinositol 3-kinase requires the serine-threonine kinase Akt/protein kinase B. *Proc. Natl. Acad. Sci. USA* 96:2077-2081.

36. Morales-Ruiz, M., Fulton, D., Sowa, G., Languino, L.R., Fujio, Y., Walsh, K., and Sessa, W.C. 2000. Vascular endothelial growth factor-stimulated actin reorganization and migration of endothelial cells is regulated via the serine/threonine kinase Akt. *Circ. Res.* 86:892-896.

37. Khwaja, A., Rodriguez-Viciana, P., Wennstrom, S., Warne, P.H., and Downward, J. 1997. Matrix adhesion and Ras transformation both activate a phosphoinositide 3-OH kinase and protein kinase B/Akt cellular survival pathway. *EMBO J.* 16:2783-2793.

38. King, W.G., Mattaliano, M.D., Chan, T.O., Tsichlis, P.N., and Brugge, J.S. 1997. Phosphatidylinositol 3-kinase is required for integrin-stimulated AKT and Raf-1/mitogen-activated protein kinase pathway activation. *Mol. Cell. Biol.* 17:4406-4418.

39. Chen, H.-C., and Guan, J.-L. 1994. Association of focal adhesion kinase with its potential substrate phosphatidylinositol 3-kinase. *Proc. Natl. Acad. Sci. U.S.A.* 91:10148-10152.

40. Downward, J. 1998. Mechanisms and consequences of activation of protein kinase B/Akt. *Curr. Opin. Cell Biol.* 10:262-267.

41. Sun, M., Wang, G., Paciga, J.E., Feldman, R.I., Yuan, Z.Q., Ma, X.L., Shelley, S.A., Jove, R., Tsichlis, P.N., Nicosia, S.V., et al. 2001. AKT1/PKBalpha kinase is frequently elevated in human cancers and its constitutive activation is required for oncogenic transformation in NIH3T3 cells. *Am. J. Pathol.* 159:431-437.

42. Paweletz, C.P., Charboneau, L., Bichsel, V.E., Simone, N.L., Chen, T., Gillespie, J.W., Emmert-Buck, M.R., Roth, M.J., Petricoin, I.E., and Liotta, L.A. 2001. Reverse phase protein microarrays which capture disease progression show activation of pro-survival pathways at the cancer invasion front. *Oncogene* 20:1981-1989.

43. Mercurio, A.M., Rabinovitz, I., and Shaw, L.M. 2001. The $\alpha_6\beta_4$ integrin and epithelial cell migration. *Current Opin. Cell Biol.* 13:541-545.

44. Maehama, T., and Dixon, J.E. 1999. PTEN: a tumour suppressor that functions as a phospholipid phosphatase. *Trends Cell Biol.* 9:125-128.

45. Tamura, M., Gu, J., Tran, H., and Yamada, K.M. 1999. PTEN gene and integrin signaling in cancer. *J. Natl. Cancer Inst.* 91:1820-1828.

46. Tamura, M., Gu, J., Matsumoto, K., Aota, S., Parsons, R., and Yamada, K.M. 1998. Inhibition of cell migration, spreading and focal adhesions by tumor suppressor PTEN. *Science* 280:1614-1617.

47. Li, J., Yen, C., Liaw, D., Podsypanina, K., Bose, S., Wang, S.I., Puc, J., Miliaresis, C., Rodgers, L., McCombie, R., et al. 1997. *PTEN*, a putative protein tyrosine phosphatase gene mutated in human brain, breast, and prostate cancer. *Science* 275:1943-1947.

48. Steck, P.A., Pershouse, M.A., Jasser, S.A., Yung, W.K., Lin, H., Ligon, A.H., Langford, L.A., Baumgard, M.L., Hattier, T., Davis, T., et al. 1997. Identification of a candidate tumour suppressor gene, MMAC1, at chromosome 10q23.3 that is mutated in multiple advanced cancers. *Nature Genet.* 15:356-362.

49. Cairns, P., Okami, K., Halachmi, S., Halachmi, N., Esteller, M., Herman, J.G., Jen, J., Isaacs, W.B., Bova, G.S., and Sidransky, D. 1997. Frequent inactivation of *PTEN/MMAC1* in primary prostate cancer. *Cancer Res.* 57:4997-5000.

50. Suzuki, H., Freije, D., Nusskern, D.R., Okami, K., Cairns, P., Sidransky, D., Isaacs, W.B., and Bova, G.S. 1998. Interfocal heterogeneity of PTEN/MMAC1 gene alterations in multiple metastatic prostate cancer tissues. *Cancer Res.* 58:204-209.

51. Teng, D.H., Hu, R., Lin, H., Davis, T., Iliev, D., Frye, C., Swedlund, B., Hansen, K.L., Vinson, V.L., Gumpper, K.L., et al. 1997. MMAC1/PTEN mutations in primary tumor specimens and tumor cell lines. *Cancer Res.* 57:5221-5225.

52. Vlietstra, R.J., van Alewijk, D.C.J.G., Hermans, K.G.L., van Steenbrugge, G.J., and Trapman, J. 1998. Frequent inactivation of *PTEN* in prostate cancer cell lines and xenografts. *Cancer Res.* 58:2720-2723.

53. McMenamin, M.E., Soung, P., Perera, S., Kaplan, I., Loda, M., and Sellers, W.R. 1999. Loss of PTEN expression in paraffin-embedded primary prostate cancer correlates with high Gleason score and advanced stage. *Cancer Res.* 59:4291-4296.

54. Campbell, S.L., Khosravi-Far, R., Rossman, K.L., Clark, G.J., and Der, C.J. 1998. Increasing complexity of Ras signaling. *Oncogene* 17:1395-1413.

55. Robinson, M.J., and Cobb, M.H. 1997. Mitogen-activated protein kinase pathways. *Current Opinion in Cell Biology* 9:180-186.

56. Chang, L., and Karin, M. 2001. Mammalian MAP kinase signalling cascades. *Nature* 410:37-40.

57. Mainiero, F., Murgia, C., Wary, K.K., Curatola, A.M., Pepe, A., Blumemberg, M., Westwick, J.K., Der, C.J., and Giancotti, F.G. 1997. The coupling of $\alpha_6\beta_4$ integrin to Ras-MAP kinase pathways mediated by Shc controls keratinocyte proliferation. *EMBO J.* 16:2365-2375.

58. Schlaepfer, D.D., and Hunter, T. 1998. Integrin signalling and tyrosine phosphorylation: just the FAKs? *Trends Cell Biol.* 8:151-157.

59. Howe, A., Aplin, A.E., Alahari, S.K., and Juliano, R.L. 1998. Integrin signaling and cell growth control. *Curr. Opin. Cell Biol.* 10:220-231.

60. Schwartz, M.A., and Assoian, R.K. 2001. Integrins and cell proliferation: regulation of cyclin-dependent kinases via cytoplasmic signaling pathways. *J. Cell Sci.* 114:2553-2560.

61. Sastry, S.K., Lakonishok, M., Wu, S., Truong, T.Q., Huttenlocher, A., Turner, C.E., and Horwitz, A.F. 1999. Quantitative changes in integrin and focal adhesion signaling regulate myoblast cell cycle withdrawal. *J. Cell Biol.* 144:1295-1309.

62. Gu, J., Tamura, M., Pankov, R., Danen, E.H., Takino, T., Matsumoto, K., and Yamada, K.M. 1999. Shc and FAK differentially regulate cell motility and directionality modulated by PTEN. *J. Cell Biol.* 146:389-403.

63. Cheresh, D.A., Leng, J., and Klemke, R.L. 1999. Regulation of cell contraction and membrane ruffling by distinct signals in migratory cells. *J. Cell Biol.* 146:1107-1116.

30

64. Frisch, S.M., and Screaton, R.A. 2001. Anoikis mechanisms. *Curr. Opin. Cell Biol.* 13:555-562.

65. Le Gall, M., Chambard, J.C., Breittmayer, J.P., Grall, D., Pouyssegur, J., and Van Obberghen-Schilling, E. 2000. The p42/p44 MAP kinase pathway prevents apoptosis induced by anchorage and serum removal. *Mol. Biol. Cell* 11:1103-1112.

66. Rosen, K., Rak, J., Leung, T., Dean, N.M., Kerbel, R.S., and Filmus, J. 2000. Activated Ras prevents downregulation of Bcl-X(L) triggered by detachment from the extracellular matrix. A mechanism of Ras-induced resistance to anoikis in intestinal epithelial cells. *J. Cell Biol.* 149:447-456.

67. Cho, S.Y., and Klemke, R.L. 2000. Extracellular-regulated kinase activation and CAS/Crk coupling regulate cell migration and suppress apoptosis during invasion of the extracellular matrix. *J. Cell Biol.* 149:223-236.

68. Gioeli, D., Mandell, J.W., Petroni, G.R., Frierson, H.F., Jr., and Weber, M.J. 1999. Activation of mitogen-activated protein kinase associated with prostate cancer progression. *Cancer Res.* 59:279-284.

69. Magi-Galluzzi, C., Mishra, R., Fiorentino, M., Montironi, R., Yao, H., Capodieci, P., Wishnow, K., Kaplan, I., Stork, P.J., and Loda, M. 1997. Mitogen-activated protein kinase phosphatase 1 is overexpressed in prostate cancers and is inversely related to apoptosis. *Lab. Invest.* 76:37-51.

70. Price, D.T., Rocca, G.D., Guo, C., Ballo, M.S., Schwinn, D.A., and Luttrell, L.M. 1999. Activation of extracellular signal-regulated kinase in human prostate cancer. *J. Urol.* 162:1537-1542.

71. Augustus, M., Moul, J.W., and Srivastava, S. 1999. *The molecular phenotype of the malignant prostate.* Washington, DC: IOS Press. 321-340 pp.

72. Parise, L.V., Lee, J., and Juliano, R.L. 2000. New aspects of integrin signaling in cancer. *Semin. Cancer Biol.* 10:407-414.

73. Vaux, D.L., Cory, S., and Adams, J.M. 1988. Bcl-2 gene promotes haemopoietic cell survival and cooperates with c- myc to immortalize pre-B cells. *Nature* 335:440-442.

74. Reed, J.C. 2000. Mechanisms of apoptosis. *Am. J. Pathol.* 157:1415-1430.

75. ang, Z., Vuori, K., Reed, J.C., and Ruoslahti, E. 1995. The $\alpha_5\beta_1$ integrin supports survival of cells on fibronectin and up-regulates Bcl-2 expression. *Proc. Natl. Acad. Sci. USA* 92:6161-6165.

76. Matter, M.L., and Ruoslahti, E. 2001. A signaling pathway from the $\alpha5\beta1$ and $\alpha V\beta3$ integrins that elevates bcl-2 transcription. *J. Biol. Chem.* 276:27757-27763.

77. Bruckheimer, E.M., Gjertsen, B.T., and McDonnell, T.J. 1999. Implications of cell death regulation in the pathogenesis and treatment of prostate cancer. *Semin. Oncol.* 26:382-398.

78. Colombel, M., Symmans, F., Gil, S., O'Toole, K.M., Chopin, D., Benson, M., Olsson, C.A., Korsmeyer, S., and Buttyan, R. 1993. Detection of the apoptosis-suppressing oncoprotein bcl-2 in hormone- refractory human prostate cancers. *Am. J. Pathol.* 143:390-400.

79. Del Bufalo, D., Biroccio, A., Leonetti, C., and Zupi, G. 1997. Bcl-2 overexpression enhances the metastatic potential of a human breast cancer line. *Faseb J.* 11:947-953.

80. Eliceiri, B.P. 2001. Integrin and growth factor receptor crosstalk. *Circ. Res.* 89:1104-1110.

81. Comoglio, P.M., Boccaccio, C., and Trusolino, L. 2003. Interactions between growth factor receptors and adhesion molecules: breaking the rules. *Curr. Opin. Cell Biol.* 15:565-571.

82. Goel, H.L., and Dey, C.S. 2002. PKC-regulated myogenesis is associated with increased tyrosine phosphorylation of FAK, Cas, and paxillin, formation of Cas-CRK complex, and JNK activation. *Differentiation* 70:257-271.

83. Howe, A.K., Aplin, A.E., and Juliano, R.L. 2002. Anchorage-dependent ERK signaling--mechanisms and consequences. *Curr. Opin. Genet. Dev.* 12:30-35.

84. Yamada, K.M., and Even-Ram, S. 2002. Integrin regulation of growth factor receptors. *Nat. Cell Biol.* 4:E75-E76.

85. Sieg, D.J., Hauck, C.R., Ilic, D., Klingbeil, C.K., Schaefer, E., Damsky, C.H., and Schlaepfer, D.D. 2000. FAK integrates growth-factor and integrin signals to promote cell migration. *Nat. Cell Biol.* 2:249-256.

86. Baron, V., Calleja, V., Ferrari, P., Alengrin, F., and Van Obberghen, E. 1998. p125Fak focal adhesion kinase is a substrate for the insulin and insulin-like growth factor-I tyrosine kinase receptors. *J. Biol. Chem.* 273:7162-7168.

87. Goel, H.L., and Dey, C.S. 2002. Insulin stimulates spreading of skeletal muscle cells involving the activation of focal adhesion kinase, phosphatidylinositol 3-kinase and extracellular signal regulated kinases. *J. Cell. Physiol.* 193:187-198.

88. Guvakova, M.A., and Surmacz, E. 1999. The activated insulin-like growth factor I receptor induces depolarization in breast epithelial cells characterized by actin filament disassembly and tyrosine dephosphorylation of FAK, Cas, and paxillin. *Exp. Cell. Res.* 251:244-255.

89. Tai, Y.T., Podar, K., Catley, L., Tseng, Y.H., Akiyama, M., Shringarpure, R., Burger, R., Hideshima, T., Chauhan, D., Mitsiades, N., et al. 2003. Insulin-like growth factor-1 induces adhesion and migration in human multiple myeloma cells via activation of β1-integrin and phosphatidylinositol 3'-kinase/AKT signaling. *Cancer Res.* 63:5850-5858.

90. Zheng, B., and Clemmons, D.R. 1998. Blocking ligand occupancy of the αVβ3 integrin inhibits insulin-like growth factor I signaling in vascular smooth muscle cells. *Proc. Natl. Acad. Sci. USA* 95:11217-11222.

91. Clemmons, D.R., and Maile, L.A. 2003. Minireview: Integral membrane proteins that function coordinately with the insulin-like growth factor I receptor to regulate intracellular signaling. *Endocrinology* 144:1664-1670.

92. Jones, J.I., Prevette, R.T., Gockerman, A., and Clemmons, D.R. 1996. Ligand occupancy of the αVβ3 integrin is necessary for smooth muscle cells to migrate in response to insulin-like growth factor I. *Proc. Natl. Acad. Sci. USA* 93:2482-2487.

93. Maile, L.A., and Clemmons, D.R. 2002. The αVβ3 integrin regulates insulin-like growth factor I (IGF-I) receptor phosphorylation by altering the rate of recruitment of the Src-homology 2-containing phosphotyrosine phosphatase-2 to the activated IGF-I receptor. *Endocrinology* 143:4259-4264.

REGULATORS OF VASCULAR ENDOTHELIAL GROWTH FACTOR EXPRESSION IN CANCER

OLIVER STOELTZING AND LEE M. ELLIS

1. INTRODUCTION

Angiogenesis is essential for solid malignancies to grow beyond 1–2 mm in size, when oxygen diffusion alone is no longer sufficient to maintain adequate tissue oxygenation [1]. The process of angiogenesis is regulated by numerous pro- and anti-angiogenic factors that function in a dynamic and coordinated manner to lead to a mature and functional vascular network. Tumor angiogenesis is the result of a net gain in pro-angiogenic stimuli through an increase in the expression of pro-angiogenic molecules, a decrease in the expression of anti-angiogenic molecules, or a combination thereof. In neoplastic lesions, this process is more accurately defined as a combination of angiogenesis and vasculogenesis in which the main blood supply is derived from preexisting vessels [2] but circulating endothelial cell (EC) precursors may contribute to the growing vascular network [3].

In general, tumor vascularization (specifically, the high vessel density within tumors) and increased expression of certain angiogenic factors in various human cancers have been associated with tumor progression and poor outcome [4]. One of the best-characterized angiogenic factors, and one that has been associated with angiogenesis and metastasis in many cancer systems [4], is vascular endothelial growth factor (VEGF), also known as vascular permeability factor [5, 6]. VEGF is also a potent EC mitogen and with its ability to mediate EC survival it facilitates angiogenesis in tumors even under adverse conditions such as hypoxia or an acidic milieu [7, 8]. In vivo, VEGF also increases vascular permeability (by 50,000 times that of histamine, the "gold standard" for induction of permeability) and may contribute to the formation of ascites or pleural effusion in patients with cancer [9]. Many other factors have been implicated in the regulation of EC proliferation and survival as well, including angiopoietins, fibroblast growth factors (FGFs), epidermal growth factor (EGF), interleukin (IL)-8, platelet-derived growth factor (PDGF), platelet-derived EC growth factor (PD-ECGF, also known as thymidine phosphorylase), and hepatocyte growth factor (HGF) (reviewed in [10]. The angiopoietins act specifically on ECs through the EC receptor Tie-2, whereas other growth factors (e.g., PDGF, EGF, or bFGF) also bind to receptors on other types of cells. Angiogenesis is not regulated by a single angiogenic molecule but rather depends on the cooperation and integration of various factors that act in a coordinated fashion (reviewed in [11]). However, the most tightly regulated angiogenic factor is VEGF. Because VEGF is of universal importance in angiogenesis, this chapter focuses on the mechanisms that regulate VEGF expression in tumor cells and discuss the potential role of these mechanisms for molecular targeted therapy.

2. THE VEGF FAMILY AND THEIR RECEPTORS

The VEGF family currently comprises 6 secreted glycoproteins designated VEGF-A, VEGF-B, VEGF-C, VEGF-D, VEGF-E, and placental growth factor (PlGF) (Figure 1) [12]. The best characterized of the VEGF family members is VEGF-A (referred to as VEGF in this chapter), a 34- to 50-kDa homodimeric glycoprotein that is expressed in various isoforms secondary to alternative exon splicing that leads to mature 121-, 165-, 189-, and 206-amino

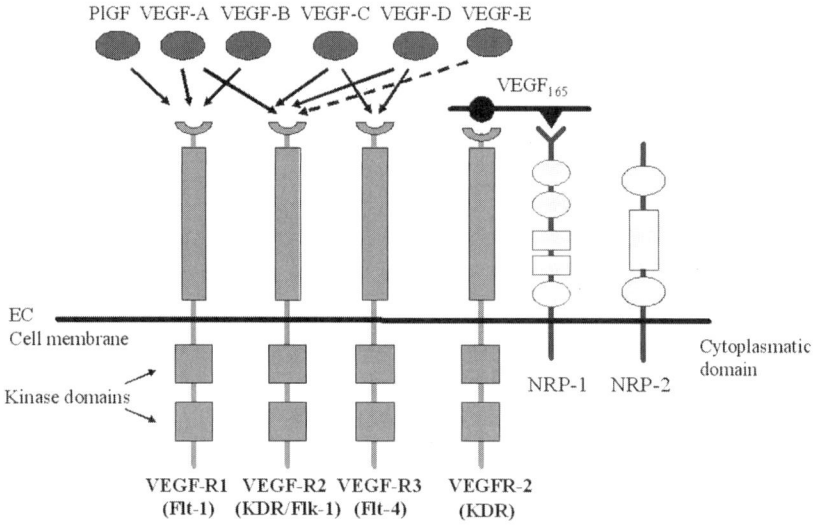

Figure 1. VEGF family and receptors. Overview of VEGFs and their principal receptors. Importantly, VEGFR-2 function can be enhanced by a simultaneous binding of a VEGF165 molecule to NRP-1 and VEGFR-2 via different binding sites (paracrine effect). The NRP-1 receptor may also be on another cell (i.e. tumor cell) presenting bound VEGF165 to an EC VEGFR-2 (juxtacrine effect). The importance of the NRP-2 receptor as co-receptor for VEGFR has not been well defined yet.

acid proteins [13]. $VEGF_{121}$ and $VEGF_{165}$ are secreted by cells, whereas the larger isoforms ($VEGF_{189}$ and $VEGF_{206}$) are associated with the cell surface and sequestered in the extracellular matrix (ECM) [13]. $VEGF_{165}$ also has intermediary properties, as a significant fraction remains bound to the cell surface and ECM [14]. $VEGF_{165}$ is a 45-kDa glycoprotein with a heparin-binding domain [15]. $VEGF_{121}$ is an acidic polypeptide that does not bind heparin, whereas $VEGF_{189}$ and $VEGF_{206}$ are highly basic and bind to heparin with high affinity. Loss of the heparin-binding domain on VEGF has been shown to result in a significant reduction of mitogenic activity [16]. The ECM-bound isoforms can be released in a diffusible form by plasmin cleavage at the C-terminus, which generates a bioactive fragment. Some less commonly expressed splice variants ($VEGF_{145}$, $VEGF_{183}$) have also been reported [17]. $VEGF_{165}$ is the predominant isoform and is commonly overexpressed in a variety of human solid tumors. Mouse embryos lacking a single VEGF allele demonstrate growth inhibition and impaired angiogenesis with degeneration of ECs, and they die between day 11 and day 12 of gestation [18]. Recent studies suggest that expression patterns of certain VEGF isoforms are tissue-specific, implying that these isoforms have specific roles in development and probably in tumor angiogenesis as well [19].

The VEGFs mediate their function via high-affinity binding to receptor tyrosine kinases that are predominantly expressed on ECs. However, recent evidence suggests that some of these receptors may also be expressed on tumor cells [20], bone-marrow–derived ECs [21], or lymphatic ECs [22]. The current nomenclature lists three VEGF receptors (VEGFRs): VEGFR-1 (also known as Flt-1), VEGFR-2 (Flk-1 or KDR; Flk-1 is the murine homolog of KDR), and VEGFR-3 (Flt-4). Two co-receptors for VEGFR-2, neuropilin-1 and neuropilin-2, have also been discovered. These neuropilin receptors are expressed on numerous cell types, including cancer cells, and have been implicated in the regulation of the binding affinity of $VEGF_{165}$ to VEGFR-2; as such, they contribute to augmenting VEGFR function without transmitting intracellular signals by themselves [23].

The ligand specificities of the above-mentioned VEGFRs differ and are illustrated in Figure 1. Upon activation, VEGFRs mediate specific functions in vitro and in vivo. In adults, VEGFR-1 and VEGFR-2 are expressed mainly in the vascular endothelium, whereas VEGFR-3 is restricted largely to the lymphatic endothelium. VEGFR-1 seems to be important for EC migration and differentiation, whereas VEGFR-2 activation enhances EC survival and proliferation and increases vascular permeability [6]. VEGFR-3 has been shown to be involved in lymphangiogenesis and lymphatic metastasis [24].

In the following sections, we describe regulators of VEGF expression in tumor cells, starting with regulators that act at the cell surface followed by signaling intermediates, and ending with the ultimate effect of these upstream signals on gene transcription.

3. REGULATORS OF VEGF EXPRESSION

3.1. Growth Factors and Their Receptors

For nearly 2 decades, growth factors and their receptors have been known to directly contribute to tumor growth, but only recently have they been found to indirectly affect tumor growth by upregulating the expression of angiogenic factors in tumor cells. Kerbel and associates demonstrated that antibodies to the ErbB (HER) receptor family members could down-regulate VEGF expression in tumor cells [25]. These studies formed the basis for other investigations of the role of other tyrosine kinase receptors in regulating VEGF, as discussed below.

3.1.1. Epidermal Growth Factor Receptor Family and Their Ligands

Among the best-studied growth factor receptor systems is that of the EGF receptor family (also known as type I receptor tyrosine kinases or ErbB receptor tyrosine kinases), which comprises four homologous receptors: the epidermal growth factor receptor (EGFR/also called HER1 or ErbB1), ErbB2 (HER2/neu), ErbB3 (HER3), and ErbB4 (HER4) (reviewed in [26]). Details regarding the epidermal growth factor family and signaling are provided by Barnes and Kumar (Chapter 1) in this book.

The importance of the EGFR (ErbB1) system in VEGF regulation and angiogenesis has been validated in several tumor systems, including colon carcinoma, pancreatic cancer, gastric cancer, glioblastoma multiforme, non-small cell lung cancer, and renal cell carcinoma (reviewed in [27]). In recent experimental models, inhibition of EGFR function by receptor-tyrosine kinase (RTK) inhibitors or selective EGFR antibodies led to a significant reduction of VEGF expression, angiogenesis, and tumor growth in mice. These effects are now being investigated in clinical trials for the treatment of solid malignancies; in these trials, VEGF plasma levels are being used as a surrogate marker of activity. We do not condone this approach. As we discuss throughout this chapter, many factors regulate VEGF, and VEGF may also serve as a marker of tumor burden; thus plasma VEGF is unlikely to be a valid surrogate marker of activity of anti-EGFR agents.

The constitutively active receptor tyrosine kinase HER2/neu (ErbB2) is also known to induce VEGF and angiogenesis in tumors [28]. Studies of NIH 3T3 fibroblasts that had been transformed with mutant HER2/neu demonstrated significant induction of VEGF expression, and the magnitude of this effect was further elevated by exposure to hypoxia [25]. Laughner and colleagues recently showed that HER2/neu overexpression in mouse 3T3 cells led to increased activity of hypoxia-inducible factor 1α (HIF-1α) (discussed further in section 3.5.1) and subsequently to increased VEGF expression in these cells [29]. Their finding that heregulin stimulation led to HER2/neu transactivation in MCF-7 breast cancer cells suggested that similar mechanisms leading to up-regulation of VEGF were also present in cancer cells [29]. As the HER2/neu receptor system has been implicated in the progression and angiogenesis of human breast cancer (and possibly lung or colon cancer as well), the development of neutralizing antibodies to HER2/neu has become a major focus in molecular cancer research. Most studies of VEGF regulation by HER2/neu, and its subsequent suppression by administration of various antibodies to HER-2/neu, have been done with breast cancer models. Petit et al. showed that treatment of the HER2/neu–positive human breast cancer cell line SKBR-3 with a specific neutralizing anti-HER2/neu monoclonal antibody (4D5) resulted in a dose-dependent reduction of VEGF protein expression in those cells [25]. HER2/neu also regulates the expression of VEGF family members other than VEGF-A in human breast cancer; specifically, Yang et al. reported that HER2/neu expression correlated with VEGF-C and VEGF-D expression in human breast cancer specimens, as measured by immunohistochemical analysis [30]. HER2/neu may therefore contribute to the development of pro-angiogenic and pro-lymphangiogenic changes in the tumor microenvironment in a variety of types of cancer in humans.

Upon being activated, various ErbB family receptors may heterodimerize, which leads to significant increases in VEGF levels in cancer cells. For example, activation of the ErbB4 receptor system by its ligand heregulin-β1 has been shown to be an important mediator of VEGF expression in breast and lung cancer cells [31]. In these experiments, Yen et al. used a panel of human breast and lung cancer cell lines that constitutively overexpressed ErbB2 or were engineered to stably overexpress the ErbB2 receptor and demonstrated that heregulin induced VEGF secretion in most cancer cell lines but not in normal human mammary or bronchial primary cells [31]. This observation supports the concept of a functional interplay between HER2 and other ErbB receptor systems such as ErbB3 and ErbB4 in cancer cells (reviewed in [32]). Also, overexpression of the HER2/neu (erbB2) receptor led to an increase in the basal level of VEGF secretion in these cells, and subsequent exposure to heregulin led to further induction of VEGF secretion [31]. These findings indicate that the HER2 system is an important target for direct and indirect therapy for cancer, and clinical trials of this approach of inhibiting HER2 are currently ongoing.

3.1.2. Insulin-like Growth Factor-I Receptor System

The insulin-like growth factor-I receptor (IGF-IR) is often overexpressed in a variety of human cancers, and its overexpression has been associated with aggressive disease and formation of metastases. Binding between IGF-IR and its major ligands, IGF-I and IGF-II, initiates intracellular signaling, mainly via the mitogen-activated protein kinase (MAPK, also known as Erk1/2) and phosphatidylinositol 3'-kinase (PI-3K)/Akt pathways and downstream activation of transcription factors such as HIF-1α and nuclear factor-kappaB (NF-κB) (Figure 2).

The main source of IGF-I is the liver, a fact of tremendous importance for the development and growth of liver metastases derived from gastrointestinal (e.g. colon or pancreas) cancers that overexpress IGF-IR [33, 34]. Recent experimental model systems have clearly demonstrated the importance of activation of the IGF-IR system in mediating angiogenesis by up-regulating VEGF expression in various neoplastic [34, 35] and non-neoplastic tissues [36]. Smith and colleagues demonstrated that interaction of IGF-I with IGF-IR increased retinal neovascularization by up-regulating VEGF expression in retinal endothelial cells through activation of the MAPK signaling pathway [36]. In this model, IGF-IR activation was necessary for VEGF to have the maximum effect in ocular angiogenesis [36]. In neoplastic systems, IGF-I–mediated induction of VEGF expression in cancer cells in vitro has been reported for breast cancer cells [37], endometrial adenocarcinoma cells [38], and colorectal cancer cells [34, 35]. However, only a few studies have addressed the role of IGF-I and IGF-IR in tumor angiogenesis (i.e., regulation of VEGF expression) in primary tumors and metastases in vivo. Reinmuth and colleagues from our laboratory showed that inhibition of IGF-IR function (by transfection with a dominant-negative receptor construct) reduced VEGF expression in vitro in two human colon carcinoma cell lines (HT29, KM12L4), which resulted in a significant reduction of angiogenesis and hepatic colorectal tumor growth in two different in vivo models [34]. Tumors from the dominant-negative IGF-IR–transfected cell lines expressed less VEGF than did tumors from control cells [34]. The anti-angiogenic and growth-inhibitory effects of inhibiting IGF-IR function seem to be even more pronounced in an orthotopic model of pancreatic cancer [33]. Because IGF-IR is highly homologous to the insulin receptor, targeting it has proven difficult; however, several agents are being tested in preclinical trials, and phase I studies are expected to be initiated in 2004.

3.1.3. Platelet-derived Growth Factors

The family of PDGFs consists of four different isoforms, PDGF-A, PDGF-B, PDGF-C, and PDGF-D. PDGF-A and -B can form homodimers and heterodimers by disulfide bonding; the recently discovered PDGF-C and PDGF-D [39] seem to be expressed only as homodimers. PDGF-A and -B bind to homo- or heterodimers of PDGF-receptor (PDGFR) -α or PDGFR-β proteins on the cell surface [40].

PDGFs promote angiogenesis in vivo by regulating EC survival and pericyte (vascular smooth muscle cell) recruitment [41, 42]. PDGFs can also regulate angiogenesis indirectly by inducing VEGF in some types of cells. Stimulation of pericytes with recombinant PDGF-BB may increase VEGF expression by up-regulating its transcription, an effect mediated by the PI-3K/Akt signaling pathway [41]. This paracrine mechanism of VEGF secretion has important implications for the survival of ECs in vivo. Overexpression of PDGF-B and PDGFR-β in human gliomas has been shown to be responsible for recruiting pericytes to vessels [42]. Overexpression of PDGF-B in glioma cells seems to enhance the formation of intracranial gliomas by stimulating VEGF expression in neovessels and by attracting vessel-associated pericytes [42]. However, direct stimulation of glioma cells with PDGF-B in those studies did not increase VEGF expression in vitro, suggesting that the observed effects were due to paracrine mediation of VEGF secretion by pericytes, as noted above.

In cells derived from neurofibromas, unlike glioma cells, PDGF-BB can directly induce VEGF expression, an effect mediated by the MAPK signaling pathway; this mechanism seems to underlie the molecular pathology of highly vascularized neurofibromas [43]. The newly described PDGF-CC isoform has also been associated with increased neovascularization in a model of corneal angiogenesis, but concrete evidence of subsequent up-regulation of endothelial (or corneal) VEGF expression upon treatment with PDGF-CC is lacking [39]. The PDGFR system has become a promising target for anti-angiogenic regimens, as shown by Shaheen and associates in 2001 [44] and Bergers et al. in 2003 [45]. These investigators demonstrated that adding inhibition of PDGFR activity to inhibitors of VEGFR activity enhanced the anti-angiogenic effect in vivo. Clinical trials of dual-kinase inhibitors to both VEGFR and PDGFR are ongoing.

3.2. Interleukin-1

Certain cytokines have been implicated in angiogenesis in malignant disorders through their contribution to raising VEGF levels in the tumor microenvironment. One such cytokine, IL-1β, has been shown to induce VEGF expression by means of its receptor, IL-1R, in human colon cancer cells [46]. This up-regulation of VEGF upon stimulation with IL-1β occurred through increased promoter activity and transcriptional regulation of the VEGF gene [46]. Yano and colleagues recently demonstrated that overexpressing IL-1β by means of stable transfection led to up-regulation of various cytokines, including VEGF, in human lung cancer cells [47]. In this experimental model of lung cancer, IL-1β-overexpressing cells formed more lung metastases and were associated with a higher degree of vascularization. These effects were abrogated by treatment with an antibody to IL-1β [47]. The IL-1/IL-1R system may contribute to VEGF expression in a variety of cell systems, and inhibition of IL-1β–induced angiogenesis by the administration of IL-1–binding antibodies is currently being investigated in other experimental models. In fact, inhibition of IL-1/IL-1R is effective in the treatment of arthritis, a disease known to depend on angiogenesis

3.3. Interleukin-6

Another cytokine that elicits angiogenic properties in vivo by modulating VEGF expression is IL-6. This function is mediated by interaction of IL-6 with the alpha chain of the IL-6 receptor (IL-6R) and subsequent tyrosine kinase activation. Interestingly, the IL-6R system also includes a signal transducer protein (gp130) that can be activated by other kinases such as Janus kinase (JAK). Sharing of a receptor subunit is a general feature of cytokine receptors and provides the molecular basis for the functional redundancy of cytokines (reviewed in [48]). In cervical carcinomas, IL-6 has been shown to promote tumor growth in vivo [49]. Wei and associates showed that this effect was caused by increased VEGF expression and increased angiogenesis in C33A cervical tumor cells [50]; using gene promoters and dominant-negative transfection, they also identified the STAT3 signaling pathway as being the exclusive pathway for mediating IL-6-induced up-regulation of VEGF in those cells. IL-6 has also been implicated in the pathogenesis and neovascularization of gastric [51] and pancreatic carcinomas [52] by up-regulating VEGF. Masui and colleagues used immunohistochemical staining to demonstrate the frequent expression of the IL-6R in human pancreatic cancer specimens and showed that stimulating the pancreatic cancer cell line CFPAC-1 with IL-6 led to up-regulation of VEGF expression in those cells [52]. However, the significance of the IL-6/IL-6R system in the regulation of angiogenic molecules such as VEGF in vivo remains to be elucidated.

3.4. Hepatocyte Growth Factor/c-MET

HGF, also known as scatter factor, has recently been shown to elicit pro-angiogenic activity by up-regulating VEGF expression in a variety of cell types and cancer systems [53]. HGF is the primary ligand for the c-Met protooncogene, a transmembrane receptor-tyrosine kinase that is primarily expressed in epithelial tissues. Numerous studies have implicated aberrant c-Met function in the progression and metastasis of human tumors such as colorectal cancer, pancreatic carcinoma, melanoma, and osteosarcoma (reviewed in [54]). However, the role of HGF as a paracrine factor in regulating the expression of angiogenesis factors by tumor cells has only recently been investigated [55]. A direct effect of HGF/c-Met signaling on VEGF expression in cancer cells has been demonstrated by several groups. In glioma cells (which are often positive for c-Met), treatment with HGF resulted in enhanced secretion of VEGF protein accompanied by increased transcription of VEGF mRNA in a dose-dependent fashion [56]. Using another tumor system, Dong and colleagues investigated the association of HGF/c-Met and VEGF regulation in serum samples from patients with head and neck squamous cell carcinoma. Their results showed that increases in serum HGF levels correlated with higher serum VEGF levels in these patients [55]. These authors explored the association between HGF/c-Met and VEGF expression further by treating various head and neck squamous cell carcinoma cell lines (which often overexpress c-Met) with recombinant HGF in vitro. These experiments led to identification of the MAPK (Erk1/2) and PI-3K/Akt signaling pathways as being critical mediators of VEGF up-regulation in these cells. The importance of the MAPK and PI-3K pathways for mediating HGF/c-Met induced VEGF expression has also been described in ECs [57]. However, to date it has been difficult to effectively impair c-Met function by using inhibitors specific for this receptor-tyrosine kinase. Therefore inhibition of the above-mentioned signaling intermediates (also discussed in section 3.5.1) might be a more promising approach for interrupting HGF-mediated angiogenesis.

3.5. Intracellular Signaling Intermediates

3.5.1. The Mitogen-Activated Protein Kinase and Phosphatidylinositol 3′ Kinase Signaling Pathways

Many of the growth factor receptors described here initiate intracellular signaling cascades that are commonly shared by those receptor systems and also by intracellular non-receptor kinases (e.g. Src) and other mediators of cellular response to stresses induced by hypoxia, acidity, cell density, and irradiation [58]. With regard to VEGF regulation in cancer cells, two signaling pathways deserve particular attention—the MAPK and PI-3K pathways, which are most often involved in the regulation and transcriptional activation of VEGF target gene expression in a variety of cancer entities (Figure 2).

MAPKs are conserved proteins that regulate cell growth, division, and death. Although activated in the cytosol, the MAPKs move into the nucleus upon activation, where they phosphorylate a large number of nuclear proteins. In studies investigating how the Ras oncogene transmits extracellular growth signals (described further below), the MAPK pathway has emerged as the crucial route between membrane-bound Ras and the nucleus. The MAPK pathway represents a cascade of phosphorylation events that includes three pivotal kinases, namely Raf, MEK, and ERK (reviewed in [59]).

The PI-3K family of enzymes is recruited upon activation of growth factor receptors and results in the production of 3'-phosphoinositide lipids. The lipid products of PI-3K act as second messengers by binding to and activating diverse cellular target proteins. This signal transduction pathway is also responsible for hypoxia-induced expression of VEGF in malignant and non-malignant cells (see below). These events constitute the start of a complex signaling cascade that ultimately results in the mediation of cellular activities such as proliferation, differentiation, chemotaxis, survival, trafficking, and glucose homeostasis (reviewed in [60]). Therefore, PI-3K has a central role in many cellular functions. Importantly, its function is under the control of the protein encoded by the tumor suppressor gene *PTEN*, which acts as a naturally occurring antagonist to PI-3K/Akt signaling [61] (Figure 2).

As noted above, up-regulation of VEGF expression in cancer cells after activation of growth factor receptors such as IGF-IR or EGFR often involves both the MAPK and PI-3K/Akt signaling pathways [62]. Akagi and colleagues recently demonstrated that inhibition of MAPK or PI-3K signaling partially inhibited EGF induced VEGF expression in gastric cancer cells in vitro. However, inhibition of the activity of another MAPK signaling intermediate, P38, almost completely abolished EGF–induced VEGF expression in these cells [62]. In contrast, in glioblastoma cells, the EGF–receptor mediated up-regulation of VEGF depended on PI-3K alone [63]. In other cancer systems, however, P38 may exclusively regulate VEGF expression [64]. Specifically, Xiong et al. showed that heregulin-1 (as a ligand for ErbB3 and ErbB4) induced VEGF expression in human breast cancer cells selectively via P38 activation [64]; inhibition of neither the MAPK nor the PI-3K pathways affected VEGF expression in this model [64]. These studies clearly suggest that there is no common rule for growth factor–induced activation of signaling intermediates such as Erk or Akt that mediate VEGF up-regulation in cancer cells. Rather, activation of signaling pathways via RTKs involved in VEGF induction depends on the cancer system being investigated and may even vary among cell lines of the same cancer entity [65].

3.5.1.1. Microenvironmental Regulation of Signaling Intermediates

Other stimuli such as hypoxia, acidic conditions, and irradiation consistently utilize specific pathways for VEGF up-regulation. A variety of studies have shown that hypoxia leads to selective activation of PI-3K/Akt and ultimately HIF-1α, one of the strongest transcriptional activators of VEGF [66]. The influence of stress caused either by serum starvation or high cell density on VEGF expression has been investigated in colon cancer cells, and the MAPK pathway has been identified as an important mediator of VEGF [58]. The same pathway, which also involves Ras, was also shown to modulate VEGF expression by means of increasing the activity of the transcriptional regulator activator protein-1 (AP-1) in human glioblastoma cells upon exposure to low-pH culture medium [67]. Goerges et al. investigated the effects of acidic pH on VEGF induction in ECs and found that this induction was mediated via activation of MAPK signaling [68]. Ionizing radiation can also cause up-regulation of VEGF via the MAPK pathway in various types of cells [69, 70]. In both astrocytoma and glioblastoma cells, MAPK activation induced by irradiation led to further downstream activation of the transcriptional regulator AP-1, which led to increases in VEGF [69, 70]. In addition to MAPKs' extrinsic activation mechanisms, activating Ras mutations and oncogenes may result in constitutive activated (phosphorylated) Erk1/2 signaling intermediates that subsequently could contribute to increased VEGF levels in tumors. Whether the signaling pathways that mediate VEGF induction (MAPK, PI-3K, or P38) could be used as molecular targets for down-regulating VEGF and angiogenesis in tumors remains uncertain at this time, as these pathways are also critically involved in multiple physiological and homeostatic processes.

3.5.2.   *Prostaglandins and Cyclooxygenase-2*

Prostaglandins (PGs) play critical roles in numerous biological processes, including the regulation of immune function, kidney development, reproductive biology, and gastrointestinal integrity. Certain PGs have recently been implicated in tumor angiogenesis through up-regulation of VEGF expression (reviewed in [71]). Prostaglandin-endoperoxide synthase (also known as cyclooxygenase or COX) is the rate-limiting enzyme involved in the oxidative transformation of arachidonic acid into PG H2, which represents the precursor of several bioactive molecules, including PGE(2), prostacyclin, and thromboxane [72]. Two COX isoforms have been well characterized to date, and they differ mainly in their pattern of expression. The role of a recently discovered third COX isoform in angiogenesis is undefined [73]. COX-1 is constitutively expressed in most tissues; COX-2 is usually absent but can be induced by numerous stimuli, including growth factors or hypoxia [74]. With respect to angiogenesis in cancer, COX-2 seems to be the most relevant enzyme in terms of its regulating PGs and potentially VEGF. Of the prostaglandins PGE(2) has been most often reported to be involved in the regulation of VEGF expression and angiogenesis [75].

Over the past decade, numerous studies have confirmed an association between COX-2 overexpression and tumor progression and increased angiogenesis (VEGF expression) in solid malignancies, including gastric cancer, colon cancer, prostate cancer, breast cancer, and pancreatic cancer (reviewed in [76]). Evidence of a direct link between PGs and increased VEGF expression comes from recent studies in which PGE(2) induced VEGF expression in cells via activation of HIF-1α (also discussed below) [77]. Fukuda et al. demonstrated that PGE(2) led to activation of MAPK and Akt in HCT116 human colon carcinoma cells, which subsequently resulted in increased VEGF expression by these cells. Interestingly, blockade of either c-src function (see below) or Erk phosphorylation inhibits VEGF mRNA and HIF-1α protein expression in response to PGE(2) [77]. Liu and associates also showed that this increase in VEGF expression through the addition of PGE(2) (also mediated through HIF-1α) could be inhibited in human prostate cancer cells by various COX-2 blocking agents [78], suggesting that COX-2 (which itself may be up-regulated by hypoxia) may directly regulate HIF-1α activity and subsequently VEGF expression in cancer cells. Our own work suggests that EGF–induced or IGF-I–induced expression of COX-2 and ultimately VEGF expression in cancer cells does not take place along a linear signaling pathway, nor did inhibition of COX-2 decrease VEGF expression in these cells. These findings may well vary among cell types. Nevertheless, clinical trials of COX-2 inhibitors are ongoing, and decreases in VEGF expression are being investigated as a secondary endpoint.

3.5.3. Signal Transducer and Activation of Transcription-3 Signaling Pathway

Non-receptor and receptor-tyrosine kinases such as Src and the EGF receptor are major inducers of VEGF expression in a variety of cells lines and tumor systems. However, tyrosine kinases signal through multiple pathways and the signaling intermediate "signal transducer and activation of transcription-3" (STAT3) is a point of convergence for many of these kinases. STAT3 is also often constitutively activated in a wide range of cancer cells [79, 80]. Niu and colleagues elegantly demonstrated that constitutive STAT3 activation can lead to overexpression of VEGF in cancer cells [79]. In those studies, transfection with an activated STAT3 mutant (STAT3C) up-regulated VEGF expression in cancer cells, which resulted in increased tumor angiogenesis. Moreover, STAT3C-induced VEGF up-regulation was abrogated when a STAT3-binding site in the VEGF promoter was mutated, and interruption of STAT3 signaling by a dominant-negative STAT3 protein or STAT3 antisense oligonucleotide subsequently down-regulated VEGF expression in tumor cells again [79]. The finding in that study showing that v-Src-mediated VEGF induction was inhibited when STAT3 signaling was blocked illustrates the role of STAT3 as point of signaling convergence [79]. STAT3 may well contribute to a pro-angiogenic microenvironment in various tumor systems. However, at this time STAT3 does not seem to be a major regulator of VEGF.

3.5.4. Oncogenes

3.5.4.1. Src

Many oncogenes have been implicated in the process of angiogenesis in solid tumors, in part because of their ability to induce pro-angiogenic growth factors such as VEGF (reviewed in [81]). This observed induction capability is further supported by the fact that various signaling inhibitors have been shown to reverse such pro-angiogenic effects, both in vitro and in vivo. The c-src protooncogene encodes a protein tyrosine kinase, pp60c-src, that has been identified as an important mediator in many signal transduction pathways. One of the pathways involving activation of the pp60c-src protein tyrosine kinase is implicated in the regulation of VEGF expression in such a way as to promote neovascularization of growing tumors. Other recent evidence suggests that several factors that regulate VEGF expression may depend in part on c-src–mediated signal transduction pathways. The tyrosine kinase activity of src is often activated in colon tumors and various colon cancer cell lines. In experimental models, inhibiting src activity by means of stable transfection with an antisense expression vector decreased VEGF expression levels in proportion to the decrease in observed src kinase activity [82]. Moreover, northern blot analyses revealed that cells with decreased src activity showed only a modest increase in VEGF expression under hypoxic conditions, but parental cells exhibited a > 50-fold increase [82]. In a subsequent in vivo experiment with nude mice, subcutaneous tumors from antisense transfectants were found to be much less vascular, suggesting that src activity regulates the expression of VEGF in colon tumor cells. Previous studies have also shown that c-src is necessary for hypoxia- and HIF-1–mediated VEGF expression in cells [83]. Thus targeting src function may be a valuable approach for inhibiting tumor angiogenesis in a variety of cancer systems. Specific inhibitors are now being tested in preclinical models.

3.5.4.2. BCR-ABL

The BCR-ABL oncogene has been identified as having a key role in the molecular pathogenesis of leukemias, which are considered angiogenesis-dependent malignancies. Recent studies in which STI-571 (Gleevec, imatinib mesylate) was used to target BCR-ABL showed that VEGF levels in BCR-ABL-positive K562 cells were reduced, in a dose-dependent fashion, by treatment with STI-571 [84]. In that same study, transfection of BCR-ABL into murine myeloid 32D cells or human megakaryocyte MO7e hematopoietic cells resulted in enhanced VEGF expression, suggesting that BCR-ABL could mediate some of its angiogenesis-dependent effects in chronic myelogenous leukemia through up-regulation of VEGF expression [84]. Similar findings on BCR-ABL–regulation of VEGF expression were reported by Janowska-Wieczorek and colleagues, who showed a four-fold increase in VEGF expression after transfecting the murine leukemia cell line FL5.12 with BCL-ABL [85]. Studies of STI-571 as an inhibitor of BCR-ABL are ongoing, but it may be difficult to sort out its anti-angiogenic effects from its direct tumoricidal effects.

3.5.4.3. Ras

Dominantly acting transforming oncogenes are generally considered to contribute to tumor development and progression by their direct effects on tumor cell proliferation and differentiation. Such oncogenes may have a much greater influence than initially realized given their indirect effects on tumor growth through induction of angiogenesis (Figure 2). An example is the induction of VEGF expression by mutant H- or K-ras oncogenes in various types of cells such as pancreatic cancer, colon cancer, and non-small cell lung cancer [86]. The expression of mutant Ras oncogenes is one of the most commonly encountered genetic changes detected in human cancer. The observation that VEGF expression levels are higher in tumors than in non-malignant tissues may be mediated in part through the Ras-Raf-MAPK signal transduction pathway, which results in activation of transcription factors such as AP-1 that leads in turn to increased transcription of the *VEGF* gene [67] (Figure 2). The presence of Ras activation in transformed epithelial cells and its association with VEGF induction was first observed by Rak, Kerbel, and others in 1995 [87]. These investigators also showed that genetic disruption of the mutant K-ras allele in human colon carcinoma cells was associated with a reduction in VEGF activity [87]. Like other signaling intermediates, the

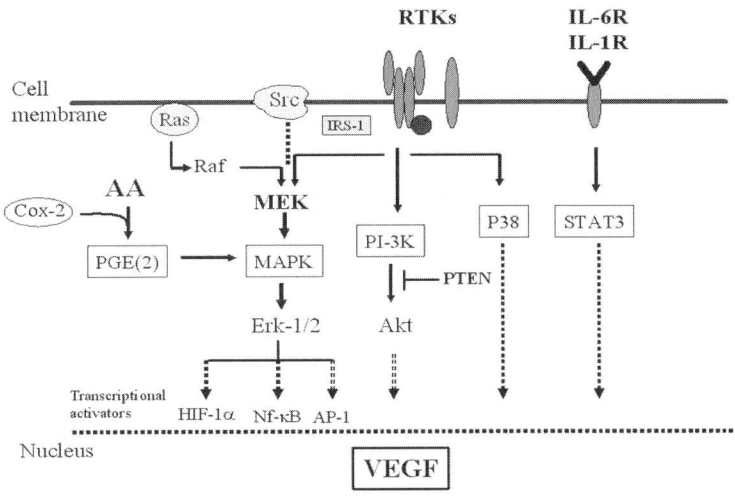

Figure 2. Activation of intracellular signaling intermediates leading to upregulation of VEGF in cancer cells. Common pathways involved in the regulation of VEGF expression in cancer cells are shown. Activation of certain growth factor receptor tyrosine kinases (RTK) or oncogenic activation (Src) may lead to phosphorylation of signaling intermediates such as MAPK, PI-3K, STAT3 and P38, respectively. In addition, Ras function and its constitutive activation has been shown to be an important mediator of a Ras-Raf-MEK signaling cascade implicated in constitutive MAPK activation and consecutive VEGF expression. MAPK mainly mediates "its" proangiogenic properties through transcription factors such as HIF-1α, Nf-κB and AP-1, which bind to certain region within the *VEGF* promoter.

activation of Ras is also part of a signaling cascade initiated by certain growth factor receptors such as EGFR [63, 88]. For example, Maity et al. found that inhibition of either EGFR or Ras reduced VEGF expression in glioblastoma cells [63]; in another study, Feldkamp et al. showed that constitutive phosphorylation of EGF-R in astrocytoma cells led to increased Ras activation and subsequent VEGF secretion [88]. This phosphorylation mechanism also seems to be important in the molecular pathogenesis of glioblastoma multiforme, which does not harbor oncogenic Ras mutations but shows functional up-regulation of Ras signaling through activation of receptor-tyrosine kinases that are overexpressed in these tumors (reviewed in [89]). Breier and colleagues underscored the importance of Ras-mediated tumor growth and angiogenesis by demonstrating that Ras-transformed mouse mammary ECs, unlike non-Ras-transformed cells, were able to promote vascularization in tumors that overexpressed transforming growth factor-β [90]. Although ras targeting with farnesyl transferase inhibitors has not been as effective as anticipated, it is possible that this approach could directly and indirectly (through inhibition of VEGF induction) inhibit tumor growth.

3.5.5. *Tumor Suppressor Genes*

3.5.5.1. *p53*

One of the most intensively studied tumor suppressor genes implicated in the molecular pathology of many solid malignancies is *p53*. Several studies have implied that p53 has a central role in the regulation of VEGF in malignant tumors. Pal et al. showed that the direct interaction of the p53 protein with the transcription factor Sp1, which is often activated upon hypoxic stimulation of cancer cells, prevented transcriptional activation of the VEGF promoter in breast cancer cells [91]. These authors also found that p53 could inhibit the hypoxic induction of Src kinase and subsequently the induction of VEGF in those cells [91]. In another study of the mechanisms by which p53 participates in angiogenesis, Ravi and colleagues showed that homozygous deletion of *p53* promoted neovascularization and growth of tumor xenografts in nude mice [92] and that the presence of wild-type *p53* promoted Mdm2-mediated ubiquitination and proteasomal degradation of HIF-1α [92]. Loss of p53 in tumor cells therefore acted to raise HIF-1α levels and consequently to augment HIF-1–dependent transcriptional activation of *VEGF* in response to hypoxia. In that same study, forced expression of HIF-1α in p53–overexpressing tumor cells increased hypoxia-induced VEGF expression and augmented the neovascularization and growth of tumor xenografts [92]. The conclusion that p53 plays a central role in VEGF regulation and angiogenesis is also supported by results of other studies in which stable transfection of endometrial carcinoma cells with wild-type p53 resulted in decreased VEGF expression [93].

3.6. *Transcription Factors Regulating VEGF mRNA Expression*

3.6.1. *Hypoxia-inducible Factor-1α*

One of the strongest natural inducers of VEGF expression in both normal and malignant tissues is hypoxia. This induction is physiologically mediated through the activity of the transcription factor HIF-1α. Under non-hypoxic conditions, HIF-1α is rapidly ubiquitinated and degraded, a process that requires a functional protein encoded by the von Hippel-Lindau (VHL) tumor suppressor gene [94]. However, upon exposure to low oxygen-tension levels,

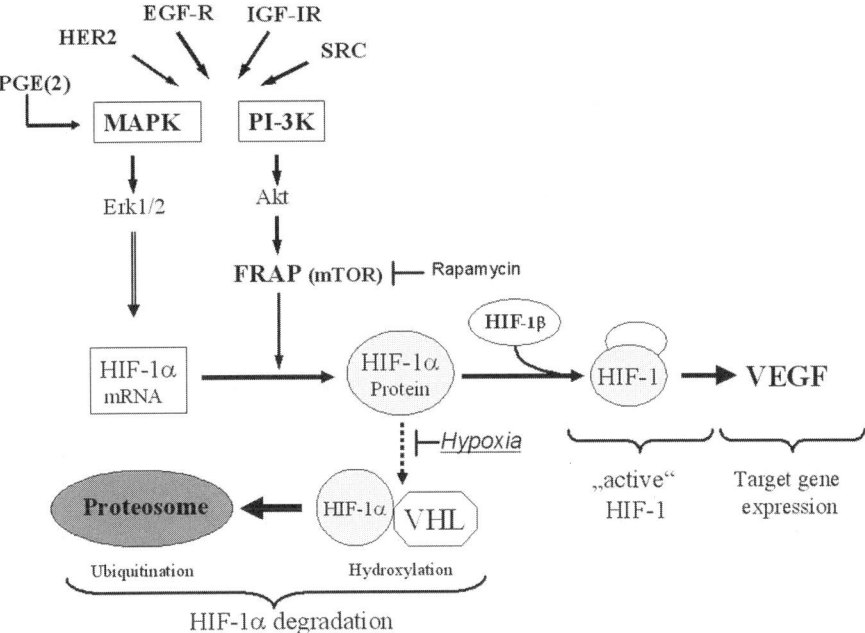

Figure 3. The HIF-1 transcription factor system in the regulation of VEGF expression. Schematic diagram of HIF-1 regulated *VEGF* expression. This system may be either activated by hypoxia and/or growth factor receptor signaling which in turn leads to increased HIF-1α expression. Under normoxic conditions, HIF-1α is rapidly degraded and ubiquitinated by the proteosome, a process that involved the VHL protein. However, under hypoxic conditions HIF-1α heterodimerizes with the constitutively expressed HIF-1β in order to form a functional HIF-1 complex, acting as a transcriptional activator. In addition to the regulation of cytoplasmatic levels of HIF-1α protein by the VHL tumor suppressor, HIF-1α mRNA transcription is critically regulated by the mammalian target of rapamycin (mTOR, FRAP).

a very common condition in tumors [95], HIF-1α forms heterodimers with the constitutively expressed HIF-1β subunit [96]. This heterodimerization is necessary for the activation of HIF-1 (the dimer of HIF-1α and HIF-1β) and its subsequent translocation into the nucleus, where it binds to the consensus sequence "hypoxia-response element" (HRE) and initiates transcriptional activity of target genes such as VEGF, erythropoietin, transferrin, endothelin-1, inducible nitric oxide synthase, and IGF-II [96] (Figure 3). HIF-1α also has a pivotal role in cancer angiogenesis and metastasis and is often overexpressed in a variety of human cancers and their metastases [97]. Constitutive HIF-1α activation in human cancer has been associated with high VEGF expression levels and development of a pro-angiogenic tumor microenvironment. However, this constitutive activation of HIF-1α can occur by either two ways—through inhibition of its degradation or through an increase in its expression, or both. In addition to hypoxic regulation of HIF-1 activation, some hypoxia-independent mechanisms have also been described. Several studies have shown that genetic alterations of tumor suppressor genes (e.g., *p53, VHL, PTEN*) and oncogenes (e.g., Src, HER2/neu, H-Ras) can induce HIF-1 activity in tumor tissues, through either its overexpression or inhibition of its degradation. Further, activation of certain growth factor receptors such as IGF-IR has been shown to up-regulate HIF-1α expression, and subsequently VEGF, in human colon carcinoma cells [98]. Studies in our laboratory validated the existence of this IGF-I/IGF-IR/HIF-1α signaling pathway in pancreatic cancer cells, where HIF-1α seems to be in part constitutively activated by an IGF-IR autocrine loop [33].Findings from these and other studies suggest that HIF-1α could be an effective molecular target for the treatment of solid malignancies. However, to date the effects of inhibiting HIF-1α on tumor growth and angiogenesis in vivo have been assessed only in mouse xenograft assays, either with genetically engineered embryonic stem cells that lack HIF-1α (HIF-1α$^{-/-}$) or with mutated constructs of other proteins that heterodimerize with HIF-1 under hypoxic conditions (e.g., p300) [99, 100]. Results from studies using these experimental models of HIF-1 inhibition have been inconsistent with regard to VEGF regulation. Specifically, inhibition of HIF-1α function was not associated with decreased VEGF expression or vessel density in tumors, but rather was associated in some cases with accelerated tumor growth. As specific inhibitors of HIF-1α are not available at this time, the effects of in vivo inhibition of HIF-1α function on tumor growth and angiogenesis in solid malignancies remain to be elucidated.

3.6.1.1. The von Hippel-Lindau Tumor Suppressor Gene

Mutation or loss of both alleles of the *VHL* tumor suppressor gene have been documented in sporadic renal cell carcinomas and neoplasms that arise in individuals with VHL syndrome. The observation that the tumors that form in VHL syndrome (e.g., hemangioblastomas, pheochromocytomas, renal cell carcinomas, pancreatic tumors, carcinoid tumors) are all well vascularized (reviewed in [101]) led to investigations of the role of VHL in the regulation of VEGF expression in VHL syndrome. Early reports showed that human renal carcinoma cells that either lacked endogenous wild-type VHL or had been transfected with an inactive mutant VHL exhibited deregulated expression of VEGF that could be reversed by the introduction of wild-type VHL [102]. A relationship between dysfunctional VHL and regulation of VEGF expression has been implicated in other disorders such as hemangioblastoma as well [103]. The mechanism by which the VHL protein regulates VEGF is as follows. VHL regulates intracellular homeostasis of HIF-1α (Figure 3) through mediating the ubiquitination of HIF-1α under normoxic conditions (reviewed in [98]). Loss of VHL function leads to accumulation of HIF-1α in the cytoplasm, which can subsequently dimerize to form active HIF-1 and initiate VEGF transcription [103].

3.6.2. *Nuclear Transcription Factors Nf-κB and Activator Protein-1*

Two transcription factors that have been identified as important downstream regulators of VEGF expression are NF-κB and AP-1. Very often, expression of the target gene (e.g., VEGF) is under the control of both factors. NF-κB is often co-activated by signaling pathways that also regulate the activity of AP-1. However, NF-κB and AP-1 are both part of signaling cascades that are often shared by activated receptors of growth factors (EGF), cytokines (interleukin-1α), or oncogenes (HER-2/neu) or that participate in cellular reactions to hypoxia [104]. Through these mechanisms, NF-κB and AP-1 can be constitutively activated and contribute to high VEGF expression levels in cancer cells [105]. Inhibition of NF-κB function by transfection with a dominant-negative construct has been shown to decrease VEGF levels and reduce angiogenesis in vivo [106]. Interestingly, inhibition of EGFR function also decreased NF-κB activity and VEGF promoter activity in head and neck squamous cell carcinoma cells [104]. Given the fact that NF-κB is regulated by multiple signaling pathways, the development of inhibitors of NF-κB may be another promising approach for lowering VEGF levels in cancer cells.

4. CONCLUSIONS

The fact that VEGF is so precisely regulated is a testament to its central role in tumor angiogenesis, progression, and metastasis. The upstream pathways that regulate VEGF expression are valid targets for therapy as long as those pathways have some role in tumor growth. Although it is theoretically possible that targeting an upstream molecule could inhibit VEGF expression, the redundancies in the pathways that regulate VEGF make the success of this approach unlikely. However, it is possible that some dominant pathway leads to VEGF induction (e.g. VHL or HIF) and that inhibition of that pathway may lead to significant reductions in VEGF levels and angiogenesis in tumors. Careful preclinical modeling and clinical trial design will provide insights into the efficacy of inhibiting upstream targets of VEGF with the intent of decreasing tumor angiogenesis.

Oliver Stoeltzing
Department of Cancer Biology, The University of Texas M. D. Anderson Cancer Center, Houston, Texas

Lee M. Ellis

Departments of Cancer Biology and Surgical Oncology, The University of Texas M. D. Anderson Cancer Center, Houston, Texas

5. REFERENCES

1. Folkman, J. The role of angiogenesis in tumor growth. Semin Cancer Biol, *3:* 65-71, 1992.
2. Fidler, I. J., Singh, R. K., Yoneda, J., Kumar, R., Xu, L., Dong, Z., Bielenberg, D. R., McCarty, M., and Ellis, L. M. Critical determinants of neoplastic angiogenesis. Cancer J, *6 Suppl 3:* S225-236, 2000.
3. Monestiroli, S., Mancuso, P., Burlini, A., Pruneri, G., Dell'Agnola, C., Gobbi, A., Martinelli, G., and Bertolini, F. Kinetics and viability of circulating endothelial cells as surrogate angiogenesis marker in an animal model of human lymphoma. Cancer Res, *61:* 4341-4344, 2001.
4. Takahashi, Y., Kitadai, Y., Bucana, C. D., Cleary, K. R., and Ellis, L. M. Expression of vascular endothelial growth factor and its receptor, KDR, correlates with vascularity, metastasis, and proliferation of human colon cancer. Cancer Res, *55:* 3964-3968, 1995.
5. Senger, D. R., Perruzzi, C. A., Feder, J., and Dvorak, H. F. A highly conserved vascular permeability factor secreted by a variety of human and rodent tumor cell lines. Cancer Res, *46:* 5629-5632, 1986.
6. Dvorak, H. F., Detmar, M., Claffey, K. P., Nagy, J. A., van de Water, L., and Senger, D. R. Vascular permeability factor/vascular endothelial growth factor: an important mediator of angiogenesis in malignancy and inflammation. Int Arch Allergy Immunol, *107:* 233-235, 1995.
7. Leung, D. W., Cachianes, G., Kuang, W. J., Goeddel, D. V., and Ferrara, N. Vascular endothelial growth factor is a secreted angiogenic mitogen. Science, *246:* 1306-1309, 1989.
8. Gerber, H. P., McMurtrey, A., Kowalski, J., Yan, M., Keyt, B. A., Dixit, V., and Ferrara, N. Vascular endothelial growth factor regulates endothelial cell survival through the phosphatidylinositol 3'-kinase/Akt signal transduction pathway. Requirement for Flk-1/KDR activation. J Biol Chem, *273:* 30336-30343, 1998.
9. Zebrowski, B. K., Yano, S., Liu, W., Shaheen, R. M., Hicklin, D. J., Putnam, J. B., Jr., and Ellis, L. M. Vascular endothelial growth factor levels and induction of permeability in malignant pleural effusions. Clin Cancer Res, *5:* 3364-3368, 1999.
10. Yancopoulos, G. D., Davis, S., Gale, N. W., Rudge, J. S., Wiegand, S. J., and Holash, J. Vascular-specific growth factors and blood vessel formation. Nature, *407:* 242-248, 2000.
11. Carmeliet, P. and Jain, R. K. Angiogenesis in cancer and other diseases. Nature, *407:* 249-257, 2000.
12. Ferrara, N. and Davis-Smyth, T. The biology of vascular endothelial growth factor. Endocr Rev, *18:* 4-25, 1997.
13. Houck, K. A., Ferrara, N., Winer, J., Cachianes, G., Li, B., and Leung, D. W. The vascular endothelial growth factor family: identification of a fourth molecular species and characterization of alternative splicing of RNA. Mol Endocrinol, *5:* 1806-1814, 1991.
14. Park, J. E., Keller, G. A., and Ferrara, N. The vascular endothelial growth factor (VEGF) isoforms: differential deposition into the subepithelial extracellular matrix and bioactivity of extracellular matrix-bound VEGF. Mol Biol Cell, *4:* 1317-1326, 1993.
15. Ferrara, N. and Henzel, W. J. Pituitary follicular cells secrete a novel heparin-binding growth factor specific for vascular endothelial cells. Biochem Biophys Res Commun, *161:* 851-858, 1989.

16. Keyt, B. A., Berleau, L. T., Nguyen, H. V., Chen, H., Heinsohn, H., Vandlen, R., and Ferrara, N. The carboxyl-terminal domain (111-165) of vascular endothelial growth factor is critical for its mitogenic potency. J Biol Chem, *271:* 7788-7795, 1996.

17. Neufeld, G., Cohen, T., Gengrinovitch, S., and Poltorak, Z. Vascular endothelial growth factor (VEGF) and its receptors. Faseb J, *13:* 9-22, 1999.

18. Ferrara, N., Carver-Moore, K., Chen, H., Dowd, M., Lu, L., O'Shea, K. S., Powell-Braxton, L., Hillan, K. J., and Moore, M. W. Heterozygous embryonic lethality induced by targeted inactivation of the VEGF gene. Nature, *380:* 439-442, 1996.

19. Cheung, N., Wong, M. P., Yuen, S. T., Leung, S. Y., and Chung, L. P. Tissue-specific expression pattern of vascular endothelial growth factor isoforms in the malignant transformation of lung and colon. Hum Pathol, *29:* 910-914, 1998.

20. Neuchrist, C., Erovic, B. M., Handisurya, A., Fischer, M. B., Steiner, G. E., Hollemann, D., Gedlicka, C., Saaristo, A., and Burian, M. Vascular endothelial growth factor C and vascular endothelial growth factor receptor 3 expression in squamous cell carcinomas of the head and neck. Head Neck, *25:* 464-474, 2003.

21. Hattori, K., Muta, M., Toi, M., Iizasa, H., Shinsei, M., Terasaki, T., Obinata, M., Ueda, M., and Nakashima, E. Establishment of bone marrow-derived endothelial cell lines from ts-SV40 T-antigen gene transgenic rats. Pharm Res, *18:* 9-15, 2001.

22. Partanen, T. A., Arola, J., Saaristo, A., Jussila, L., Ora, A., Miettinen, M., Stacker, S. A., Achen, M. G., and Alitalo, K. VEGF-C and VEGF-D expression in neuroendocrine cells and their receptor, VEGFR-3, in fenestrated blood vessels in human tissues. Faseb J, *14:* 2087-2096, 2000.

23. Soker, S., Miao, H. Q., Nomi, M., Takashima, S., and Klagsbrun, M. VEGF165 mediates formation of complexes containing VEGFR-2 and neuropilin-1 that enhance VEGF165-receptor binding. J Cell Biochem, *85:* 357-368, 2002.

24. Achen, M. G., Jeltsch, M., Kukk, E., Makinen, T., Vitali, A., Wilks, A. F., Alitalo, K., and Stacker, S. A. Vascular endothelial growth factor D (VEGF-D) is a ligand for the tyrosine kinases VEGF receptor 2 (Flk1) and VEGF receptor 3 (Flt4). Proc Natl Acad Sci U S A, *95:* 548-553, 1998.

25. Petit, A. M., Rak, J., Hung, M. C., Rockwell, P., Goldstein, N., Fendly, B., and Kerbel, R. S. Neutralizing antibodies against epidermal growth factor and ErbB-2/neu receptor tyrosine kinases down-regulate vascular endothelial growth factor production by tumor cells in vitro and in vivo: angiogenic implications for signal transduction therapy of solid tumors. Am J Pathol, *151:* 1523-1530, 1997.

26. Klapper, L. N., Kirschbaum, M. H., Sela, M., and Yarden, Y. Biochemical and clinical implications of the ErbB/HER signaling network of growth factor receptors. Adv Cancer Res, *77:* 25-79, 2000.

27. Arteaga, C. Targeting HER1/EGFR: a molecular approach to cancer therapy. Semin Oncol, *30:* 3-14, 2003.

28. Kumar, R. and Yarmand-Bagheri, R. The role of HER2 in angiogenesis. Semin Oncol, *28:* 27-32, 2001.

29. Laughner, E., Taghavi, P., Chiles, K., Mahon, P. C., and Semenza, G. L. HER2 (neu) signaling increases the rate of hypoxia-inducible factor 1alpha (HIF-1alpha) synthesis: novel mechanism for HIF-1-mediated vascular endothelial growth factor expression. Mol Cell Biol, *21:* 3995-4004, 2001.

30. Yang, W., Klos, K., Yang, Y., Smith, T. L., Shi, D., and Yu, D. ErbB2 overexpression correlates with increased expression of vascular endothelial growth factors A, C, and D in human breast carcinoma. Cancer, *94:* 2855-2861, 2002.

31. Yen, L., You, X. L., Al Moustafa, A. E., Batist, G., Hynes, N. E., Mader, S., Meloche, S., and Alaoui-Jamali, M. A. Heregulin selectively upregulates vascular endothelial growth factor secretion in cancer cells and stimulates angiogenesis. Oncogene, *19:* 3460-3469, 2000.

32. Arteaga, C. L., Chinratanalab, W., and Carter, M. B. Inhibitors of HER2/neu (erbB-2) signal transduction. Semin Oncol, 28: 30-35, 2001.

33. Stoeltzing, O., Liu, W., Reinmuth, N., Fan, F., Parikh, A. A., Bucana, C. D., Evans, D. B., Semenza, G. L., and Ellis, L. M. Regulation of hypoxia-inducible factor-1α, vascular endothelial growth factor, and angiogenesis by an insulin-like growth factor-I receptor autocrine loop in human pancreatic cancer. Am J Pathol, 163: (in print), 2003.

34. Reinmuth, N., Fan, F., Liu, W., Parikh, A. A., Stoeltzing, O., Jung, Y. D., Bucana, C. D., Radinsky, R., Gallick, G. E., and Ellis, L. M. Impact of insulin-like growth factor receptor-I function on angiogenesis, growth, and metastasis of colon cancer. Lab Invest, 82: 1377-1389, 2002.

35. Warren, R. S., Yuan, H., Matli, M. R., Ferrara, N., and Donner, D. B. Induction of vascular endothelial growth factor by insulin-like growth factor 1 in colorectal carcinoma. J Biol Chem, 271: 29483-29488, 1996.

36. Smith, L. E., Shen, W., Perruzzi, C., Soker, S., Kinose, F., Xu, X., Robinson, G., Driver, S., Bischoff, J., Zhang, B., Schaeffer, J. M., and Senger, D. R. Regulation of vascular endothelial growth factor-dependent retinal neovascularization by insulin-like growth factor-1 receptor. Nat Med, 5: 1390-1395, 1999.

37. Oh, J. S., Kucab, J. E., Bushel, P. R., Martin, K., Bennett, L., Collins, J., DiAugustine, R. P., Barrett, J. C., Afshari, C. A., and Dunn, S. E. Insulin-like growth factor-1 inscribes a gene expression profile for angiogenic factors and cancer progression in breast epithelial cells. Neoplasia, 4: 204-217, 2002.

38. Bermont, L., Lamielle, F., Fauconnet, S., Esumi, H., Weisz, A., and Adessi, G. L. Regulation of vascular endothelial growth factor expression by insulin-like growth factor-I in endometrial adenocarcinoma cells. Int J Cancer, 85: 117-123, 2000.

39. Cao, R., Brakenhielm, E., Li, X., Pietras, K., Widenfalk, J., Ostman, A., Eriksson, U., and Cao, Y. Angiogenesis stimulated by PDGF-CC, a novel member in the PDGF family, involves activation of PDGFR-alphaalpha and -alphabeta receptors. Faseb J, 16: 1575-1583, 2002.

40. Bergsten, E., Uutela, M., Li, X., Pietras, K., Ostman, A., Heldin, C. H., Alitalo, K., and Eriksson, U. PDGF-D is a specific, protease-activated ligand for the PDGF beta-receptor. Nat Cell Biol, 3: 512-516, 2001.

41. Reinmuth, N., Liu, W., Jung, Y. D., Ahmad, S. A., Shaheen, R. M., Fan, F., Bucana, C. D., McMahon, G., Gallick, G. E., and Ellis, L. M. Induction of VEGF in perivascular cells defines a potential paracrine mechanism for endothelial cell survival. Faseb J, 15: 1239-1241, 2001.

42. Guo, P., Hu, B., Gu, W., Xu, L., Wang, D., Huang, H. J., Cavenee, W. K., and Cheng, S. Y. Platelet-derived growth factor-B enhances glioma angiogenesis by stimulating vascular endothelial growth factor expression in tumor endothelia and by promoting pericyte recruitment. Am J Pathol, 162: 1083-1093, 2003.

43. Kotsuji-Maruyama, T., Imakado, S., Kawachi, Y., and Otsuka, F. PDGF-BB induces MAP kinase phosphorylation and VEGF expression in neurofibroma-derived cultured cells from patients with neurofibromatosis 1. J Dermatol, 29: 713-717, 2002.

44. Shaheen, R. M., Tseng, W. W., Davis, D. W., Liu, W., Reinmuth, N., Vellagas, R., Wieczorek, A. A., Ogura, Y., McConkey, D. J., Drazan, K. E., Bucana, C. D., McMahon, G., and Ellis, L. M. Tyrosine kinase inhibition of multiple angiogenic growth factor receptors improves survival in mice bearing colon cancer liver metastases by inhibition of endothelial cell survival mechanisms. Cancer Res, 61: 1464-1468, 2001.

45. Bergers, G., Song, S., Meyer-Morse, N., Bergsland, E., and Hanahan, D. Benefits of targeting both pericytes and endothelial cells in the tumor vasculature with kinase inhibitors. J Clin Invest, 111: 1287-1295, 2003.

46. Akagi, Y., Liu, W., Xie, K., Zebrowski, B., Shaheen, R. M., and Ellis, L. M. Regulation of vascular endothelial growth factor expression in human colon cancer by interleukin-1beta. Br J Cancer, *80:* 1506-1511, 1999.
47. Yano, S., Nokihara, H., Yamamoto, A., Goto, H., Ogawa, H., Kanematsu, T., Miki, T., Uehara, H., Saijo, Y., Nukiwa, T., and Sone, S. Multifunctional interleukin-1beta promotes metastasis of human lung cancer cells in SCID mice via enhanced expression of adhesion-, invasion- and angiogenesis-related molecules. Cancer Sci, *94:* 244-252, 2003.
48. Ozaki, K. and Leonard, W. J. Cytokine and cytokine receptor pleiotropy and redundancy. J Biol Chem, *277:* 29355-29358, 2002.
49. Wei, L. H., Kuo, M. L., Chen, C. A., Cheng, W. F., Cheng, S. P., Hsieh, F. J., and Hsieh, C. Y. Interleukin-6 in cervical cancer: the relationship with vascular endothelial growth factor. Gynecol Oncol, *82:* 49-56, 2001.
50. Wei, L. H., Kuo, M. L., Chen, C. A., Chou, C. H., Lai, K. B., Lee, C. N., and Hsieh, C. Y. Interleukin-6 promotes cervical tumor growth by VEGF-dependent angiogenesis via a STAT3 pathway. Oncogene, *22:* 1517-1527, 2003.
51. Huang, S. P., Wu, M. S., Wang, H. P., Yang, C. S., Kuo, M. L., and Lin, J. T. Correlation between serum levels of interleukin-6 and vascular endothelial growth factor in gastric carcinoma. J Gastroenterol Hepatol, *17:* 1165-1169, 2002.
52. Masui, T., Hosotani, R., Doi, R., Miyamoto, Y., Tsuji, S., Nakajima, S., Kobayashi, H., Koizumi, M., Toyoda, E., Tulachan, S. S., and Imamura, M. Expression of IL-6 receptor in pancreatic cancer: involvement in VEGF induction. Anticancer Res, *22:* 4093-4100, 2002.
53. Sengupta, S., Gherardi, E., Sellers, L. A., Wood, J. M., Sasisekharan, R., and Fan, T. P. Hepatocyte growth factor/scatter factor can induce angiogenesis independently of vascular endothelial growth factor. Arterioscler Thromb Vasc Biol, *23:* 69-75, 2003.
54. Danilkovitch-Miagkova, A. and Zbar, B. Dysregulation of Met receptor tyrosine kinase activity in invasive tumors. J Clin Invest, *109:* 863-867, 2002.
55. Dong, G., Chen, Z., Li, Z. Y., Yeh, N. T., Bancroft, C. C., and Van Waes, C. Hepatocyte growth factor/scatter factor-induced activation of MEK and PI3K signal pathways contributes to expression of proangiogenic cytokines interleukin-8 and vascular endothelial growth factor in head and neck squamous cell carcinoma. Cancer Res, *61:* 5911-5918, 2001.
56. Moriyama, T., Kataoka, H., Hamasuna, R., Yokogami, K., Uehara, H., Kawano, H., Goya, T., Tsubouchi, H., Koono, M., and Wakisaka, S. Up-regulation of vascular endothelial growth factor induced by hepatocyte growth factor/scatter factor stimulation in human glioma cells. Biochem Biophys Res Commun, *249:* 73-77, 1998.
57. Sengupta, S., Sellers, L. A., Li, R. C., Gherardi, E., Zhao, G., Watson, N., Sasisekharan, R., and Fan, T. P. Targeting of mitogen-activated protein kinases and phosphatidylinositol 3 kinase inhibits hepatocyte growth factor/scatter factor-induced angiogenesis. Circulation, *107:* 2955-2961, 2003.
58. Jung, Y. D., Nakano, K., Liu, W., Gallick, G. E., and Ellis, L. M. Extracellular signal-regulated kinase activation is required for up-regulation of vascular endothelial growth factor by serum starvation in human colon carcinoma cells. Cancer Res, *59:* 4804-4807, 1999.
59. Seger, R. and Krebs, E. G. The MAPK signaling cascade. Faseb J, *9:* 726-735, 1995.
60. Cantley, L. C. The phosphoinositide 3-kinase pathway. Science, *296:* 1655-1657, 2002.
61. Davies, M. A., Koul, D., Dhesi, H., Berman, R., McDonnell, T. J., McConkey, D., Yung, W. K., and Steck, P. A. Regulation of Akt/PKB activity, cellular growth, and apoptosis in prostate carcinoma cells by MMAC/PTEN. Cancer Res, *59:* 2551-2556, 1999.
62. Akagi, M., Kawaguchi, M., Liu, W., McCarty, M. F., Takeda, A., Fan, F., Stoeltzing, O., Parikh, A. A., Jung, Y. D., Bucana, C. D., Mansfield, P. F., Hicklin, D. J., and Ellis, L. M.

Induction of neuropilin-1 and vascular endothelial growth factor by epidermal growth factor in human gastric cancer cells. Br J Cancer, *88:* 796-802, 2003.

63. Maity, A., Pore, N., Lee, J., Solomon, D., and O'Rourke, D. M. Epidermal growth factor receptor transcriptionally up-regulates vascular endothelial growth factor expression in human glioblastoma cells via a pathway involving phosphatidylinositol 3'-kinase and distinct from that induced by hypoxia. Cancer Res, *60:* 5879-5886, 2000.

64. Xiong, S., Grijalva, R., Zhang, L., Nguyen, N. T., Pisters, P. W., Pollock, R. E., and Yu, D. Up-regulation of vascular endothelial growth factor in breast cancer cells by the heregulin-beta1-activated p38 signaling pathway enhances endothelial cell migration. Cancer Res, *61:* 1727-1732, 2001.

65. Woods, S. A., McGlade, C. J., and Guha, A. Phosphatidylinositol 3'-kinase and MAPK/ERK kinase 1/2 differentially regulate expression of vascular endothelial growth factor in human malignant astrocytoma cells. Neuro-oncol, *4:* 242-252, 2002.

66. Jiang, B. H., Jiang, G., Zheng, J. Z., Lu, Z., Hunter, T., and Vogt, P. K. Phosphatidylinositol 3-kinase signaling controls levels of hypoxia-inducible factor 1. Cell Growth Differ, *12:* 363-369, 2001.

67. Xu, L., Fukumura, D., and Jain, R. K. Acidic extracellular pH induces vascular endothelial growth factor (VEGF) in human glioblastoma cells via ERK1/2 MAPK signaling pathway: mechanism of low pH-induced VEGF. J Biol Chem, *277:* 11368-11374, 2002.

68. Goerges, A. L. and Nugent, M. A. Regulation of vascular endothelial growth factor binding and activity by extracellular pH. J Biol Chem, *278:* 19518-19525, 2003.

69. Park, J. S., Qiao, L., Su, Z. Z., Hinman, D., Willoughby, K., McKinstry, R., Yacoub, A., Duigou, G. J., Young, C. S., Grant, S., Hagan, M. P., Ellis, E., Fisher, P. B., and Dent, P. Ionizing radiation modulates vascular endothelial growth factor (VEGF) expression through multiple mitogen activated protein kinase dependent pathways. Oncogene, *20:* 3266-3280, 2001.

70. Mori, K., Tani, M., Kamata, K., Kawamura, H., Urata, Y., Goto, S., Kuwano, M., Shibata, S., and Kondo, T. Mitogen-activated protein kinase, ERK1/2, is essential for the induction of vascular endothelial growth factor by ionizing radiation mediated by activator protein-1 in human glioblastoma cells. Free Radic Res, *33:* 157-166, 2000.

71. Gately, S. The contributions of cyclooxygenase-2 to tumor angiogenesis. Cancer Metastasis Rev, *19:* 19-27, 2000.

72. Needleman, P., Turk, J., Jakschik, B. A., Morrison, A. R., and Lefkowith, J. B. Arachidonic acid metabolism. Annu Rev Biochem, *55:* 69-102, 1986.

73. Chandrasekharan, N. V., Dai, H., Roos, K. L., Evanson, N. K., Tomsik, J., Elton, T. S., and Simmons, D. L. COX-3, a cyclooxygenase-1 variant inhibited by acetaminophen and other analgesic/antipyretic drugs: cloning, structure, and expression. Proc Natl Acad Sci U S A, *99:* 13926-13931, 2002.

74. Herschman, H., Gilbert, R., Reddy, S., and Xie, W. L. Coordinate regulation of the inducible forms of prostaglandin synthase and nitric oxide synthase in fibroblasts and macrophages. Adv Exp Med Biol, *400A:* 177-182, 1997.

75. Amano, H., Hayashi, I., Endo, H., Kitasato, H., Yamashina, S., Maruyama, T., Kobayashi, M., Satoh, K., Narita, M., Sugimoto, Y., Murata, T., Yoshimura, H., Narumiya, S., and Majima, M. Host prostaglandin E(2)-EP3 signaling regulates tumor-associated angiogenesis and tumor growth. J Exp Med, *197:* 221-232, 2003.

76. Masferrer, J. Approach to angiogenesis inhibition based on cyclooxygenase-2. Cancer J, *7 Suppl 3:* S144-150, 2001.

77. Fukuda, R., Kelly, B., and Semenza, G. L. Vascular endothelial growth factor gene expression in colon cancer cells exposed to prostaglandin E2 is mediated by hypoxia-inducible factor 1. Cancer Res, *63:* 2330-2334, 2003.

78. Liu, X. H., Kirschenbaum, A., Lu, M., Yao, S., Dosoretz, A., Holland, J. F., and Levine, A. C. Prostaglandin E2 induces hypoxia-inducible factor-1alpha stabilization and nuclear localization in a human prostate cancer cell line. J Biol Chem, *277:* 50081-50086, 2002.

79. Niu, G., Wright, K. L., Huang, M., Song, L., Haura, E., Turkson, J., Zhang, S., Wang, T., Sinibaldi, D., Coppola, D., Heller, R., Ellis, L. M., Karras, J., Bromberg, J., Pardoll, D., Jove, R., and Yu, H. Constitutive Stat3 activity up-regulates VEGF expression and tumor angiogenesis. Oncogene, *21:* 2000-2008, 2002.

80. Mora, L. B., Buettner, R., Seigne, J., Diaz, J., Ahmad, N., Garcia, R., Bowman, T., Falcone, R., Fairclough, R., Cantor, A., Muro-Cacho, C., Livingston, S., Karras, J., Pow-Sang, J., and Jove, R. Constitutive activation of Stat3 in human prostate tumors and cell lines: direct inhibition of Stat3 signaling induces apoptosis of prostate cancer cells. Cancer Res, *62:* 6659-6666, 2002.

81. Rak, J., Mitsuhashi, Y., Sheehan, C., Tamir, A., Viloria-Petit, A., Filmus, J., Mansour, S. J., Ahn, N. G., and Kerbel, R. S. Oncogenes and tumor angiogenesis: differential modes of vascular endothelial growth factor up-regulation in ras-transformed epithelial cells and fibroblasts. Cancer Res, *60:* 490-498, 2000.

82. Ellis, L. M., Staley, C. A., Liu, W., Fleming, R. Y., Parikh, N. U., Bucana, C. D., and Gallick, G. E. Down-regulation of vascular endothelial growth factor in a human colon carcinoma cell line transfected with an antisense expression vector specific for c-src. J Biol Chem, *273:* 1052-1057, 1998.

83. Mukhopadhyay, D., Tsiokas, L., Zhou, X. M., Foster, D., Brugge, J. S., and Sukhatme, V. P. Hypoxic induction of human vascular endothelial growth factor expression through c-Src activation. Nature, *375:* 577-581, 1995.

84. Ebos, J. M., Tran, J., Master, Z., Dumont, D., Melo, J. V., Buchdunger, E., and Kerbel, R. S. Imatinib mesylate (STI-571) reduces Bcr-Abl-mediated vascular endothelial growth factor secretion in chronic myelogenous leukemia. Mol Cancer Res, *1:* 89-95, 2002.

85. Janowska-Wieczorek, A., Majka, M., Marquez-Curtis, L., Wertheim, J. A., Turner, A. R., and Ratajczak, M. Z. Bcr-abl-positive cells secrete angiogenic factors including matrix metalloproteinases and stimulate angiogenesis in vivo in Matrigel implants. Leukemia, *16:* 1160-1166, 2002.

86. Zhang, X., Gaspard, J. P., and Chung, D. C. Regulation of vascular endothelial growth factor by the Wnt and K-ras pathways in colonic neoplasia. Cancer Res, *61:* 6050-6054, 2001.

87. Rak, J., Mitsuhashi, Y., Bayko, L., Filmus, J., Shirasawa, S., Sasazuki, T., and Kerbel, R. S. Mutant ras oncogenes upregulate VEGF/VPF expression: implications for induction and inhibition of tumor angiogenesis. Cancer Res, *55:* 4575-4580, 1995.

88. Feldkamp, M. M., Lau, N., Rak, J., Kerbel, R. S., and Guha, A. Normoxic and hypoxic regulation of vascular endothelial growth factor (VEGF) by astrocytoma cells is mediated by Ras. Int J Cancer, *81:* 118-124, 1999.

89. Mischel, P. S. and Cloughesy, T. F. Targeted molecular therapy of GBM. Brain Pathol, *13:* 52-61, 2003.

90. Breier, G., Blum, S., Peli, J., Groot, M., Wild, C., Risau, W., and Reichmann, E. Transforming growth factor-beta and Ras regulate the VEGF/VEGF-receptor system during tumor angiogenesis. Int J Cancer, *97:* 142-148, 2002.

91. Pal, S., Datta, K., and Mukhopadhyay, D. Central role of p53 on regulation of vascular permeability factor/vascular endothelial growth factor (VPF/VEGF) expression in mammary carcinoma. Cancer Res, *61:* 6952-6957, 2001.

92. Ravi, R., Mookerjee, B., Bhujwalla, Z. M., Sutter, C. H., Artemov, D., Zeng, Q., Dillehay, L. E., Madan, A., Semenza, G. L., and Bedi, A. Regulation of tumor angiogenesis by p53-induced degradation of hypoxia-inducible factor 1alpha. Genes Dev, *14:* 34-44, 2000.

58

93. Fujisawa, T., Watanabe, J., Kamata, Y., Hamano, M., Hata, H., and Kuramoto, H. Effect of p53 gene transfection on vascular endothelial growth factor expression in endometrial cancer cells. Exp Mol Pathol, *74:* 276-281, 2003.

94. Krieg, M., Haas, R., Brauch, H., Acker, T., Flamme, I., and Plate, K. H. Up-regulation of hypoxia-inducible factors HIF-1alpha and HIF-2alpha under normoxic conditions in renal carcinoma cells by von Hippel-Lindau tumor suppressor gene loss of function. Oncogene, *19:* 5435-5443, 2000.

95. Helmlinger, G., Yuan, F., Dellian, M., and Jain, R. K. Interstitial pH and pO2 gradients in solid tumors in vivo: high-resolution measurements reveal a lack of correlation. Nat Med, *3:* 177-182, 1997.

96. Wang, G. L., Jiang, B. H., Rue, E. A., and Semenza, G. L. Hypoxia-inducible factor 1 is a basic-helix-loop-helix-PAS heterodimer regulated by cellular O2 tension. Proc Natl Acad Sci U S A, *92:* 5510-5514, 1995.

97. Zhong, H., De Marzo, A. M., Laughner, E., Lim, M., Hilton, D. A., Zagzag, D., Buechler, P., Isaacs, W. B., Semenza, G. L., and Simons, J. W. Overexpression of hypoxia-inducible factor 1alpha in common human cancers and their metastases. Cancer Res, *59:* 5830-5835, 1999.

98. Fukuda, R., Hirota, K., Fan, F., Jung, Y. D., Ellis, L. M., and Semenza, G. L. Insulin-like growth factor 1 induces hypoxia-inducible factor 1-mediated vascular endothelial growth factor expression, which is dependent on MAP kinase and phosphatidylinositol 3-kinase signaling in colon cancer cells. J Biol Chem, *277:* 38205-38211, 2002.

99. Carmeliet, P., Dor, Y., Herbert, J. M., Fukumura, D., Brusselmans, K., Dewerchin, M., Neeman, M., Bono, F., Abramovitch, R., Maxwell, P., Koch, C. J., Ratcliffe, P., Moons, L., Jain, R. K., Collen, D., Keshert, E., and Keshet, E. Role of HIF-1alpha in hypoxia-mediated apoptosis, cell proliferation and tumour angiogenesis. Nature, *394:* 485-490, 1998.

100. Kung, A. L., Wang, S., Klco, J. M., Kaelin, W. G., and Livingston, D. M. Suppression of tumor growth through disruption of hypoxia-inducible transcription. Nat Med, *6:* 1335-1340, 2000.

101. Lonser, R. R., Glenn, G. M., Walther, M., Chew, E. Y., Libutti, S. K., Linehan, W. M., and Oldfield, E. H. von Hippel-Lindau disease. Lancet, *361:* 2059-2067, 2003.

102. Siemeister, G., Weindel, K., Mohrs, K., Barleon, B., Martiny-Baron, G., and Marme, D. Reversion of deregulated expression of vascular endothelial growth factor in human renal carcinoma cells by von Hippel-Lindau tumor suppressor protein. Cancer Res, *56:* 2299-2301, 1996.

103. Krieg, M., Marti, H. H., and Plate, K. H. Coexpression of erythropoietin and vascular endothelial growth factor in nervous system tumors associated with von Hippel-Lindau tumor suppressor gene loss of function. Blood, *92:* 3388-3393, 1998.

104. Bancroft, C. C., Chen, Z., Yeh, J., Sunwoo, J. B., Yeh, N. T., Jackson, S., Jackson, C., and Van Waes, C. Effects of pharmacologic antagonists of epidermal growth factor receptor, PI3K and MEK signal kinases on NF-kappaB and AP-1 activation and IL-8 and VEGF expression in human head and neck squamous cell carcinoma lines. Int J Cancer, *99:* 538-548, 2002.

105. Bancroft, C. C., Chen, Z., Dong, G., Sunwoo, J. B., Yeh, N., Park, C., and Van Waes, C. Coexpression of proangiogenic factors IL-8 and VEGF by human head and neck squamous cell carcinoma involves coactivation by MEK-MAPK and IKK-NF-kappaB signal pathways. Clin Cancer Res, *7:* 435-442, 2001.

106. Shibata, A., Nagaya, T., Imai, T., Funahashi, H., Nakao, A., and Seo, H. Inhibition of NF-kappaB activity decreases the VEGF mRNA expression in MDA-MB-231 breast cancer cells. Breast Cancer Res Treat, *73:* 237-243, 2002.

INTEGRIN-LINKED KINASE (ILK) IN COMBINATION MOLECULAR TARGETING

LINCOLN A EDWARDS, JENNIFER A SHABBITS,
MARCEL BALLY AND SHOUKAT DEDHAR

1. INTRODUCTION

Cancer is a complex multi-step process involving the activation and/or abrogation of signal transduction pathways resulting in a variety of cellular activities. Cancer, in effect, is the loss or disregard of normal cell communication, which engenders unregulated cell proliferation, migration, and an inhibition of apoptotic arrest. Although the complexity of cellular communication is intimidating, knowledge of cell signal communication is to a stage where we can consider strategies designed to specifically augment or inhibit these pathways in a reasonably specific manner. At the root of cellular communication is the discovery of protein kinases; the consequences of which now allow for the possibility of identifying small molecules which block kinase activity as well as molecular strategies which can block or elicit production of these enzymes. While chemotherapy remains one of the primary ways to treat cancer, conventional chemotherapeutic agents are rather blunt instruments that are easily circumvented by rapid drug metabolism and tumor drug resistance. This problem is further exacerbated by the lack of specificity of chemotherapy and the associated systemic toxicity on the cancer patient. Specific and selective inhibitors of kinases offer a more surgical approach to management of cancer, particularly when the role of the enzyme is well understood in normal and diseased cells. For these reasons there has been a concerted effort into gaining a better understanding of protein kinases, not only for the intimate role they play in proliferative diseases, but also as molecular targets in cancer therapy.

Integrin-linked kinase (ILK) defines such a molecular target, as the overexpression of ILK leads to various oncogenic-related events including cell migration, invasion and an inhibition of apoptosis. Increased ILK activity has been correlated to several tumor malignancies including colon, prostate, breast and brain cancer. Given the important role of ILK in these diseases it can be argued that it will be an important therapeutic target, however, it is recognized that although necessary, its inhibition may not be sufficient in light of other cancer-associated cell signaling pathways that may render monotherapy ineffective. This chapter seeks to describe signaling pathways activated by ILK, the use of ILK inhibitors in rational targeted drug combination therapy, how to translate these combinations from an in vitro to an in vivo setting, and how appropriate targeted drug combinations may lead to synergistic or additive drug responses. A drug targeting approach that is more selective and specific may provide a much needed alternative to conventional chemotherapeutic agents.

2. ILK AND ITS IMPORTANCE IN CANCER

2.1. Overview

Integrin linked kinase (ILK), was first discovered by yeast two hybrid screening trying to identify proteins that interact with the cytoplasmic domains of the cell surface receptors known as integrins [1]. Co-immunoprecipitation studies verified that ILK interacts with the β1 integrins and was later found to interact with the β3 components of integrins [1]. Integrins are a large family of receptors that exist as heterodimers composed of α and β subunits that can interact with extracellular matrix proteins such as collagen and fibronectin [2]. The result of an integrin-ILK interaction identifies not only how signaling proteins connect and communicate with the extracellular matrix, but also how intracellular cytoskeletal proteins are also activated. Evidence of intracellular cytoskeletal activity can be seen with several proteins that interact with ILK and connect to the actin cytoskeleton. Yeast two hybrid analysis with ILK as bait led to the identification of CH-ILKBP also known as α-parvin (found in mice) and actopaxin. CH-ILKBP, which contains two calponin homology (CH) domains, and mediates ILK association through its CH2 domain, at ILKs COOH terminal domain, [3] also connects to the actin cytoskeleton. Similarly, two other proteins, affixin (the human orthologue of mouse β-parvin) which shares significant sequence similarity to CH-ILKBP within the CH2 domain, can interact with ILK at ILKs COOH terminal domain and paxillin, which interacts with ILK at ILKs COOH terminal domain through an LD1 motif can, like CH-ILKBP, connect to the actin cytoskeleton [4, 5]. This new class of ILK binding proteins which consists of CH-ILKBP-actopaxin-α-paxin, affixin-β-parvin and paxillin are thought to act as one of the major bridging molecules, linking ILK to the actin filaments.

Overexpression of the CH2 binding domain of CH-ILKBP-actopaxin-α-parvin in cells results in loss of actin stress fiber formation and cell shape change [5, 3]. In addition, *C. elegans* with mutations in either ILK or CH-ILKBP-actopaxin-α-paxin show defects in the assemblage of muscle dense bodies that attach actin filaments to sarcolemma (Lin, X., and B.S.Williams. 2000. 40th American Society for Cell Biology Annual Meeting abstract). The regulated interaction of these proteins with ILK mediate the dynamic interaction of integrins with the actin cytoskeleton during cell attachment, spreading and migration.

ILK can initiate a signal transduction cascade from its interaction with integrins, and it is more than likely that ILKs ability to phosphorylate β1 integrins at Ser790 regulates localization of β1 integrins to focal adhesions. ILK has also been shown to localize to fibrillar adhesions [6]. ILKs involvement in both focal adhesions and fibrillar adhesion is consistent with a role of ILK in cell adhesion, spreading and fibronectin matrix assembly [1, 7, 8]. In addition, ILKs interaction with integrins at its C-terminus leading to cell signaling is only part of ILKs ability to initiate cell communication. The N-terminus of ILK, along with having a phosphoinositide-binding motif, contains four ankyrin repeats that allow for protein-protein interactions. PINCH is a LIM domain-only protein, that can interact with ILK at the ILK N-terminus, through the LIM1 domain of PINCH. The adaptor protein Nck-2 (also known as Nckβ or Grb4), interacts with PINCH through the PINCH LIM4 domain and the Nck-2 SH3 domain, Nck-2 can then be recruited by growth factor receptors such as EGFR. Moreover, Nck-2 could conceivably bring other components of the growth factor and small GTPase signaling pathways into proximity. The coupling of ILK to integrins and growth factor receptors indicates that ILK has a role in both integrin and growth factor receptor signaling, making ILK very versatile and potentially able to initiate other cell signaling cascades. ILK appears to be deeply rooted in the PI3K pathway, as ILK is positively regulated by PI3K and negatively regulated by the protein phosphatase PTEN. Specifically, through its phosphoinositide binding motif at its N-terminus, ILK can bind PIP_3 resulting in ILK activation, PTENs ability to dephosphorylate PIP_3 can therefore inhibit ILK activity. Similarly, PKB/AKT is known to be an important regulator of cell cycle progression and cell survival, which is also affected by PIP_3 levels regulated by PTEN [9]. ILK has been shown to regulate the cell cycle by increasing expression of cyclin D1 and cyclin A [10, 11], and increased ILK activity corresponds with increased PKB/AKT expression in a PI3K dependent manner [12, 7]. Full activation of PKB/AKT is dictated by phosphorylation of PKB/AKT at two sites: Thr-308 and Ser473. The phosphorylation of Thr-308 is carried out by phosphatidylinositol 3-kinase dependent kinase-1 (PDK1) and for complete activation of PKB/AKT, ILK stimulates phosphorylation at the Ser473 site [13].

There is little doubt that ILK can regulate PKB/AKT phosphorylation however, the precise mechanism remains unclear. ILK causes PKB/AKT phosphorylation via an indirect mechanism. On the other hand, ILK has been shown to directly phosphorylate several proteins in vitro including PKB/AKT [4, 14, 15]. Recently, RNA interference (siRNA) targeted at ILK resulted in a significant decrease in phosphorylation on Ser473 of PKB/AKT that could be rescued by kinase-active ILK. Further, conditional knock-out of ILK using the Cre-Lox system resulted in inhibition of phosphorylation on Ser473 of PKB/AKT. Since the activity of ILK and PKB/AKT is positively regulated by PI3K and is negatively regulated by PTEN, and ILK mediates PKB/AKT activity, ILK appears to occupy a nodal position in the regulation of PKB/AKT activity (see Fig 1).

2.2. The Role of ILK in Tumorigenesis and Invasion

The downstream targets of ILK, the most prominent being PKB/AKT, appear to be involved in several oncogenic related processes including anchorage-independent growth, invasion, migration and prolonged cell survival when ILK is overexpressed or constitutively active. Wu and colleagues showed that ILK can act as a proto-oncogene by demonstrating that overexpression of ILK in epithelial cells induces tumorigenicity in nude mice [16]. Reduced sensitivity to anoikis is also observed when ILK is overexpressed in anoikis-sensitive SCP2 cells causing inhibition of apoptosis in these cells. On the other hand, downregulation of ILK activity by dominant-negative ILK or dominant negative PKB/AKT reverses anoikis sensitivity in SCP2 cells, indicating again ILKs role in the regulation of PKB/AKT [17]. Prolonged cell survival results from ILKs activation of PKB/AKT leading to the inactivation of downstream pro-apoptotic factors such as BAD [18] and caspase-9 [19]. ILKs ability to induce an invasive phenotype is observed with ILK overexpression in mammary and intestinal epithelial cells (IECs). ILK overexpression is associated with increased expression of matrix metalloproteinase (MMP-9). Soft agar studies of IECs with ILK overexpression revealed cell growth, essentially promoting anchorage-independent growth [20]. Further, ILK overexpression leads to the loss of cell-cell adhesion due to suppression of E-cadherin expression via activation of the E-cadherin repressor Snail, again promoting a migratory/invasive phenotype [21]. The activities of ILK in tumorigenesis and cell invasion are not limited to just the in vitro setting. Transgenic mice that overexpressed ILK via a mouse mammary tumor virus (MMTV/ILK) were associated with the development of papillary adenocarcinomas and spindle cell tumors [22]. Lately, ILK expression is becoming increasingly identified with patient survival and is a potential prognostic indicator of cancer. Tissue microarray and immunohistochemistry of 67 primary melanomas, showed that ILK expression was found to increase with melanoma invasion and progression and that 5-year patient survival was inversely correlated to ILK expression.

2.3. Relevance of Integrin-linked Kinase to Human Cancer

Evaluation of ILK in primary prostate tumor tissue revealed that ILK expression increased in high grade tumors (i.e. prostatic adenocarcinoma) but did not increase in benign prostatic hyperplasias and low-grade prostatic adenocarcinomas [8]. Importantly, ILK expression was also inversely related to 5-year patient survival, 81% of patients whose tumors overexpressed ILK failed to survive beyond 5 years [8]. This suggests that ILK could be a prognostic marker for prostate cancer. The consistency of ILK expression and its potential as a prognostic marker is even greater in Ewing's Sarcoma and primitive ectodermal tumors, as 100% of the cases analyzed showed ILK expression [23]. The activity of ILK in several cancers and its potential as a prognostic marker is further seen in ovarian cancer where recently seventy-three tissue samples (10 normal, 10 benign, 14 borderline, 17 grade I/II, and 22 grade III) were analyzed for ILK expression by immunohistochemistry. Immunoreactive ILK was detected in 53 specimens where intensity of ILK staining correlated with tumor grade. Normal ovarian tissue showed no immunoreactive ILK [24]. In addition, peritoneal tumor fluid increased ILK expression in ovarian cancer cell lines, but had no effect on normal ovarian surface epithelial cell lines, indicating that peritoneal tumor fluid may maintain sustained ILK expression in ovarian cancer cell lines [24].

Although the main focus of ILK activity has been on PKB/AKT, ILK is involved in the phosphorylation of other cell signaling proteins that are associated with a cancerous phenotype. One such cell signaling protein is glycogen synthase kinase-3 (GSK-3). ILK overexpression negatively regulates a phosphorylation mediated inhibition of GSK-3. Inactivation of GSK-3 results in a loss of cell-cell adhesion due to loss of E-cadherin expression. Since metastasis requires the separation of tumor cells from the bulk tumor this evidence suggests the involvement of ILK, GSK-3 and E-cadherins in metastatic activity. Further, loss of E-cadherin expression is highly correlated with breast cancer invasion and metastasis [25].

ILK expression has also been demonstrated to be highly elevated in precancerous colon polyps from patients with familial adenomatous polyposis and colon carcinomas [26]. Mutations in proteins that act in a PI3K dependent manner could potentially, like PTEN, be associated with cancer progression. Recently, besides PTEN, two potential tumor suppressor proteins, disabled homolog 2/differentially expressed protein 2 (DOC-2) and stomach cancer-associated protein-tyrosine phosphatase-1 (SAP-1) have been shown to negatively regulate ILK activity and are associated with breast and stomach cancer respectively [27-28]. Another negative regulator of ILK has been identified, ILKAP, a protein phosphatase which selectively inhibits ILK

64

activity on the Wnt signaling pathway [29], however, evidence of tumor activity has yet to be associated with ILKAP mutation.

2.4. ILK as a Therapeutic Molecular Target

It is clear that overexpression of ILK results in malignant progression. In normal cells, overexpression leads to tumorigenicity, anchorage-independence, cell cycle progression and tumor invasiveness [1, 30, 11, 16]. Inhibition of ILK activity has also been shown to suppress the growth of human colon carcinoma cells in SCID mice [21].

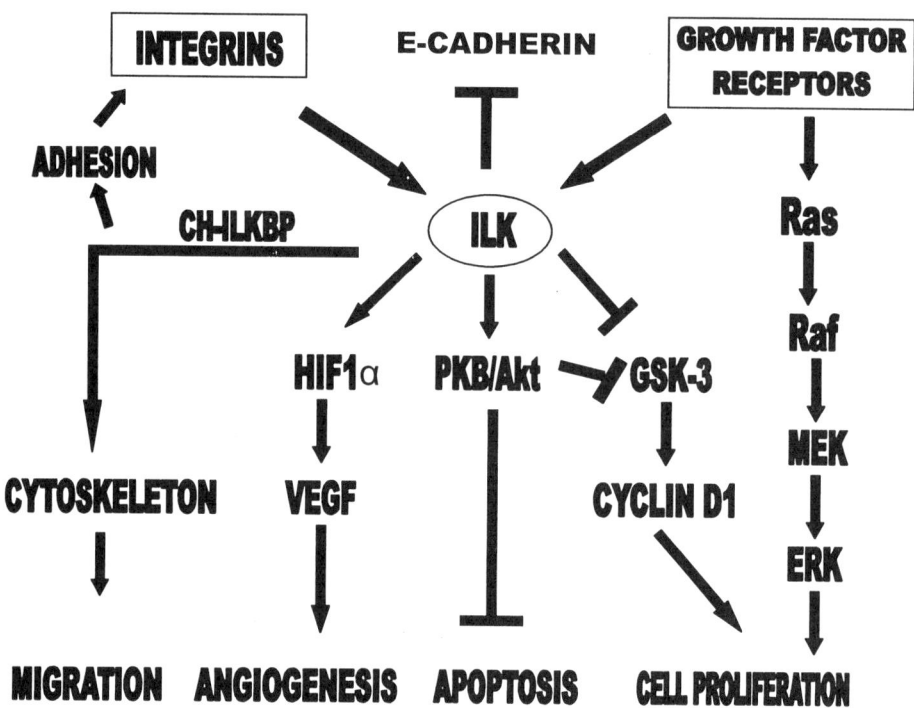

Figure 1. Cell signaling pathways activated by ILK.. ILK is a major effector serine/threonine kinase that links integrins and receptor tyrosine kinases to the cytoskeleton and downstream signaling molecules.

In human colon cancer, increased ILK activity has been associated with downstream cell signaling molecules such as GSK-3 and as much as a 9-fold increase in MBP phosphotransferase activity [31]. In addition, ILK overexpression is observed in malignant crypts from both primary and metastatic lesions. Marotta and colleagues relate that selective inhibition of ILK alone or in combination would be an effective means of treating this cancer [31]. Recent studies have demonstrated a critical role of ILK in tumor angiogenesis, whereby ILK regulates VEGF expression in tumor cells and

VEGF mediated vascular morphogenesis (Dedhar, S., unpublished). Since the function of ILK surrounds the activation of key downstream PI3K regulators such as PKB/AKT, GSK-3 [30, 32] and various cellular functions including angiogenesis, apoptosis, and cell proliferation, (see Figure 1) and given that inactivation of ILK by dominant negative ILK, kinase deficient ILK, and ILK siRNA can block ILK activity and cell transformation, ILK is a very good candidate for drug targeting.

3. MOLECULARLY TARGETED DRUG COMBINATIONS

3.1. Rationale for Combination Approaches

The increased genetic instability typically exhibited by cancer cells predisposes them to acquire numerous molecular alterations or mutations, and indeed the vast majority of cancers consist of multiple molecular alterations. Consequently, therapies directed against a single cellular or molecular target are unlikely to result in a significant response. Combinations of conventional chemotherapy drugs have been used as one approach; however, the emergence of drug resistance remains a considerable limiting factor. Since cell signaling is largely responsible for the apoptotic and cytostatic effects of conventional chemotherapy and radiation-based therapeutics, the addition of molecular target inhibition to these established treatment approaches has the potential to enhance the effectiveness of both established and emerging therapies. Combination strategies that attack multiple cellular targets and/or signaling in parallel or sequential pathways known to contribute to the overall transformed state will likely yield the most dramatic therapeutic benefit.

3.2. Why Molecular Targeting Involving ILK?

ILK has been identified as a plausible target for the treatment of cancer in which the PI3K pathway is involved. In addition, the Ras/Raf/MAPK pathway has also been shown to be involved in several cancers. Although these pathways in cancer are typically presented as separate entities, the reality is that there is a great deal of cross-talk between these pathways, including at the level of ILK regulation. Molecular combination therapy could exploit approaches that target specific pathways in conjunction with established treatment practices or approaches that target multiple independent pathways each known to influence cancer progression and survival. Before any attempt can be made to discuss molecular drug combinations however, it is reasonable to test whether the combinations have the potential to interact in a synergistic, additive or antagonistic manner as judged by cell based screening assays. These data provide evidence that support further evaluations of selected drug combinations but typically provide no information on mechanisms governing the combined drugs or, importantly, how to use the drugs *in vivo* in a manner that insure optimal synergistic effects are maintained. The Chou and Talalay median-effect approach [33] is one of the most prevalent methods to assess drug combination effects and no alternative approaches will be discussed here. For a comprehensive review of approaches to assess drug interaction effects, please refer to the excellent review by Greco et al.[34].

3.3. *The Evaluation of Targeted Molecular Drug Combinations (in vitro)*

To evaluate drug combinations the multiple drug effect equation (Eq 1) of Chou was introduced [35-36]. The multiple drug effect equation that defines the additive effect only, indicates that any effect greater than the additive effect is a synergistic one.

$$f_a/f_u = (D/D_m)^m$$

Eq 1

From equation 1, where D= the dose of drug, D_m= the median-effect dose indicating the potency (determined from the x-intercept of the graph of the median effect plot), f_a=the fraction affected by the dose, f_u=the fraction unaffected, where f_u=1-f_a and finally m=an exponent showing the shape of the dose-effect curve determined by the median effect plot. From this equation the combination index (CI) equation was formed (Chou & Talalay 1983) [40]. The benefit of the CI equation for drug combinations in the treatment of cancer is that the CI gives a quantitative determination of synergy, antagonism and additivity (see Eq 2). Chou and Talalay proposed that CI=1 represents the additive effect, CI < 1 indicates a synergistic effect and that a CI > 1 indicates an antagonistic effect [37].

$$CI = (D)_1/(D_x)_1 + (D)_2/(D_x)_2$$

Eq 2

For each combination dose, the data can be plotted as the percent of cells inhibited (i.e. the fraction affected-Fa) and the CI value determined for each combination, which results in a Fa versus a CI plot. Equations 1 and 2 and the graphs that can be generated from drug combination data have been used in the development of a user friendly software program CalcuSyn (BioSoft, Ferguson MO, USA). A representative graph generated by CalcuSyn is shown using the conventional chemotherapy combination of BCNU and Carboplatin, which are used in the treatment of malignant glioma, and in our hands has been shown as a positive control of a synergistic relationship (Figure 2).

These data illustrate an important point when considering combination effects by this approach. CI varies as a function of the fraction of affected cells, a parameter that is obviously related to the drug dose (eq 1). Thus CI values indicating antagonism are estimated at doses inducing a measured effect (e.g. % cytotoxicity) of less than 20% while the CI is indicative of synergistic interactions when these drug doses used achieve effect levels of >0.5.

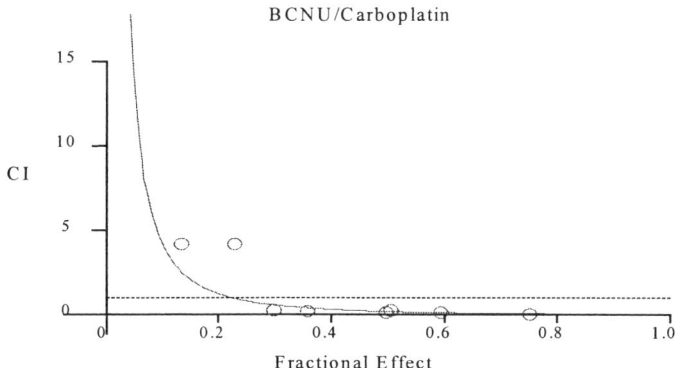

Figure 2. Representative plot showing a synergistic effect of the conventional chemotherapeutic alkylating agents BCNU and Carboplatin. Note that each data point (black open circles) moving from left to right is an increasing ratio of the drug combination of BCNU and Carboplatin. The data points that fall below the dotted horizontal line which represents the CI value 1 indicates a synergistic relationship.

The point of molecular targeted drug combinations is not only to limit monotherapy failure, as has been indicated in some Gleevec ™ patients [38, 39], but also to increase the therapeutic benefit of the drug combination. This can occur at dose levels less than that required by either one of the agents used alone. Provided that toxicity to normal cells is not enhanced by the use of the combination, then the synergistic combination should achieve improved therapeutic results. In addition, a multi-targeted attack on cell signaling proteins should reduce or limit the emergence of drug resistance [40]. Although *in vitro* assays defining synergistic interactions, may provide little information as to how this synergy can be obtained *in vivo*. It can be argued that an approach to insuring synergy is achieved *in vivo* could be developed on the basis of the concentration of each drug required to achieve synergistic effects and relating this information to pharmacokinetic parameters. Moreover, it should be noted that in any drug combination approach there will be a balance between efficacy and toxicity. For example, targeting ILK and MEK1/2 protein may provide a good response with an acceptable level of toxicity, while targeting ILK and the EGFR may give a better efficacy response with comparable toxicity. The importance of choosing the "right" targets will, therefore, be only one of the parameters that must be considered when developing drug combinations for clinical use. Other issues such as those effecting biodistribution, metabolism and toxicity must also be considered if we are going to define what will prevent cancer cells from circumventing the targeted pathways.

3.4. Rationalized Molecular Drug Combinations

Cell signaling gone awry in cancer allows not only for the unregulated cellular activities that define cancer, but also allow cancer cells to evade or escape attempts at single agent therapy. For example, the G-protein Ras stimulates the Ras/Raf/MAPK pathway, (a cell proliferation pathway) however, this stimulation can be inhibited by wortmannin, which is a PI3K pathway (a cell survival pathway) inhibitor. In contrast, stimulation of the Ras/Raf/MAPK pathway via receptor tyrosine kinases such as EGFR is insensitive to wortmannin [41]. Further, EGFR overexpression which has been implicated in breast and ovarian cancer, was inhibited by the EGFR inhibitor CI-1033 in MDA-MB-435 breast cancer cells. Even though reduced levels of PKB/AKT were seen in the MDA-MB-453 breast cancer cells, this was not sufficient to induce apoptosis in these cells. In addition, exposure of gemcitabine to MDA-MB-453 cells resulted in the activation of Erk1/2 with low levels of apoptosis. However, the combination of CI-1033 and gemcitabine 24hrs later in the MDA-MB-435 cells resulted in the suppression of PKB/AKT and Erk1/2 with a significant increase in apoptosis over either of the two agents alone [42]. These examples illustrate the importance of knocking out multiple targets especially in light of the cross-talk that can exist between intracellular pathways. In the context of drugs targeting ILK, studies from our lab have demonstrated that the combination of the EGFR inhibitor PD153035 and an ILK inhibitor (KP392) [43] and the combination of PD153035 and the ILK inhibitor (KP307-2) [44] can result in an additive and synergistic drug combination effect on SF-188 and U87MG glioblastoma cells respectively (see Figure 3).

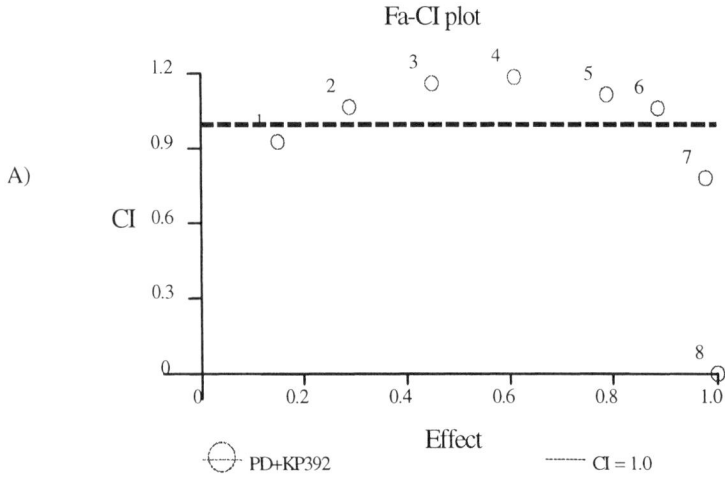

A)

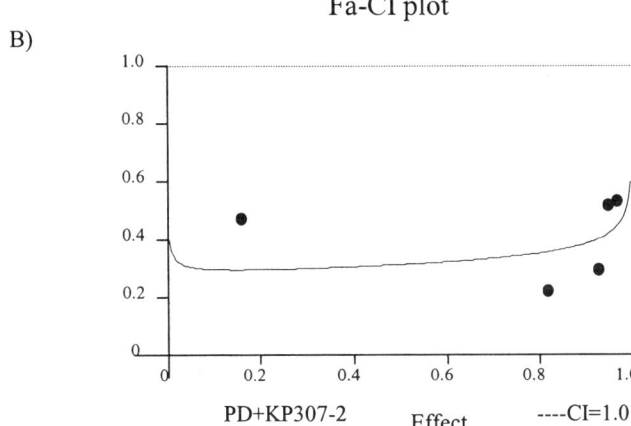

Fa-CI plot

B)

PD+KP307-2   Effect   ----CI=1.0

Figure 3. (A) fraction affected (Fa) versus combination index (CI) plot where the dotted horizontal line represents CI=1, indicating an additive effect of the EGFR inhibitor PD153035 and the ILK inhibitor KP392 in combination against SF-188 glioblastoma cells. Data points (open circles) moving from left to right represent increasing drug combination concentrations. (B) A similar Fa versus CI plot this time indicating a strong synergistic effect of the EGFR inhibitor PD153035 and the ILK inhibitor KP307-2 against U87MG glioblastoma cells .Data points (closed circles) moving from left to right represent increasing drug combination concentrations. Both graphs were generated using the software program CalcuSyn.

Interestingly, studies have also shown a synergistic interaction can be obtained with an ILK inhibitor and the EGFR inhibitor PD153035 when tested in vitro against U87MG cells. This is illustrated in Figure 4 through the use of a dose reduction index (DRI) plot obtained from data generated by CalcuSyn. The dose reduction index determines how much of each drug in a synergistic combination may be reduced at a given effect level versus the doses for each agent alone. In this case the plot was generated at an effect level in which 90% cell kill is achieved i.e. Fa=0.9 (see Figure 4). The same level of cytotoxicity can be achieved by using a drug combination at a dose that is 5-fold lower for the ILK inhibition and 6-fold lower for the EGFR inhibitor. The DRI really highlights one of the primary outcomes associated with use of
a synergistic drug combination; where reduced amounts of drug in combination will achieve the same effect (an affect on cell viability) but at a lower dose. It can be argued that this synergy may, in turn, lead to effect control of tumor growth at dose levels that would exhibit reduced toxicity.

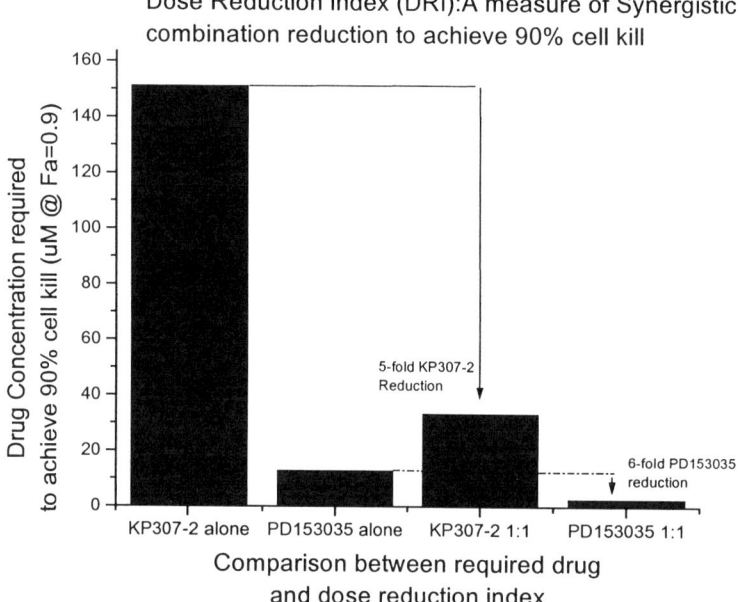

Figure 4. The DRI of the ILK inhibitor KP307-2 and the EGFR inhibitor PD153035 that results in a synergistic drug combination upon exposure to the U87MG glioblastoma cells.

3.5. The Evaluation of Targeted Molecular Drug Combinations (in vivo)

The evaluation of drug combinations via the Chou and Talalay CalcuSyn program is an important tool in defining the potential of molecularly targeted drug combinations. The ability of this program to give a quantitative assessment of a synergistic, antagonistic or an additive drug effect is a crucial first step in the development of novel drug combinations for the treatment of cancer. However, this evaluation is primarily limited to analysis of *in vitro* based model systems where cell culture assays typically provide a measure of activity based on drug concentrations required to inhibit cell proliferation or induce cell toxicity. Unfortunately, even the most promising results obtained using this approach may ultimately prove ineffective in more advanced pre-clinical and clinical models. This problem is highlighted dramatically by the results of *in vitro* assessments to define an optimal therapeutic drug combinations. While it is possible to regulate and maintain the desired concentration of combined drugs in the highly controlled environment of an in vitro assay, this level of control is immediately lost in the in vivo setting, where drugs are rarely delivered to target cells in a site-specific manner.

Further, drug biodistribution and metabolism become a major factor influencing the level of the drugs at the target site. Thus it is anticipated that two drugs displaying synergistic interactions *in vitro* may not display synergy *in vivo* due to the independent pharmacokinetic, biodistribution, and metabolic behavior of the individual drug. Although the evaluation of conventional chemotherapeutic agents from *in vitro* to *in vivo* has resulted in the current chemotherapeutic agents commonly used for the treatment of cancer; it is clear that emerging therapies aimed at combining drugs which block kinase activity by way of kinase inhibitors, or the successful delivery of therapeutic antisense, or peptide molecules, will require the development of novel approaches to ensure optimal delivery of both drugs prior to clinical use.

Several investigators have outlined the complexities associated with developing these novel therapeutics for use as clinically relevant drugs [45-48]. Certainly one of the most important issues in the use of these novel therapeutics is to define better clinical endpoints, such as modulation of the target protein activity or level and to incorporate proxy endpoints that may aid in tailoring therapeutic dosing and dose schedules [49]. Indeed, from a pharmacokinetic perspective it is essential that therapeutically effective concentrations of the agents are achieved in the plasma and/or at the tumor site. There must be sufficient drug in the body, at the right site(s), in a bioavailable form and maintained for sufficient duration in order to see a therapeutic response. The failure of novel therapeutics often arises at the level of dosing, where ineffective plasma concentrations result from a lack of understanding of appropriate dose-response effects. Perhaps a more pharmacological rather than toxicological approach should be employed, with greater emphasis placed on defining the optimal in vivo biological dose rather than the maximum tolerated dose. More rigorous dosing schedules such as infusions or the use of loading doses may be required to obtain a therapeutic response in vivo. Yet another level of complexity arises when pharmacokinetic interactions are observed between the drug combinations. However, molecularly targeted approaches aimed at target proteins/molecules in different pathways may be less susceptible to this drawback, giving it another advantage over conventional drug combination approaches.

4. CONCLUSIONS

Understanding cell signaling is central to further understanding the mechanisms that go awry in the development and progression of cancer. In addition, knowledge of cell signaling is important for defining potential molecular targets that make up the cell signaling machinery and to improve upon existing cancer treatment. In determining molecular targets that are important for cancer progression, cell signaling cross-talk must be considered. The possibility of pathway cross-talk which may allow for the continuation of

72

cancer progression even after effective inhibition of one pathway is achieved with a single agent is an argument for molecular targeted drug combinations. An evaluation of drug combinations for the treatment of cancer using the methods like those developed by Chou and Talalay can provide important drug combination effect information. In turn, investigators and clinicians are hoping that significant improvements in efficacy and reduced drug induced toxicity will be achieved targeting cell signaling proteins in combination with existing drugs as well as novel agents.

Inhibition of integrin linked kinase (ILK) will be of therapeutic value in the treatment of cancer. Particularly when it is used in combination with established chemotherapeutic agents as well as other novel agents effecting parallel dysregulated pathways involving receptor tyrosine kinases. There is sufficient evidence that ILK is a valid molecular target, playing a central role not only for the formation, progression and maintenance of cancer, but is also a key component in the cross-talk that can exist between pathways (i.e. the PI3K/PKB/AKT and Ras/Raf/MAPK pathway). The preclinical development of agents known to control the activity and/or expression of ILK seems warranted. The potential for further drug combinations that involve ILK are not limited to the treatment of glioblastoma, ovarian and breast cancer, but for potentially all cancers in which these cell signaling pathways are central for that cancer's formation, progression and survival. The leap to clinically relevant combinations will require new therapeutic criteria than what has been seen for conventional chemotherapy. Our group is currently focused on generating preclinical data with the specific aim of providing compelling evidence to support the clinical development of agents targeting ILK activity. With specific targets such as ILK in mind, and encouraged by the success of Gleevec ^{TM}, the improvements in cancer treatment and response using this approach should be substantial.

5. ACKNOWLEDGEMENTS

This work was supported by the National Cancer Institute of Canada in the form of operating grants (SD), a Terry Fox New Frontiers Initiative Grant and a grant from the Breast Cancer Research Initiative of the NCIC.

Lincoln A. Edwards
Advanced Therapeutics, BC Cancer Agency and Research Centre, Department of Pathology & Laboratory Medicine, The University of British Columbia

Jennifer Shabbits
Advanced Therapeutics, BC Cancer Agency and Research Centre

Marcel Bally , Advanced Therapeutics, BC Cancer Agency and Research Centre, Celator Technologies, Department of Pathology & Laboratory Medicine, The University of British Columbia

Shoukat Dedhar,
BC Cancer Agency; Cancer Biology Program, Department of Biochemistry and Molecular Biology, The University of British Columbia, Jack Bell Research Centre

6. REFERENCES

1. Hannigan, GE., Leung-Hagesteijn, C., Fitz-Gibbon, L., Coppolino, MG., Radeva, G., Filmus, J., Bell, JC., Dedhar, S. (1996). Regulation of cell adhesion and anchorage-dependent growth by a new beta 1-integrin-linked protein kinase. *Nature* 379: 91-96.
2. Giancotti, FG & Ruoslahti, E. (1999). Integrin signaling. *Science* 285: 1028-1032.
3. Tu, Y., Huang, Y., Zhang, Z., Hua, Y., Wu, C. (2001). A new focal adhesion protein that interacts with integrin-linked kinase and regulates cell adhesion and spreading. *J Cell Biol* 153: 585-598.
4. Yamaji, S., Suzuki, A., Sugiyama, Y., Koide, Y., Yoshida, M., Kanamori, H., Mohri, H., Ohno, S., Ishigatsubo, Y. (2001). A novel integrin-linked kinase-binding protein, affixin, is involved in the early stage of cell-substrate interaction. *J Cell Biol* 153: 1251-1264.
5. Nikolopoulos, SN., Turner, CE. (2001). Integrin-linked kinase (ILK) binding to paxillin LD1 motif regulates ILK localization to focal adhesions. *J Biol Chem* 276: 23499-23505.
6. Guo, L., Wu, C. (2002). Regulation of fibronectin matrix deposition and cell proliferation by the PINCH-ILK-CH-ILKBP complex. *FASEB J* 16: 1298-1300.
7. Lynch DK., Ellis, CA., Edwards, PA, Hiles, ID (1999). Integrin-linked kinase regulates phosphorylation of serine 473 of protein kinase B by an indirect mechanism. *Oncogene* 18: 8024-8032.
8. Graff, JR., Deddens, JA., Konicek, BW, Colligan, BM., Hurst, BM., Carter, HW, Carter, JH. (2001). Integrin-linked kinase expression increases with prostate tumor grad. *Clin Can Res* 7: 1987-1991.
9. Stambolic, V., Suzuki, A., de la Pompa, JL., Brothers, GM., Mirtsos, C., Sasaki, T., Ruland, J., Penninger, JM., Siderovski, DP., Mak, TW. (1998). Negative regulation of PKB/AKT-dependent cell survival by the tumor suppressor PTEN. *Cell* 95: 29-39.
10. D'Amico, M., Hulit, J., Amanatullah, DF., Zafonte, BT., Albanese, C., K-I., Moon, RT., Davis, R, Lisanti, MP., Shtutman, M., Zhurinsky, J., Ben-Ze'ev, A., Troussard, AA., Dedhar, S., Pestell, RG. (2000). The integrin-linked kinase regulates the cyclin D1 gene through glycogen synthase kinase 3beta and cAMP-responsive element-binding protein-dependent pathways. *J Biol Chem* 275: 32649-32657.
11. Radeva, G., Petrocelli, T., Behrend, E., Leung-Hagesteijn, C., Filmus, J., Slingerland, J., Dedhar, S. (1997). Overexpression of the integrin-linked kinase promotes anchorage-independent cell cycle progression. *J Biol Chem* 272: 13937-13944.
12. Nicholson, KM., Anderson, NG. (2002). *Cell Signal.* 14: 381-395.
13. Persad, S., Attwell, S., Gray, V., Mawji, N., Deng, JT., Leung, D., Yan, J., Sanghera, J., Walsh, MP., Dedhar, S. (2001). Regulation of Protein Kinase B/Akt-Serine 473 Phosphorylation by Integrin-linked Kinase. *J Biol Chem* 276(29): 27462-27469.
14. Kiss, E., Muranyi, A., Csortos, C., Gergely, P., Ito, M., Hartshorne, DJ, Erdodi, F .(2002). Integrin-linked kinase phosphorylates the myosin phosphatase target subunit at the inhibitory site in platelet cytoskeleton. *Biochem J* 365: 79-87.
15. Persad, S., Attwell, S., Gray, V., Delcommenne, M., Troussard, A., Sanghera, J., Dedhar, S. (2000). Inhibition of integrin-linked kinase (ILK) suppresses activation of protein

74

kinase B/Akt and induces cell cycle arrest and apoptosis of PTEN-mutant prostate cancer cells. *Proc Natl Acad Sci USA* 97: 3207-1322.

16. Wu, C., Keightley, SY., Leung-Hagesteijn, C., Radeva, G., Coppolino, M., Goicoechea, S., McDonald, JA., Dedhar, S. (1998). Integrin-linked protein kinase regulates fibronectrin matrix assembly, E-cadherin expression, and tumorigenicity. *J Biol Chem* 273: 528-536.

17. Attwell, S., Roskelley, C., Dedhar, S. (2000). The integrin-linked kinase (ILK) suppresses anoikis. *Oncogene* 19: 3811-3815.

18. Downward, J (1998). Mechanisms and consequences of activation of protein kinase B/Akt. *Curr Opin Cell Biol* 10: 262-267.

19. Cardone, MH., Roy, N., Stennicke, HR et al. (1998). Regulation of cel death protease caspase-9 by phosphorylation. *Science* 282: 1318-1321.

20. Troussard, AA., Tan, C., Yoganathan, TN., Dedhar, S (1999). Cell-extracellular matrix interactions stimulate the AP-1 transcription factor in an integrin-linked kinase- and glycogen synthase kinase 3-dependent manner. *Mol Cell Biol* 19: 7420-7427.

21. Tan, C., Costello, P., Sanghera, J., Dominguez, D., Baulida, J., de Herreros, AG., Dedhar, S. (2001) Inhibition of integrin linked kinase (ILK) suppresses beta-catenin-Lef/Tcf-dependent transcription and expression of the E-cadherin repressor, snail, in APC-/-human colon carcinoma cells. *Oncogene* 20: 133-140.

22. White, DE., Cardiff, RD., Dedhar, S., Muller, WJ. (2001). Mammary epithelial-specific expression of the integrin-linked kinase (ILK) results in the induction of mammary gland Hyperplasias and tumors in transgenic mice. *Oncogene* 20: 7064-7072.

23. Chung, DH., Lee, JI., Kook, MC., Kim, JR., Kim, SH., Choi, EY., Park, SH., Song, HG. (1998). ILK (beta 1 integrin-linked protein kinase): A novel immunohistochemical marker for Ewing's sarcoma and primitive neuroectodermal tumor. *Virchows Arch* 433: 113-117.

24. Ahmed, N., Riley, C., Olica, K., Stutt, E., Rice, GE., Quinn, MA. (2003). Integrin-linked kinase expression increases with ovarian tumour grade and is sustained by peritoneal tumour fluid. The Journal of Pathology (In press)

25. Berx, G., Van Roy, F. (2001). The E-cadherin/catenin complex: An important gatekeeper in breast cancer tumorigenesis and malignant proression. *Breast Cancer Res* 3: 289-293.

26. Marotta, A., Tan, C., Gray, V., Malik, S., Gallinger, S., Sanghera, J., Dupuis, B., Owen, D., Dedhar, S., & Salh, B. (2001). Dysregulation of integrin-linked kinase (ILK) signaling in colonic polyposis. *Oncogene* 20: 6250-6257.

27. Wang, SC., Makino, K., Xia, W., Kim, JS., Im, SA., Peng, H., Mok, SC., Singletary, SE., Jung, MC. (2001). DOC-2/hDab-2 inhibits ILK activity and induces anoikis in breast cancer cells through an Akt-independent pathway. *Oncogene* 20: 6960-6964.

28. Takada, T., Noguchi, T., Inagaki, K., Hosooka, T., Fukunaga, K., Yamao, T., Ogawa, W., Matozaki, T., Kasuga, M. (2002). Induction of apoptosis by stomach cancer-associated protein-tyrosine phosphatase-1 (SAP-1). *J Biol Chem* 277: 34359-34366.

29. Leung-Hagesteijn, C., Mahendra, A., Naruszewicz, I., Hannigan, GE. (2001). Modulation of integrin signal transduction by ILKAP, a protein phosphatase 2C associating with the integrin-linked kinase, ILK1. *EMBO J* 20(9): 2160-2170.

30. Dedhar, S. (2000). Cell-substrate interactions and signaling through ILK. *Curr Opin Cell Biol* 12: 250-256.

31. Marotta, A., Parhar, K., Owen, D., Dedhar, S., Salh, B. (2003). Characterisation of integrin-linked kinase siganlling in sporadic human colon cancer. *BJC* 88: 1755-1762.

32. Delecommenne, M., Tan C., Gray, V., Rue, L., Woodgett, J., Dedhar, S. (1998). Phosphoinositide-3-OH kinase-dependent regulation of glycogen synthase kinase 3 and protein kinase B/AKT by the integrin-linked kinase. *Proc Natl Acad Sci USA* 95: 11211-11216.

33. Chou, T.-C. and Talalay, P. (1984). Generalized equations for the analysis of inhibitors of Michaelis-Menten and higher order kinetic systems with two or more mutually exclusive and nonexclusive inhibitors. *Europ J Biochem* 115:207-217.

34. Greco, WR., Bravo, G., Parsons, JC. (1995). The Search for Synergy: A Critical Review from a Response Surface Perspective. *The Amer Soc Pharm Exp Ther* 47(2): 332-3382.

35. Chou, T-C. (1974). Relationships between inhibition constants and fractional inhibitions in enzyme-catalyzed reactions with different numbers of reactants, different reaction mechanisms, and different types of mechanisms of inhibition. *Mol Pharmacol* 10: 235-247.
36. Chou, T.C. (1976). Derivation and properties of Michaelis-Menten type and Hill type equeations for reference ligands. *J Theoret Biol* 65: 345-356.
37. Chou, T-C., and Talalay, P. (1983). Analysis of combined drug effects: A new look at a very old problem. *Trends Pharmacol Sci* 4: 450-454.
38. Gorre, ME., Mohammed, M., Ellwood, K., Hsu, N., Paquette, R., Rao, N., Sawyers, CL (2001). Clinical Resistance to STI-571 Cancer Therapy Caused by BCR-ABL Gene Mutation or Amplification. *Science* 293: 876-880.
39. Barthe, C., Cony-Makhoul, P., Melo, JV., Reiffers Francois-Xavier Mahon, J., Hochhaus, A., Kreil, S., Corbin, A., Rosee, PL., Lahaye, T., Berger, U., Cross, NCP., Linkesch, W., Druker, BJ., Hehlmann, R., Passerini, CG., Corneo, G., D'Incalci, M., Gorre, M., Shah, N., Ellwood, K., Nicoll, J., Sawyers, CL. (2001). Roots of Clinical Resistance to STI-571 Cancer Therapy. *Science* 21: 2163.
40. Chou, T-C., Rideout, D., Chou, J., Bertino, JR. (1991). Chemotherapeutiv Synergism, Potentiation and Antagonism. *Encyclopedia of Human Biology* 2: 371-379.
41. Lopez-Ilasaca, M., Crespo, P., Pellici, PG., Gutkind, JS., Wetzker, R. (1997). Linkage of G protein-coupled receptors to the MAPK signaling pathway through PI 3-kinase γ. *Science* 275: 394-397.
42. Nelson, JM., Fry, D. (2001). Akt, MAPK (Erk1/2), and p38 Act in Concert to Promote Apoptosis Response to ErbB Receptor Family Inhibition. *JBC* 276(18): 14842-14847.
43. Tan, C., Mui, A., Dedhar, S (2002). Integrin linked kinase regulates Inducible Nitric Oxide Synthase and Cyclooxygenase-2 (COX-2) Expression in a NF-κappa B- Dependent Manner *JBC* 277(5): 3109-16.
44. Yoganathan, N., Yee, A., Zhang, Z., Leung, D., Yan, J., Fazli, L., Kojic, DL., Costello, PC., Jabali, M., Dedhar, S., Sanghera, J (2002). Integrin-linked kinase, a promising cancer therapeutic target biochemical and biological properties. *Pharm & Therapeut* 93: 233-242.
45. Workman, P. (2000). *Curr Opin Oncol Endocrin Metab Invest Drugs* 2: 21-25.
46. Sausville, EA. (2000). *Anti-Cancer Drug Design* 15: 1-2.
47. Hudes, G. (1999). *J Clin Oncol* 17: 1093-1094.
48. Eisenhauer, EA. (1998). *Annals Oncol* 9: 1047-1052.
49. Morin, MJ. (2000). From oncogene to drug: development of small molecule tyrosine kinase inhibitors as anti-tumor and anti-angiogenic agents. *Oncogene* 19: 6574-6583.

P21-ACTIVATED KINASE 1: AN EMERGING THERAPEUTIC TARGET

RATNA K. VADLAMUDI AND RAKESH KUMAR

1. INTRODUCTION

The p21-activated kinase 1 (Pak1), is a serine/threonine kinase needed for a various cellular functions, including morphogenesis, motility, mitosis, angiogenesis and survival [1-4]. Pak1 has been identified as a target of the activated Rho GTPases Cdc42 and Rac1, which stimulate Pak1 autophosphorylation and activity [5]. Stimulation of Pak1 activity results in several phenotypic changes reminiscent of those produced by Cdc42 and Rac1 [6, 7]. Pak1 is widely expressed in numerous tissues and is activated by several polypeptide factors and extracellular signals in both a GTPase-dependent (through CDC42 and Rac1) and a GTPase-independent manner through its localization to membrane or focal adhesions [8-11]. Pak1 is also activated by lipids [9], tyrosine kinases [12-14], and novel substrates such as filamin [15] and G-proteins [16].

Pak1 regulates cytoskeletal reorganization in a kinase-dependent and -independent manner by protein-protein interactions. A number of interacting or substrate proteins have been identified, including actin, filamin, myosin light-chain kinase, LIM kinase, stathmin, vimentin, and desmin [3]. In addition, Pak1 stimulates a number of signaling pathways, including the mitogen-activated protein kinase MAPKs-p42/44 and -p38, Jun N-terminal kinase, and nuclear factor kappa B pathways, thus indirectly contributing to activation of gene transcription [17-19]. Pak1 was also shown to phosphorylate the C-terminal binding protein, triggering its cellular redistribution and blocking its corepressor functions, thus indirectly activating transcription of its repressed target genes [20]. The complexity of the Pak1 signaling pathway provides multiple opportunities to design interfering drugs or small molecules to modulate selective pathways that are hyperactivated in pathologic conditions. In this review we will summarize the key evidence that suggests the novel functions of Pak1 and discuss the possibility of targeting the unique Pak1-mediated signaling pathway in drug development.

2. PAK1 AS A THERAPEUTIC TARGET

2.1. Pak1Role in Cell Motility

Actin cytoskeleton reorganization is a fundamental component of several cellular processes, including determination of cellular polarity, shape, and motility, all of which are controlled by spatial and temporal polymerization and depolymerization of actin [21, 22]. Pak1 phosphorylate myosin light chain kinase, and this phosphorylation inhibits activation of the later kinase, thus reducing the phosphorylation of myosin light chain [23]. Pak1 also stimulates the activity of LIM kinase 1 and inturn, increases the phosphorylation and inactivation of cofilin, leading to reduced actin filament depolymerization [24]. Activation of Pak1 has been shown to be accompanied by the disassembly of stress fibers and focal adhesion complexes [25, 26] and for the maintenance of the integrity of the cells leading edge [6, 7] necessary for motility, which itself is a fundamental requirement of a productive cellular invasion. Adam *et al.* [25] showed a mechanistic role for Pak1 activation in the increased cellular invasion of breast cancer cells by heregulin. Furthermore, expression of a catalytically inactive Pak1 mutant in the highly invasive breast cancer cell lines MDA-MBA-435 and MDA-231 led to stabilization of stress fibers, enhanced cell spreading, reduction in the activation of the Jun N-terminal kinase/AP1 pathway, and reduced invasiveness [27]. Conversely, conditional expression of catalytically active T423E Pak1 in the noninvasive breast cancer cell line MCF-7 promoted cellular migration and anchorage-independent growth [26]. Pak1 activity also has been shown to play a role in directional motility [7, 28]. Pak1-mediated phosphorylation and inactivation of the NF2, a tumor-suppressor gene, play a role in tumor cell spreading and metastasis [29]. Pak1 signaling also plays a role in intermediate filament and microtubule reorganization. Pak1 phosphorylates desmin and inhibits its ability to bind intermediate filaments [30] aswell as regulating the reorganization of vimentin filaments through phosphorylation [31]. Pak1 phosphorylates stathmin at Serine 16 and thus plays a role in stabilizing microtubules [32]. The ability of Pak1 to phosphorylate multiple substrates involved in cytoskeletal reorganization and the effect of dominant-negative Pak1 to substantially reduce the migratory potential of cancer cells suggest that targeting Pak1 may be useful in modulating cancer cell motility.

2.2. Role in Cell Cycle

An earlier study of Pak homologues in *Saccharomyces cervisiae* (Ste20) showed that Paks may have a role in cytokinesis and mitosis [33]. In breast cancer cells, regulated expression of kinase-active Pak1 resulted in abnormal organization of mitotic spindles, characterized by the appearance of multiple spindle orientations [26]. Later studies have also shown the presence of Pak1 in the nuclear compartment with a putative role in chromatin remodeling. At the onset of mitosis, Pak1 becomes activated and translocates to the nucleus. Histone H3 was identified as a specific interacting substrate of Pak1, that phosphorylates histone H3 on Serine 10 both in vitro and in vivo [34]. Deregulation of Pak1 signaling thus may influence DNA ploidy and may contribute to anchorage-independent growth and other chromosomal abnormalities observed in tumor cells. According to one study Pak1 is phosphorylated during mitosis in mammalian fibroblasts [35]; another study suggested that cyclin B1/Cdc2 phosphorylates Pak1 in cells undergoing mitosis [36]. Collectively these studies suggest that Pak1 play an important role in cell cycle progression and that targeting Pak1 activity or blocking its localization to the nuclear compartment may have therapeutic value.

2.3. Role in Cancer

Several recent studies have suggested Pak1's involvement in cancer. Comparative genomic hybridization array and tumor tissue microarrays studies showed increased DNA copy number gains in Pak1. The 11q13-q14 amplicon, represented by six oncogenes (CCND1, FGF4, FGF3, EMS1, GARP, and Pak1) revealed preferential gene copy number gains of Pak1, which is located at 11q13.5-q14. Pak1 copy number gains were observed in 30% of the ovarian carcinomas and PakI protein was expressed in 85% of the tumors [37]. Pak1 expression is deregulated in breast cancers and increased Pak1 activity correlates to some extent with the invasiveness of human breast cancer cells and tumors [26, 27]. Increased expression of Pak1 in breast cancer cells stimulated cyclin D1 promoter activity, elevated levels of cyclin D1 mRNA and protein, and nuclear accumulation of cyclin D1. Conversely, Pak1 inhibition by an auto-inhibitory peptide ie., amino acids 83-149 or Pak1 knockdown by short-interference RNA markedly reduced the expression and nuclear levels of cyclin D1, suggesting the requirement of a functional Pak1 pathway for optimal expression of cyclin D1 [38].

Tumor growth and progression are intimately linked with the process of angiogenesis. Pak1 signaling is required for vascular endothelial growth factor expression and, consequently for its function and angiogenesis [39]. Pak1 has been shown to be a potent modulator of endotheial cell migration [40, 41]. Heregulin regulates angiogenesis by means of upregulation of vascular endothelial growth factor expression through activation of Pak1 [39]. Accordingly, Kiosses and collegues showed that treatment of cells with a Pak N-terminal peptide specifically inhibits endothelial cell migration and contractility, in a manner similar to that of treatment with a full-length dominant-negative Pak, suggesting a role for Paks in angiogenesis [41].

The expression of a catalytically active Pak1 mutant stimulates anchorage-independent growth of breast cancer cells in soft agar in a preferential MAPK-sensitive manner [26]. In another study, researchers observed that inhibition of protein kinase A allows anchorage-independent stimulation of the MAPK cascade by growth factors through Pak1 [42]. Pak1 effects the cell survival pathway by directly phosphorylating and inactivating the proapoptotic functions of Bad . Overexpression of a constitutively-active Pak1 T423E mutant promotes the survival of NIH 3T3 murine fibroblasts, while overexpression of the autoinhibitory domain of Pak1 ie., amino acids 83 to 149 enhances apoptosis [43].

Pak1 activation has been shown to have a role in cellular transformation by ras, an oncogene activated in 30% of human cancers. Evidence suggests that activation of Raf-1 kinase by ras is achieved through a combination of both physical interaction and indirect mechanisms involving the activation of a second ras effector, PI 3-kinase, which directs Pak-mediated regulatory phosphorylation of Raf-1 [44]. ras effectors also cooperate with the PI 3-kinase to inactivate Pak activity. However, results from studies with rat fibroblasts suggest that Pak activation may be necessary, but not sufficient, for cooperative transformation of Rat-1 fibroblasts by ras, or Rac, and other Rho GTPases [45, 46]. Rat-1 cell lines expressing Pak1 (K299R) were highly resistant to ras transformation, while cells expressing wild-type Pak1 were efficiently transformed by ras. Pak1 (L83, L86, R299), a mutant that fails to bind either Rac or Cdc42, also was resistant to ras transformation [45].

The PI-3 kinase-Pak1-p38MAPK pathway has been shown to be hyperactivated in breast tumors. Furthermore, PI-3 kinase and Pak1 are co-overexpressed in breast tumors with Pak1 having higher activity in grade II breast tumors [47]. Etk/Bmx, a member of the Tec family of nonreceptor protein-tyrosine kinases, directly associates with Pak1 by means of its N-terminal pleckstrin homology domain and also phosphorylates Pak1 on tyrosine residues [13]. The constitutively active mutant epidermal growth factor receptor induces tyrosine-phosphorylated forms of Pak1. The catalytic activity of the Pak family of serine/threonine kinase is stimulated by tyrosine phosphorylation [12]. Pak1-mediated phosphorylation and inactivation of NF2 may play a role in tumor cell spreading and metastasis [29].

Abnormalities in the expression and signaling pathways downstream of the epidermal growth factor receptor (EGFR) contribute to malignant transformation in human cancers. The blockade of HER2 receptor by an anti-HER2 monoclonal antibody results in the inhibition of HRG-mediated stimulation of the PI-3 kinase-Pak pathway and also the formation of motile actin cytoskeleton structures [25]. Platelet-derived growth factor (PDGF)-induced p38MAPK activation is also blocked by expression of a mutant form of the Pak1 binding proteins β-pix SH3m (W43K) and β-pix DHm (L238R,L239R) suggesting a role for the Pak1-β-Pix pathway in PDGF -mediated cytoskeletal reorganization [48]. Hepatocyte growth factor, the ligand for the Met receptor tyrosine kinase, is a potent modulator of epithelial-mesenchymal transition and dispersal of epithelial cells, processes that play crucial roles in tumor development, invasion, and metastasis. Hepatocyte growth factor induces activation of the Cdc42/Rac-regulated Pak and its translocation to membrane ruffles and is shown to play a role in inducing epithelial cell spreading [49]. PDGF mediated activation of Pak1 family kinases and cell migration of fibroblasts requires transactivation of EGFR [50]. Overexpression of mutant Pak1 blocked PDGF-induced chemotactic cell migration, suggesting that PDGF may require Pak1-mediated signaling to the p38MAPK and other pathways [51]

The process of tumor progression requires, increased transformation, directional migration, and enhanced cell survival among other steps. Several agents including EGFR tyrosine kinase inhibitor ZD1839 was shown notonly suppress of EGF-induced stimulation of EGFR autophosphorylation but also inhibit Pak1 activity, block EGF-induced cytoskeleton remodeling, cell growth, and in vitro invasiveness of cancer cells; and induce a differentiated squamous cell phenotype. These studies suggest that the EGFR-tyrosine kinase inhibitor ZD1839 may cause potent inhibition of the EGFR and Pak1 pathways, resulting in attenuation of transformed cell phenotypes and induced differentiation in human cancer cells deregulated in these growth factor receptor pathways [52]. These results suggest that targeting upstream activators of Pak1 is also useful in modulating Pak1-mediated functions in cancer.

2.4. Role in Hormonal Resistance

Amplification of loci present on band q13 of human chromosome 11 is a feature of a subset of estrogen receptor-positive breast carcinomas prone to metastasic. Pak1 was identified as one of the genes in these loci [53]. Pak1 also interacts with the ER and functions as a coactivator of ER transcriptional activity. Recently, kinase-active T423E Pak1 transgenic mouse studies have established the role of Pak1 signaling in the development of hyperplasia in mammary epithelium, with an underlying mechanism that involved direct phosphorylation of the ER at Serine 305 and subsequent transactivation independent of estrogen [54]. Estrogen rapidly activates Pak1 activity in a phosphatidylinositol 3-kinase-insensitive manner. Estrogen also induced phosphorylation and perinuclear localization of the cell survival forkhead transcription factor FKHR in the cytoplasm in a Pak1-dependent manner. In addition, Pak1 directly interacted with FKHR and phosphorylated it. Together, these results identify a novel signaling pathway linking estrogen action to Pak1 signaling and Pak1 to FKHR, suggesting that Pak1 is an important mediator of estrogen's cell survival functions [55]. The ability of Pak1 to interact with steroid hormonal receptors suggests that this kinase may play an important role in the cross-talk between steroid hormone receptors and Rho GTPase-mediated signal transduction pathways, which could influence the hormonal independence of tumors. In addition the ability of

Pak1 to induce cyclin D1 expression may also indirectly contribute to hormonal independence.

2.5. Role in Pathogen Entry

Macropinocytosis, a ruffling-driven process that allows the capture of marco material, is an essential aspect of normal cell function [56]. Vesicles are formed when membrane ruffles fuse to generate macropinosomes, which then move along the tip of the actin. Macropinocytosis is dramatically but transiently stimulated by different stimuli, such as phorbol esters or growth factors [57]. Pak1, an effector of Cdc42/Rac1, was shown to specifically regulate macropinocytosis but not clathrin- or receptor-mediated endocytosis [58]. Although the mechanism by which Pak1 regulates macropinocytosis is unknown, earlier studies have suggested that Pak1 activity is essential in the process [58-60]

Modulation of cellular signal transduction pathways is widely used by viruses and other pathogens for entering host cells. Several viruses manipulate the macropinocytosis pathway and use it as a route of entry. Herpes simplex virus type 2, Vaccinia virus, and Myxoma virus use the Pak1 pathway to promote viral replication and entry [61-63]. Salmonella and shigella infections, which are accompanied by profuse macropinocytosis, require Pak1 for their productive infection [64-66]. Althrough the molecular mechanisms by which pathogen-mediated deregulation of Pak facilitates their entry is not clear, the suspicion is that Pak activation facilitates cytoskeletal reorganization and ruffling and thus may help pathogens to enter cells.

Currently available antiviral drugs target essential HIV enzymes but the success of therapy is limited by both drug resistance resulting from mutations and the adverse effects of the drugs used [67]. To overcome these problems, recent efforts have been directed to targeting virus entry, including receptor-mediated entry, using CD4 [68] and chemokine receptors [69]. New evidence also has suggested that HIV uses macropinocytosis to enter macrophages [70] and brain microvascular endothelia [71]. Drugs such as Chlroquine, that block acidification of lysosomes dramatically increased HIV infectivity in one study, suggesting that HIV entry via macropinocytosis may be a factor in the prevalence of AIDS in underdeveloped countries where these drugs are widely used [72]. Interestingly, the HIV and simian immunodeficiency virus accessory protein Nef was shown to interact with and activate Pak, mechanism shown to be important in HIV infection [73-76]. Collectively, these findings point to Pak1 as a potential physiologic target to intervene against viral entry and replication.

3. CONCLUSIONS

The impressive amount of basic research data available on Pak1, the possibility that Pak1 is amplified or overexpressed in tumors and Pak1 ability to modulate motility and cell survival functions have generated considerable anticipation in the field of cancer research that strategies targeting Pak1 may lead to the design of target specific drugs. The identification of unique interacting proteins and the availability of the crystal structure of Pak1 provide an ideal environment to develop novel small molecules or peptides to interfere with Pak1 functions. To date no specific drug targeting Pak1 activity has been developed, althrough several companies are currently attempting to produce such drugs that inhibit Pak1 kinase activity. Similarly, althrough siRNA methods of blocking Pak1 expression is shown to work, the clinical application is still in the very early phases of development. Recent data have suggested that blocking of Pak1 upstream activators may also be benefecial by indirectly blocking Pak activity. Growth factor-induced activation of Pak1 can be inhibited by the PDGF-R-specific inhibitor AG1478 or by the EGF receptor specific inhibitors (Iressa), monoclonal antibody 225 or the HER2 blocker trastuzumab (Herceptin). Therefore these compounds may be clinically useful in regulating Pak activity in growth factor-regulated tumors [50]. Small peptide molecules blocking Pak1 interactions with its substrates could also be used to inhibit Pak1 functions. The feasibility of this approach was demonstrated using a small peptide that blocked the adaptor protein NCK from binding to Pak1 and thus interfered with angiogenesis [41]. Because Pak1 promotes cell survival, and its expression is deregulated in tumors, inhibition of its functions could induce apoptosis in cancer cells. The difference in the levels of expression of the Pak1 between normal and cancer cells could provide a therapeutic opportunity to use drugs to selectively inhibit cancer cells. In addition, drugs causing Pak1 inhibition in combination with other commonly used chemotherapy drugs, may have synergistic effects. The ability of Pak1 to phosphorylate estrogen receptor [54] and induce expression of cyclin D1 [38], both of which activities are widely deregulated in breast cancer, suggests that pak1 inhibitors may be useful in managing or at least delaying hormonal resistance seen in breast cancer. Becasue Pak1 plays an important role in modulating cytoskeleton and because many pathogens use cytoskeleton to enter cells and translocate there, Pak inhibitors may have a therapeutic role in interfering with these pathogenic viral infections. Emerging studies unraveling the novel roles of Pak1 signaling and the possible clinical potential of targeting Pak1 will continue to stimulate efforts to develop Pak1 inhibitors or blockers.

4. ACKNOWLEDGEMENTS

This work was supported in part by NIH grants CA90970, CA65746, and CA80066.

Ratna K. Vadlamudi and Rakesh Kumar
Molecular and Cellular Oncology, The University of Texas M. D. Anderson Cancer Center, Houston, Texas

5. REFERENCES

1. Bokoch, G. M. (2003) Biology of the p21-Activated Kinases. *Annual Review of Biochemistry*, 72:743-781.
2. Jaffer, Z. M., Chernoff, J. (2002) p21-activated kinases: three more join the Pak. *Int J Biochem Cell Biol* 34, 713-717
3. Kumar, R., Vadlamudi, R. K. (2002) Emerging functions of p21-activated kinases in human cancer cells. *J Cell Physiol* 193, 133-144
4. Vadlamudi, R. K., Kumar (2003) p21-activated kinases in human cancer. *Cancer and metastasis Rev* 22:385-393
5. Manser, E., Leung, T., Salihuddin, H., Zhao, Z. S., Lim, L. (1994) A brain serine/threonine protein kinase activated by Cdc42 and Rac1. *Nature* 367, 40-46
6. Sells, M. A., Knaus, U. G., Bagrodia, S., Ambrose, D. M., Bokoch, G. M., Chernoff, J. (1997) Human p21-activated kinase (Pak1) regulates actin organization in mammalian cells. *Curr Biol* 7, 202-210
7. Sells, M. A., Boyd, J. T., Chernoff, J. (1999) p21-activated kinase 1 (Pak1) regulates cell motility in mammalian fibroblasts. *J Cell Biol* 145, 837-849
8. Bokoch, G. M., Wang, Y., Bohl, B. P., Sells, M. A., Quilliam, L. A., Knaus, U. G. (1996) Interaction of the Nck adapter protein with p21-activated kinase (PAK1). *J Biol Chem* 271, 25746-25749
9. Bokoch, G. M., Reilly, A. M., Daniels, R. H., King, C. C., Olivera, A., Spiegel, S., Knaus, U. G. (1998) A GTPase-independent mechanism of p21-activated kinase activation. Regulation by sphingosine and other biologically active lipids. *J Biol Chem* 273, 8137-8144
10. Bagrodia, S., Bailey, D., Lenard, Z., Hart, M., Guan, J. L., Premont, R. T., Taylor, S. J., Cerione, R. A. (1999) A tyrosine-phosphorylated protein that binds to an important regulatory region on the cool family of p21-activated kinase-binding proteins. *J Biol Chem* 274, 22393-22400
11. Zhao, Z. S., Manser, E., Loo, T. H., Lim, L. (2000) Coupling of PAK-interacting exchange factor PIX to GIT1 promotes focal complex disassembly. *Mol Cell Biol* 20, 6354-6363
12. McManus, M. J., Boerner, J. L., Danielsen, A. J., Wang, Z., Matsumura, F., Maihle, N. J. (2000) An oncogenic epidermal growth factor receptor signals via a p21-activated kinase-caldesmon-myosin phosphotyrosine complex. *J Biol Chem* 275, 35328-35334
13. Bagheri, Y., Mandal, M., Taludker, A. H., Wang, R. A., Vadlamudi, R. K., Kung, H. J., Kumar, R. (2001) Etk/Bmx tyrosine kinase activates Pak1 and regulates tumorigenicity of breast cancer cells. *J Biol Chem* 276, 29403-29409
14. Bagheri, Y., Mandal, M., Taludker, A. H., Wang, R. A., Vadlamudi, R. K., Kung, H. J., Kumar, R. (2001) Etk/Bmx tyrosine kinase activates Pak1 and regulates tumorigenicity of breast cancer cells. *J Biol Chem* 276, 29403-29409

15. Vadlamudi, R. K., Li, F., Adam, L., Nguyen, D., Ohta, Y., Stossel, T. P., Kumar, R. (2002) Filamin is essential in actin cytoskeletal assembly mediated by p21-activated kinase 1. *Nat Cell Biol* 4, 681-690

16. Lian, J. P., Crossley, L., Zhan, Q., Huang, R., Coffer, P., Toker, A., Robinson, D., Badwey, J. A. (2001) Antagonists of calcium fluxes and calmodulin block activation of the p21-activated protein kinases in neutrophils. *J Immunol* 166, 2643-2650

17. Zhang, S., Han, J., Sells, M. A., Chernoff, J., Knaus, U. G., Ulevitch, R. J., Bokoch, G. M. (1995) Rho family GTPases regulate p38 mitogen-activated protein kinase through the downstream mediator Pak1. *J Biol Chem* 270, 23934-23936

18. Frost, J. A., Xu, S., Hutchison, M. R., Marcus, S., Cobb, M. H. (1996) Actions of Rho family small G proteins and p21-activated protein kinases on mitogen-activated protein kinase family members. *Mol Cell Biol* 16, 3707-3713

19. Frost, J. A., Swantek, J. L., Stippec, S., Yin, M. J., Gaynor, R., Cobb, M. H. (2000) Stimulation of NFkappa B activity by multiple signaling pathways requires PAK1. *J Biol Chem* 275, 19693-19699

20. Barnes, C. J., Vadlamudi, R. K., Mishra, S. K., Jacobson, R. H., Li, F., Kumar, R. (2003) Functional inactivation of a transcriptional corepressor by a signaling kinase. *Nat.Struct.Biol.* 10, 622-628

21. Hall, A. (1998) Rho GTPases and the actin cytoskeleton. *Science* 279, 509-514

22. Ridley, A. J., Paterson, H. F., Johnston, C. L., Diekmann, D., Hall, A. (1992) The small GTP-binding protein rac regulates growth factor-induced membrane ruffling. *Cell* 70, 401-410

23. Sanders, L. C., Matsumura, F., Bokoch, G. M., de, L. (1999) Inhibition of myosin light chain kinase by p21-activated kinase. *Science* 283, 2083-2085

24. Edwards, D. C., Sanders, L. C., Bokoch, G. M., Gill, G. N. (1999) Activation of LIM-kinase by Pak1 couples Rac/Cdc42 GTPase signalling to actin cytoskeletal dynamics. *Nat Cell Biol* 1, 253-259

25. Adam, L., Vadlamudi, R., Kondapaka, S. B., Chernoff, J., Mendelsohn, J., Kumar, R. (1998) Heregulin regulates cytoskeletal reorganization and cell migration through the p21-activated kinase-1 via phosphatidylinositol-3 kinase. *J Biol Chem* 273, 28238-28246

26. Vadlamudi, R. K., Adam, L., Wang, R. A., Mandal, M., Nguyen, D., Sahin, A., Chernoff, J., Hung, M. C., Kumar, R. (2000) Regulatable expression of p21-activated kinase-1 promotes anchorage-independent growth and abnormal organization of mitotic spindles in human epithelial breast cancer cells. *J Biol Chem* 275, 36238-36244

27. Adam, L., Vadlamudi, R., Mandal, M., Chernoff, J., Kumar, R. (2000) Regulation of microfilament reorganization and invasiveness of breast cancer cells by kinase dead p21-activated kinase-1. *J Biol Chem* 275, 12041-12050

28. Sells, M. A., Pfaff, A., Chernoff, J. (2000) Temporal and spatial distribution of activated Pak1 in fibroblasts. *J Cell Biol* 151, 1449-1458

29. Xiao, G. H., Beeser, A., Chernoff, J., Testa, J. R. (2002) p21-activated kinase links Rac/Cdc42 signaling to merlin. *J Biol Chem* 277, 883-886

30. Ohtakara, K., Inada, H., Goto, H., Taki, W., Manser, E., Lim, L., Izawa, I., Inagaki, M. (2000) p21-activated kinase PAK phosphorylates desmin at sites different from those for Rho-associated kinase. *Biochem Biophys Res Commun* 272, 712-716

31. Goto, H., Tanabe, K., Manser, E., Lim, L., Yasui, Y., Inagaki, M. (2002) Phosphorylation and reorganization of vimentin by p21-activated kinase (PAK). *Genes Cells* 7, 91-97

32. Daub, H., Gevaert, K., Vandekerckhove, J., Sobel, A., Hall, A. (2001) Rac/Cdc42 and p65PAK regulate the microtubule-destabilizing protein stathmin through phosphorylation at serine 16. *J Biol Chem* 276, 1677-1680

33. Faure, S., Vigneron, S., Doree, M., Morin, N. (1997) A member of the Ste20/PAK family of protein kinases is involved in both arrest of Xenopus oocytes at G2/prophase of the first meiotic cell cycle and in prevention of apoptosis. *EMBO J* 16, 5550-5561

34. Li, F., Adam, L., Vadlamudi, R. K., Zhou, H., Sen, S., Chernoff, J., Mandal, M., Kumar, R. (2002) p21-activated kinase 1 interacts with and phosphorylates histone H3 in breast cancer cells. *EMBO Rep* 3, 767-773

35. Thiel, D., Reeder, M., Pfaff, A., Coleman, T., Sells, M., Chernoff, J. (2002) Cell Cycle-Regulated Phosphorylation of p21-Activated Kinase 1. *Curr Biol* 12, 1227
36. Banerjee, M., Worth, D., Prowse, D., Nikolic, M. (2002) Pak1 phosphorylation on t212 affects microtubules in cells undergoing mitosis. *Curr Biol* 12, 1233
37. Schraml, P., Schwerdtfeger, G., Burkhalter, F., Raggi, A., Schmidt, D., Ruffalo, T., King, W., Wilber, K., Mihatsch, M. J., Moch, H. (2003) Combined array comparative genomic hybridization and tissue microarray analysis suggest PAK1 at 11q13.5-q14 as a critical oncogene target in ovarian carcinoma. *Am.J.Pathol.* 163, 985-992
38. Balasenthil, S., Sahin, A. A., Barnes, C. J., Wang, R. A., Pestell, R. G., Vadlamudi, R. K., Kumar, R. (2003) P21-activated kinase-1 signaling mediates cyclin D1 expression in mammary epithelial and cancer cells. *J.Biol.Chem.* .,
39. Bagheri, Y., Vadlamudi, R. K., Wang, R. A., Mendelsohn, J., Kumar, R. (2000) Vascular endothelial growth factor up-regulation via p21-activated kinase-1 signaling regulates heregulin-beta1-mediated angiogenesis. *J Biol Chem* 275, 39451-39457
40. Kiosses, W. B., Daniels, R. H., Otey, C., Bokoch, G. M., Schwartz, M. A. (1999) A role for p21-activated kinase in endothelial cell migration. *J Cell Biol* 147, 831-844
41. Kiosses, W. B., Hood, J., Yang, S., Gerritsen, M. E., Cheresh, D. A., Alderson, N., Schwartz, M. A. (2002) A dominant-negative p65 PAK peptide inhibits angiogenesis. *Circ Res* 90, 697-702
42. Howe, A. K., Juliano, R. L. (2000) Regulation of anchorage-dependent signal transduction by protein kinase A and p21-activated kinase. *Nat Cell Biol* 2, 593-600
43. Schurmann, A., Mooney, A. F., Sanders, L. C., Sells, M. A., Wang, H. G., Reed, J. C., Bokoch, G. M. (2000) p21-activated kinase 1 phosphorylates the death agonist Bad and protects cells from apoptosis. *Mol Cell Biol* 20, 453-461
44. Sun, H., King, A. J., Diaz, H. B., Marshall, M. S. (2000) Regulation of the protein kinase Raf-1 by oncogenic Ras through phosphatidylinositol 3-kinase, Cdc42/Rac and Pak. *Curr Biol* 10, 281-284
45. Tang, Y., Chen, Z., Ambrose, D., Liu, J., Gibbs, J. B., Chernoff, J., Field, J. (1997) Kinase-deficient Pak1 mutants inhibit Ras transformation of Rat-1 fibroblasts. *Mol Cell Biol* 17, 4454-4464
46. Tang, Y., Yu, J., Field, J. (1999) Signals from the Ras, Rac, and Rho GTPases converge on the Pak protein kinase in Rat-1 fibroblasts. *Mol Cell Biol* 19, 1881-1891
47. Salh, B., Marotta, A., Wagey, R., Sayed, M., Pelech, S. (2002) Dysregulation of phosphatidylinositol 3-kinase and downstream effectors in human breast cancer. *Int J Cancer* 98, 148-154
48. Lee, S. H., Eom, M., Lee, S. J., Kim, S., Park, H. J., Park, D. (2001) BetaPix-enhanced p38 activation by Cdc42/Rac/PAK/MKK3/6-mediated pathway. Implication in the regulation of membrane ruffling. *J Biol Chem* 276, 25066-25072
49. Royal, I., Lamarche, V., Lamorte, L., Kaibuchi, K., Park, M. (2000) Activation of cdc42, rac, PAK, and rho-kinase in response to hepatocyte growth factor differentially regulates epithelial cell colony spreading and dissociation. *Mol Biol Cell* 11, 1709-1725
50. He, H., Levitzki, A., Zhu, H. J., Walker, F., Burgess, A., Maruta, H. (2001) Platelet-derived growth factor requires epidermal growth factor receptor to activate p21-activated kinase family kinases. *J Biol Chem* 276, 26741-26744
51. Dechert, M. A., Holder, J. M., Gerthoffer, W. T. (2001) p21-activated kinase 1 participates in tracheal smooth muscle cell migration by signaling to p38 Mapk. *Am J Physiol Cell Physiol* 281, C123-C132
52. Barnes, C. J., Bagheri-Yarmand, R., Mandal, M., Yang, Z., Clayman, G. L., Hong, W. K., Kumar, R. (2003) Suppression of Epidermal Growth Factor Receptor, Mitogen-activated Protein Kinase, and Pak1 Pathways and Invasiveness of Human Cutaneous Squamous Cancer Cells by the Tyrosine Kinase Inhibitor ZD1839 (Iressa). *Mol.Cancer Ther.* 2, 345-351
53. Bekri, S., Adelaide, J., Merscher, S., Grosgeorge, J., Caroli, B., Perucca, L., Kelley, P. M., Pebusque, M. J., Theillet, C., Birnbaum, D., Gaudray, P. (1997) Detailed map of a region

commonly amplified at 11q13-->q14 in human breast carcinoma. *Cytogenet Cell Genet* 79, 125-131

54. Wang, R. A., Mazumdar, A., Vadlamudi, R. K., Kumar, R. (2002) P21-activated kinase-1 phosphorylates and transactivates estrogen receptor-alpha and promotes hyperplasia in mammary epithelium. *EMBO J* 21, 5437-5447

55. Mazumdar, A., Kumar, R. (2003) Estrogen regulation of Pak1 and FKHR pathways in breast cancer cells. *FEBS Lett.* 535, 6-10

56. Conner, SD., and Schmid, SL. (2003) Regulated portals of entry into the cell Nature 422:37-44.

57. Swanson, J. A., Watts, C. (2002) Macropinocytosis. Trends Cell Biol 5, 424-428.

58. Dharmawardhane, S., Sanders, L. C., Martin, S. S., Daniels, R. H., Bokoch, G. M. (1997) Localization of p21-activated kinase 1 (PAK1) to pinocytic vesicles and cortical actin structures in stimulated cells. *J Cell Biol* 138, 1265-1278

59. Dharmawardhane, S., Brownson, D., Lennartz, M., Bokoch, G. M. (1999) Localization of p21-activated kinase 1 (PAK1) to pseudopodia, membrane ruffles, and phagocytic cups in activated human neutrophils. *J Leukoc Biol* 66, 521-527

60. Dharmawardhane, S., Schurmann, A., Sells, M. A., Chernoff, J., Schmid, S. L., Bokoch, G. M. (2000) Regulation of macropinocytosis by p21-activated kinase-1. *Mol Biol Cell* 11, 3341-3352

61. Sieczkarski, S. B., Whittaker, G. R. (2002) Dissecting virus entry via endocytosis. *J.Gen.Virol.* 83, 1535-1545

62. Murata, T., Goshima, F., Daikoku, T., Takakuwa, H., Nishiyama, Y. (2000) Expression of herpes simplex virus type 2 US3 affects the Cdc42/Rac pathway and attenuates c-Jun N-terminal kinase activation. *Genes Cells* 5, 1017-1027

63. Johnston, J. B., Barrett, J. W., Chang, W., Chung, C. S., Zeng, W., Masters, J., Mann, M., Wang, F., Cao, J., McFadden, G. (2003) Role of the serine-threonine kinase PAK-1 in myxoma virus replication. *J.Virol.* 77, 5877-5888

64. Yoshida, S., Katayama, E., Kuwae, A., Mimuro, H., Suzuki, T., Sasakawa, C. (2002) Shigella deliver an effector protein to trigger host microtubule destabilization, which promotes Rac1 activity and efficient bacterial internalization. *EMBO J.* 21, 2923-2935

65. Rudrabhatla, R. S., Sukumaran, S. K., Bokoch, G. M., Prasadarao, N. V. (2003) Modulation of myosin light-chain phosphorylation by p21-activated kinase 1 in Escherichia coli invasion of human brain microvascular endothelial cells. *Infect.Immun.* 71, 2787-2797

66. Chen, L. M., Bagrodia, S., Cerione, R. A., Galan, J. E. (1999) Requirement of p21-activated kinase (PAK) for Salmonella typhimurium-induced nuclear responses. *J.Exp.Med.* 189, 1479-1488

67. Kilby, J. M., Eron, J. J. (2003) Novel therapies based on mechanisms of HIV-1 cell entry. *N Engl J Med* 348, 2228-2238

68. Dalgleish, A. G., Beverley, P. C., Clapham, P. R., Crawford, D. H., Greaves, M. F., Weiss, R. A. The CD4 (T4) antigen is an essential component of the receptor for the AIDS retrovirus. *Nature* 312, 763-767

69. Rubbert, A., Combadiere, C., Ostrowski, M., Arthos, J., Dybul, M., Machado, E., Cohn, M. A., Hoxie, J. A., Murphy, P. M., Fauci, A. S., Weissman, D. (1998) Dendritic cells express multiple chemokine receptors used as coreceptors for HIV entry. *J Immunol* 160, 3933-3941.

70. Marechal, V., Prevost, M. C., Petit, C., Perret, E., Heard, J. M., Schwartz, O. (2001) Human immunodeficiency virus type 1 entry into macrophages mediated by macropinocytosis. *J Virol* 75, 11166-11177

71. Liu, N. Q., Lossinsky, A. S., Popik, W., Li, X., Gujuluva, C., Kriederman, B., Roberts, J., Pushkarsky, T., Bukrinsky, M., Witte, M., Weinand, M., Fiala, M. (2002) Human immunodeficiency virus type 1 enters brain microvascular endothelia by macropinocytosis dependent on lipid rafts and the mitogen-activated protein kinase signaling pathway. *J Virol* 76, 6689-6700.

72. Fredericksen, B. L., Wei, B. L., Yao, J., Luo, T., Garcia, J. V. (2002) Inhibition of endosomal/lysosomal degradation increases the infectivity of human immunodeficiency virus. *J Virol* 76, 11440-11446.

73. Wolf, D., Witte, V., Laffert, B., Blume, K., Stromer, E., Trapp, S., Aloja, P., Schurmann, A., Baur, A. S. (2001) HIV-1 Nef associated PAK and PI3-kinases stimulate Akt-independent Bad-phosphorylation to induce anti-apoptotic signals. *Nat Med* 7, 1217-1224.

74. Nunn, M. F., Marsh, J. W. (1996) Human immunodeficiency virus type 1 Nef associates with a member of the p21-activated kinase family. *J Virol* 70, 6157-6161.

75. Brown, A., Wang, X., Sawai, E., Cheng, M. (1999) Activation of the PAK-related kinase by human immunodeficiency virus type 1 Nef in primary human peripheral blood lymphocytes and macrophages leads to phosphorylation of a PIX-p95 complex. *J Virol* 73, 9899-9907.

76. Schaeffer, E., Geleziunas, R., Greene, W. C. (2001) Human immunodeficiency virus type 1 Nef functions at the level of virus entry by enhancing cytoplasmic delivery of virions. *J Virol* 75, 2993-3000

BASIS AND IMPORTANCE OF SRC AS A TARGET IN CANCER

VICTOR A. LEVIN

1. INTRODUCTION

This chapter will attempt to explain why Src is a potentially important target for cancer therapy. To this end, I will seek to answer four key questions: What is Src and what does it do in the cell? Why is Src important in the propagation of cancer? How could inhibiting the function of Src impact cancer cell proliferation, invasion, and tumorgenicity? Which therapeutic strategies have been developed to inhibit Src activity?

Src, the first known cellular homologue of a viral oncoprotein [1], holds a unique place in the history of molecular biology and in my academic life. Nearly 100 years ago, in 1911, Peyton Rous discovered the virus that was to bear his name and provide new tools to study the process of malignant transformation [2]. Over 25 years ago, Src was heralded as a proto-oncogene and helped initiate a revolution in thinking about cancer biology and genetics [1, 3, 4]. Shortly thereafter it was determined that Src was a protein tyrosine kinase (PTK)[5, 7].

In the early 1980s I organized informal bimonthly meetings to discuss possible approaches for discovering and developing new anticancer therapeutic strategies. The meetings were held under the auspices of the Northern California Cancer Program and brought together interested scientists and clinicians from the University of California San Francisco (UCSF) and Stanford University. At one of the meetings, George Stark, who was then a professor at Stanford University, suggested that it might be timely to develop drugs to inhibit proto-oncogenes such as Src. Thus began a long and, at times, frustrating trip of hope, failure, dogged discovery, and finally success that traversed 15 years of NIH National Cooperative Drug Discovery Group grant support, pharmaceutical support, the support of two academic institutions, (UCSF and The University of Texas M. D. Anderson Cancer Center), and lastly the creation of Signase Inc, a start-up company that, working closely with Tripos Inc., succeeded in discovering and developing selective small molecule inhibitors of Src. Today, at least three pharmaceutical companies are posed to initiate phase I clinical trials with inhibitors of Src tyrosine kinase activity.

When our program first began, there was only modest evidence that Src could be an efficacious target for cancer therapy [6, 8]. Today, substantiating evidence for that idea is considerable and compelling. Concomitantly with knowing a great deal about Src and how it interacts with various cellular signaling moieties and processes, total elucidation is elusive along with our understanding of numerous other signaling molecules that intersect with Src signaling.

2. WHAT IS SRC AND WHAT DOES IT DO IN THE CELL?

Src is one of 32 currently identified cytoplasmic protein tyrosine kinases (PTKs) and the first known member of what today is now called the Src-family of PTKs. Src encodes a non-receptor, membrane-associated PTK that has a special place in oncogenesis [3, 8-11]. Other members of the Src-family are Yes, Fyn, Lyn, Lck, Hck, Fgr, and Blk: as an aggregate they are among the most studied oncoproteins [12]. Src, Yes and Fyn are ubiquitously expressed, whereas the other family members tend to be expressed in a tissue-specific manner. Because of the ubiquitous expression, its potent tyrosine kinase activity relative to other family members, and its frequently increased activity in human tumors, Src is by far the most studied member of the PTK family.

To understand the role of Src in cancer it is important to appreciate that its activation, and its ability to phosphorylate tyrosine (Tyr) residues on proteins, is tightly controlled through phosphorylation and intermolecular interactions. Src is a phosphoprotein, the activation of which increases its ability to phosphorylate tyrosine-containing substrates [13, 14]. In many circumstances, Src amplifies signals emanating from a wide spectrum of cell surface receptors, including G proteins, cytokines, calcium channels, growth factors, integrins, and immune response receptors (Table 1). In addition, Src serves as a scaffolding protein involved in recruiting many signaling enzymes and complexes. Less commonly, Src interacts with receptor protein kinases (RTKs) to downregulate RTK function. Thus, Src plays a central role in a myriad of growth-regulatory cascades.

Table 1. Substrates for Src and role in cancer cell function.

PROTEIN	PATHWAY	ROLE IN CELL	REFERENCES
ACK	I, M	Integrin, ACKs are non-receptor PTK effectors for the small GTPase, Cdc42, but	[15]

PROTEIN	PATHWAY	ROLE IN CELL	REFERENCES
		signaling by these proteins remains poorly defined.	
annexin II	A	Phospholipid binding	[16]
ASAP1	I, M	Arf GTPase-activating protein, ASAP1 localizes to focal adhesions and cycles with focal adhesion proteins when cells are stimulated to move	[17, 18]
caveolin, caveolin-1	R	Caveolar structure or function, bind to integrin-α subunits and recruit Fyn	[19]
connexin 43	C, A	Gap function	[20, 21]
cortactin	M	Cortical actin binding, important in migration	[22, 23]
EGFR	R	Important in mitogenesis	[24]
p120^{CAS}	I, M	Adaptor protein, catenin, cell-cell adhesion	[25]
p125FAK	I, M	Tyrosine kinase, integrin binding and mitotic signaling	[26, 27]
p130^{CAS}	I, M	Adaptor molecule, integrin signaling, Cas functions as a molecule promoting cell movement, cell migration, and cell spreading	[28]
p190RhoGAP	T	GTPase activator of Rho, Src activation decreases in RhoA activity, early integrin signaling induces activation and tyrosine phosphorylation of p190RhoGAP via a mechanism that requires Src, transcription	[29, 30]
p85 PI3 kinase	C	Cell cycle progression	[31]
Paxillin	I	Integrin signaling	[32, 33]
PDGFR	R	Promotes mitogenesis but inhibits chemotaxis and actin reorganization	[17, 18]
PKC	R	Actin reorganization is dependent on PKC-induced Src activation, receptor response	[29]
PLCγ		PI-specific phospholipase C	[34]
RasGAP	C, T	GTPase activator of Ras	[25]
Sam68		RNA binding	[34]
Shc	R, G, I, M	Binds Grb2 and can activate Ras/Raf/Erk signal cascades, important in receptor signaling from wide range of growth factor, G-protein, and integrin receptors	[35, 36]
Stat3	T, A	Stat3 is a transcription factor with Src SH2 domains that are activated by Tyr phosphorylation in response to cytokines and growth factors, protects against apoptosis	[37, 38]
Stat5	T	Stat5 is a transcription factor with Src SH2 domains that are activated by Tyr phosphorylation in response to cytokines and growth factors	[39, 40]
Talin, tensin	I, M	Actin-binding, integrin signaling	[27]
Vinculin	I, M	Actin-binding, integrin signaling	[41]
Integrins	I, M	Cell-substrate adhesion and signaling	
β–catenin,	I, M	Cadherin binding, cell-cell adhesion,	[42, 43]

PROTEIN	PATHWAY	ROLE IN CELL	REFERENCES
γ–catenin		migration	

Pathways: A = apoptosis, C = cell cycle signaling, G = G-protein receptors, I = integrin signaling, M = cell migration, R = cytokine/growth factor receptors, T = transcription

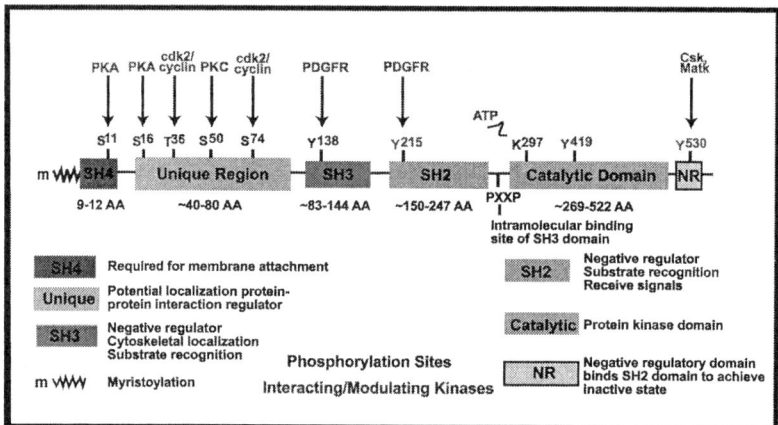

Figure 1. Schematic representation of pp60-src (Src) domains and functionality (adapted with permission from BioSource International, Camarillo, CA).

Src is a 60 kDa protein that is myristoylated at its N-terminus. It has four distinct domains and a linker region. *Figure 1* is a schematic representation of our understanding of Src's structure and the functional interactions implicated in its activation. In its inactive state, phosphorylation of Src's C-terminal Tyr530 residue (527 in mouse and 530 in human) [44-46] maintains a structural interaction with the SH2 domain, and interaction between the Src SH3 domain and the linker region between the SH2 domain and the N-terminal kinase region [47, 48]. The intramolecular interactions between the SH2 and SH3 domains repress kinase activity by steric hindrance of the active catalytic site, which is conformationally blocked by the positioning of its activation loop. Removal of Tyr530 phosphorylation leads to a change in the tertiary structure of Src and its activation state. In another mode of activation, binding through its SH2 domain to specific tyrosine autophosphorylation sites in ligand-stimulated receptor protein tyrosine kinases (RTKs) can result in SH2 displacement from phosphorylated Tyr530 or binding of the SH3 domain to specific proline motifs in target proteins [48, 49]. This mechanism results in autophosphorylation of the Tyr419 (416 in mouse and 419 in human) activation loop and stabilization of the active conformation. *Figure 2* depicts the change in configuration from inactive to activated Src.

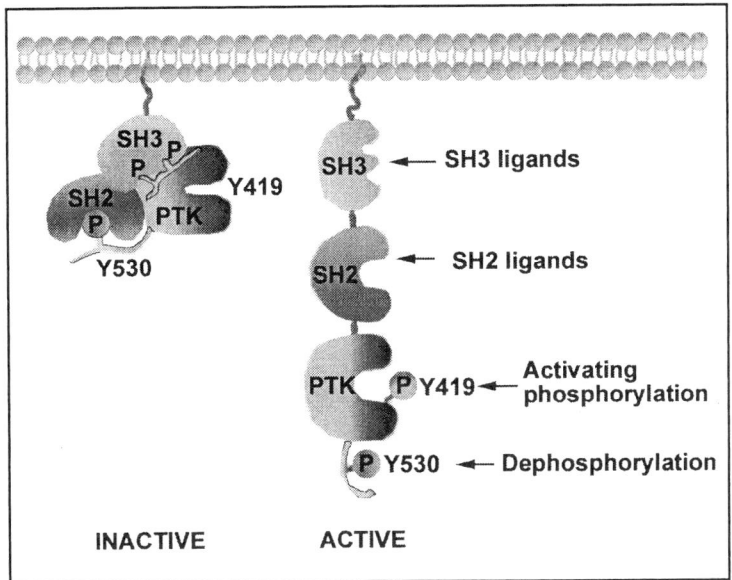

Figure 2. Schematic representation of inactive (closed) and active (open) configurations of human Src (adapted from Blume-Jensen).

In humans, Tyr419 can be a target for intermolecular autophosphorylation [50] and a target for other tyrosine kinases, resulting in Src activation [51]. Phosphorylation of Tyr419 can also prompt Src activity when simultaneous phosphorylation at Tyr530 occurs [13, 52]. This phenomenon suggests that, conversely, when active Src is phosphorylated at Tyr419, it cannot be fully inactivated by phosphorylation at Tyr530 alone, but requires the simultaneous dephosphorylation of Tyr419. This is consistent with the observation that inactive Src family members could be the target of kinases that are capable of phosphorylating Tyr419, resulting in their activation [51]. For instance, during mitosis, phosphorylation of Thr34, Thr46, and Ser72 [53] can destabilize SH2/Tyr530 interactions and sensitize Src to the action of a protein tyrosine phosphatase (PTP) that dephosphorylates Tyr530 [54-57] and inactivates Src. In addition, elevated activity or expression of selective PTPs can dephosphorylate Src Tyr530, including receptor-like PTPα, PTPλ and RPTPε, and the cytoplasmic PTP1B, SHP-1 and SHP-2, which leads to increased Src activity [58]. We also know that kinases, in particular Csk and Matk, can phosphorylate Src at Y530, thereby inhibiting Src PTK activity [57, 59].

Src is localized primarily to the cytosol and the perinuclear region of the cell and, upon activation, is translocated to the cell periphery and the external cell membrane [60-62]. Translocation can serve either to bring Src into direct contact with potential substrates and/or relocate Src, where it becomes a substrate for another signaling molecule. With this dual capability,

Src is able to participate in complex interactions with different proteins in the regulation of cell function. This is demonstrated by the fact that Src-binding proteins such as platelet-derived growth factor (PDGFR) [63, 64] and focal adhesion protein (FAK) [65, 66] can compete for binding to the SH2 domain and elicit Src activation. In addition, p130Cas, a possible docking protein, can bind to Src's SH2 and SH3 domains and facilitate its activation [67].

Src can also be a target for RTK. When the PDGFR is activated, Src can act as a substrate and can be phosphorylated within its SH3 domain at Y138 [68] and within its SH2 domain at Y213 [69], possibly disrupting the intramolecular interactions that stabilize its inactive conformation. Although the phosphorylation of Y138 can activate Src and appears necessary for PDGF-induced DNA synthesis [68], it does not fully correlate with Src activation following PDGF stimulation of cells. Y213, a site phosphorylated in vivo upon PDGF stimulation [69], also appears capable of activating Src. Src can also target receptor tyrosine kinases. Both EGFR and HER2/neu, among others, can be phosphorylated by Src.

Regulation by localization is complex, as exemplified by Src binding to FAK which necessitates the autophosphorylation of FAK at Y397 to create a binding site for the Src SH2 binding domain [70, 71], in turn allowing Src to phosphorylate FAK and increase FAK tyrosine kinase activity [72]. As is typical of many intracellular signaling events, the consequences of the formation of the FAK complex is a downstream activation, in this case of the Ras-MEK-ERK cascade [73], and phosphorylation of the cytoskeletal adapter protein, paxillin [74]. Paxillin is a cytoplasmic protein that localizes to focal adhesions, discrete sites of cell attachment to the extracellular matrix, and can interact with structural and signaling proteins such as vinculin, FAK, PYK2, Src, and Crk [32].

3. WHY IS SRC IMPORTANT IN THE PROPAGATION OF CANCER?

The survival of cancer cells in the body requires them to respond quickly to compete for survival in an environment in which their unrestrained intracellular signaling leads to robust cell proliferation that results in competition for oxygen and nutrients with normal surrounding cells and other cancer cells. In this somewhat hostile environment, the cancer cell relies on information from its external milieu to make adjustments necessary for survival. To this end, Src can function to amplify signals being initiated by cell surface receptors and to regulate receptor and non-receptor function. Primarily through the modulation of receptor signals, Src variably influences tumor growth, invasion, angiogenesis, metastases, and apoptosis [75-79].

Operationally, cellular stimuli that lead to Src activation result in an increased association between Src and the cytoskeleton. As a result, Src mediates the phosphorylation of many intracellular substrates such as EGFR, FAK, PYK2, paxillin, Stat3 and cyclin D [24, 33, 80-84]. Biologically, these responses affect cell motility and adhesion, cell cycle progression, and apoptosis. Thus, Src modulates a diverse assortment of physiologic responses and functions that support the propagation of cancer cells and enable responses to regional hypoxia, limited nutrients, and internal cellular efforts to self-destruct.

3.1. Src Activation

Although the increased specific activity of Src is found transiently in many normal cells in response to myriad physiologic conditions, Src activation in malignant cells results in a much higher proportion of Src molecules that have a high level of specific activity with a concomitant inability to downregulate such activity. Thus, "malignant activation" can be distinguished from "normal activation" both in tumor cell lines and tumors compared with cultured cells and normal tissue. Malignant activation of Src, as determined by constitutively high enzymatic levels, is seen in many human cancers: breast cancer [85, 86], gastric cancer [87], colon cancer [14, 88-94], pancreatic cancer [95], hairy cell leukemia/subgroup of B-cell lymphomas [96], low-grade bladder carcinoma [97], neuroblastoma [89, 98, 99], non-small cell lung carcinoma [100], and gliomas [98, 100, 101]. Significantly, Src activation in colon polyps can differentiate polyps destined to become malignant [102], and increased Src activity may be a harbinger of colon carcinoma development in patients with ulcerative colitis [103, 104]. Also, increases in specific Src activity are often associated with the metastatic phenotype [93, 105, 106] and activity appears, in the case of colon carcinoma, to be predictive of a poor prognosis [93, 107]. The high enzymatic levels of Src found in such a wide array of malignancies highlights its importance and utility as a target for therapeutic intervention.

3.2. Src Mutations

Src has been mapped to two chromosomal bands, 1p36 and 20q13, of which both are involved in molecular genetic rearrangements in human tumors [108-111]. A mutant Src with an activating truncation of the C terminus, ending with Tyr530, has been observed in 12% of advanced human colon cancer [112]. The importance of Src mutations in cancer may not be important as other groups were unable to find mutations [113-115]. Thus, it would appear that mutations in Src contributing to metastasis and tumor progression are more supportable than being the basis for Src activation in human colon cancer cases. From a therapeutic perspective, it is significant that in colon and breast cancer Src activation occurs more frequently than mutations of other oncogenes or inactivation of tumor suppressor genes currently observed in these tumors. Of further interest, Src activation has not been associated with the amount of Src protein present.

3.3. *Activation and Redundancy in Src Function*

Src, Yes, and Fyn are the most widely expressed Src-family members and share the greatest homology. This attribute helps explain signal path redundancy of inhibition of Src function in many normal cells. It also leads to compensatory activation of two other Src-family members, Yes and/or Fyn, with little biologic consequence to the cell [12, 76, 116]. Interestingly, in mouse systems [117] and human tumor cell lines [93], Src activation appears to be more critical than activation of Yes or Fyn for tumorigenic growth of cells in culture and for tumor growth in rodent models, as evidenced when Src protein or activity are reduced [118-120]. This is also true in mouse systems in which polyoma middle T transgenics, driven from an MMTV promoter, yield mammary carcinomas. When crossed with the src-/- mouse, few tumors arise; yet when crossed with the yes-/- mouse, tumors are still formed. Thus, an apparent lack of functional redundancy in some human cancers may turn out to be the tumor's Achilles heel, but an opportunity for creating successful directed therapies.

4. HOW WOULD INHIBITING SRC FUNCTION IMPACT CANCER CELL PROLIFERATION, INVASION, AND TUMORGENICITY?

Understanding Src function in normal cells and observations of Src function in cancer cells leads one to believe that inhibiting Src PTK activity and/or its interactions as part of a protein complex would impact many important signaling pathways that cancer cells find necessary for survival. If one looks at cell systems, Src also impacts proteins involved in integrin function and migration, enhancement of receptor signaling, regulation of transcription, cell-cycle control, and protection against apoptosis. The sections that follow provide more detailed examples of these interactions.

4.1. *Integrin and Focal Adhesion Function*

Integrins are transmembrane receptors that mediate cell attachment through their cytoplasmic contacts with cytoskeletal proteins, leading to focal adhesions with structural and signaling importance. Focal adhesion proteins associated with Src are paxillin, FAK, and p130CAS; their phosphorylation is regulatory for cell spreading and migration. *Figure 3* schematically represents some aspects of this signaling process. Inhibiting Src PTK activity would be expected to down-regulate the functionality of focal adhesions and the ability of tumor cells to invade and metastasize.

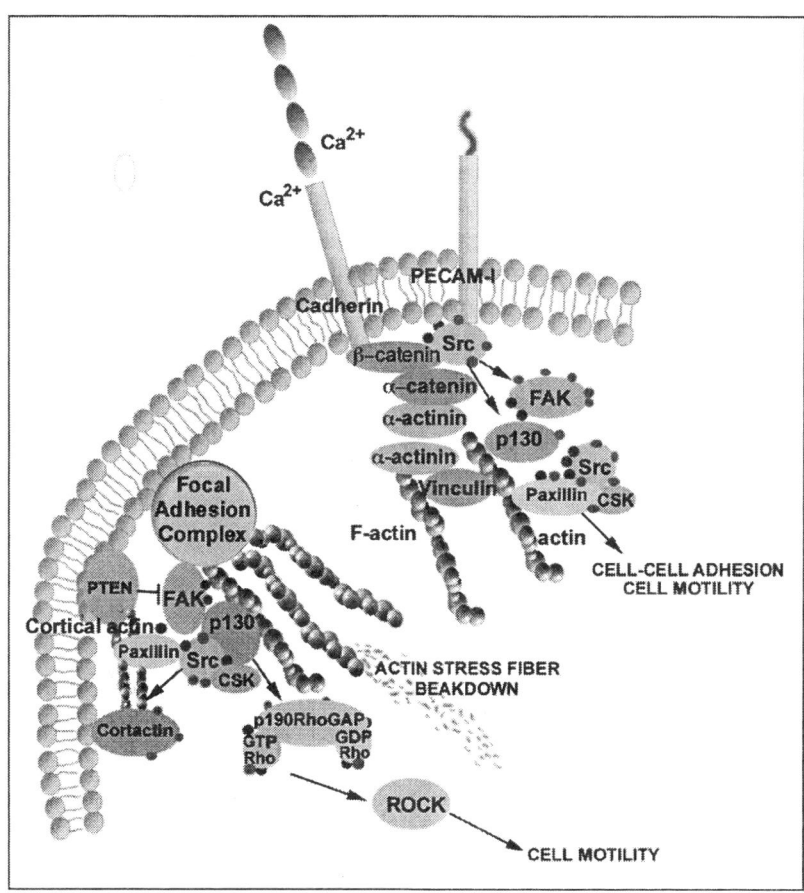

Figure 3. Schematic cartoon of our understanding of Src signaling associated with FAK, cadherin, and PECAM-1 function (adapted with permission from BioSource International, Camarillo, CA).

4.2. c-Met Function and Hepatocyte Growth Factor (HGF)/Scatter Factor (SF) Production

There is increasing evidence that a relationship exists between c-Met, the receptor for HGF/SF, and Src in carcinomas. c-Met is overexpressed early in prostatic intraepithelial neoplasia [121, 122] and its upregulation correlates with a poor prognosis. In metastatic colon cancer c-Met preferentially activates Src [123], and c-Met-dependent colony growth of mammary carcinoma cells was blocked by a dominant negative Src [124]. In addition, activated Src in breast carcinoma cells can stimulate HGF/SF expression, possibly contributing to bone resorption and lysis in metastatic cells [125]. Importantly, c-Met activates several pathways, including PI3 kinase and Akt, thereby promoting cellular survival. Inhibiting Src activity would thus be expected to decrease the production of HGF/SF and inhibit Src functioning downstream of c-Met. Activation of Src by c-Met is also critical to continued growth of prostate tumor cells in an orthotopic nude mouse model [126].

4.3. EGFR Function

There are many points of interaction between EGFR and Src that have the potential to modulate and put at risk tumor cell survival by virtue of Src inhibition. *Figure 4* is a schematic representation of some aspects of the signaling interaction between Src and EGFR, integrins, and FAK. Some examples follow. UV-induced ERK activation leads to Tyr phosphorylation of EGFR. Because both responses were completely abolished in the presence of a selective EGFR inhibitor (AG1478) or the Src inhibitor PP2 and by the expression of a kinase-dead Src mutant, it appears that UV-induced ERK activation, which provides a survival signal against stress-induced apoptosis, is mediated through Src-dependent Tyr phosphorylation of EGF receptors [127]. Src and EGFR interact with phosphorylation, occasionally cycling from one to another: Src can be activated by EGFR and Src can phosphorylate EGFR on Y845 in an EGF-dependent manner [128]. This type of interaction between receptor and non-receptor PTK is not common, and certainly not unique.

EGF stimulation of T47D brain cancer cells activates the Src PTK; the Src kinase inhibitor PP1 blocks EGF-induced phosphorylation of c-Cbl, the protein product of the proto-oncogene c-Cbl, but not the activation/ phosphorylation of the EGF receptor [129]. PP1 treatment blocks EGF-induced activation of the anti-apoptotic protein kinase Akt, suggesting that Src regulates the activation of Akt, perhaps by phosphorylating c-Cbl, which leads to PI3-K phosphorylation and activation of Akt. In addition, ErbB signaling, which has a role in transcription, requires the engagement of a novel Src-dependent route to MAPK. MAPK is thus quickly activated, and subsequently efficiently stimulates transcription [130]. Additionally, both v-Src and the EGFR can phosphorylate human GAP on Tyr-460, which suggests Src's importance in regulating mitogenic Ras signaling pathways [131] that can directly influence cell proliferation, a hallmark of oncogenesis.

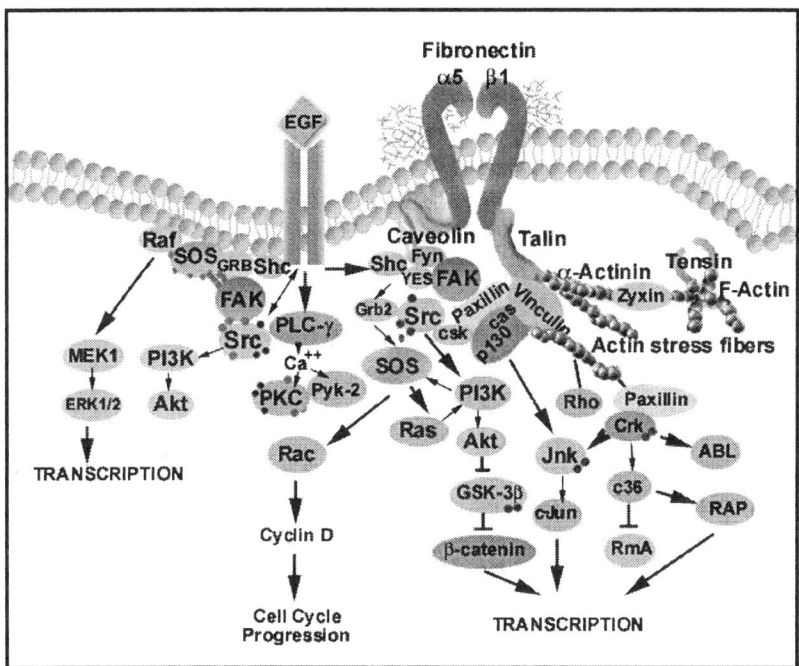

Figure 4. Schematic cartoon of our understanding of Src signaling
associated with integrin and receptor tyrosine kinase function
(adapted with permission from BioSource International, Camarillo, CA).

4.4. PDGFR Function

Src has been implicated in the control of cell division, the production of autocrine growth factors, cell survival responses, and cell motility. Src PTK activity appears to be necessary for cells to enter the cell cycle when treated with mitogens such as PDGF [132]. The observation that PDGF increases the catalytic activity of Src family members suggests that they contribute to PDGF-dependent responses. In some cells the mechanism used by the PDGFR to activate the mitogen-activated-protein kinase (p(42)/p(44) MAPK) pathway appears to be achieved through the regulation of Src and Grb-2/PI3K, which are intermediates in the p42/p44 MAPK cascade [133].

It is known that PDGF activates intracellular signal transduction cascades that can regulate adhesion, spreading, and migration, and inhibition of Src kinase activity can reduce PDGF-activated cell migration [134]. While the Src family tyrosine kinases are considered to be co-transducers of signals emanating from the activated PDGFR, the specific Src family kinase substrates that are involved in PDGF-induced signaling have not been elucidated. A 29 kDa protein was recently identified in caveolae, which is phosphorylated in response to PDGF stimulation and is prominent in cells overexpressing Fyn, suggesting its function as a Fyn substrate in PDGF-stimulated signaling pathways [135]. c-Abl is an important effector of Src for PDGF-induced DNA synthesis. Src-induced c-Abl activation involves phosphorylation of Y245 and Y412, two residues required for c-Abl mitogenic function [136]. Also, since c-Abl function was dispensable in cells deficient in active p53 and inhibition of c-Abl reduced mitogen-induced c-myc expression, it has been proposed that a PTK signaling cascade of PDGFR/Src/c-Abl is important for mitogenesis [136]. It has also been shown that PDGF can exert a mitogenic effect on vascular formation and that expression of Src can exert regulatory effects on endothelial proliferation, size of cultures, and cytoskeletal organization in two-dimensional culture, and on the formation of a differentiated multicellular network in three-dimensional culture [137]); all of these attributes are associated with tumorigenicity and the malignant phenotype. *Figure 5* is a schematic representation of some of the interactions between Src and PDGFR, IGF-IR, EGFR, and VEGFR.

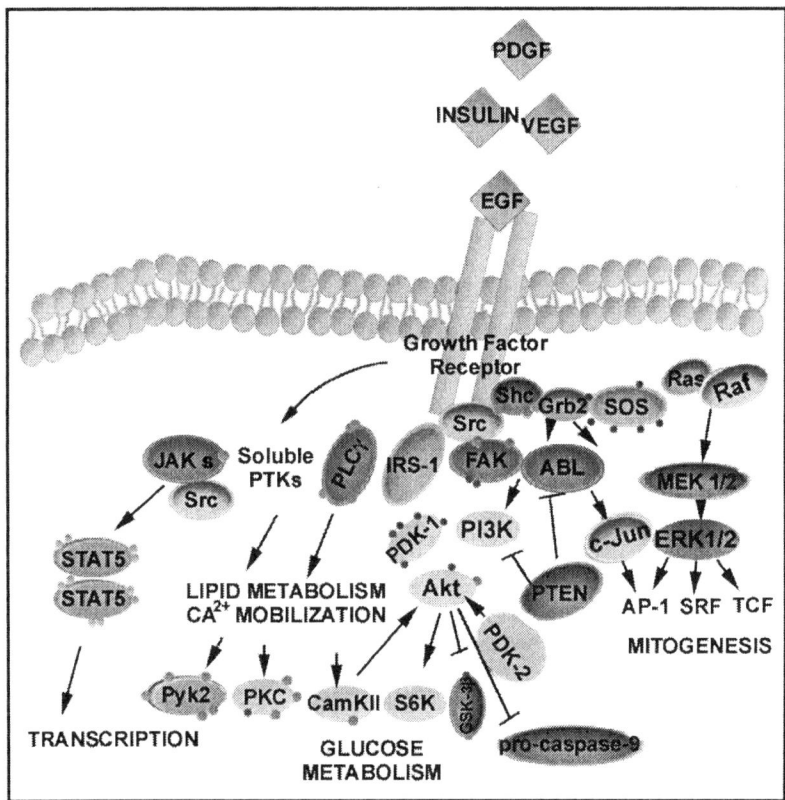

Figure 5. Schematic cartoon of our understanding of Src signaling associated with receptor tyrosine kinases such as EGFR, IGF-IR, PDGFR, and VEGFR (adapted with permission from BioSource International, Camarillo, CA).

4.5. Insulin-like Growth Factor I Receptor (IGF-IR) Function

The IGF-IR has been shown to mediate mitogenesis and the suppression of apoptosis [138] and Src has been associated with the IGF-1R. For instance, overexpression of kinase-activated SrcY527F significantly increased IGF-1-dependent cell proliferation in a pancreatic carcinoma line, PANC-1 [139]. It was also shown that while IGF-IR protein levels were higher in PANC-1 cells transfected with constitutively active Akt than in cells transfected with active Src, Akt kinases are triggered by active Src and inhibited by dominant negative Src or PTEN, as shown in in vitro kinase assays [140]. Additionally, Akt-induced IGF-IR expression was downregulated by dominant-negative Src or PTEN. These data provide evidence of a link between Akt signaling and the regulation of IGF-1R expression and demonstrate that active Akt promotes the invasiveness of pancreatic cancer cells by upregulating the expression of IGF-1R. Analogous activities involving IGF-IR, Src, Akt, and PTEN in high-grade gliomas, suggest the existence of associations similar to those found in pancreatic tumors.

4.6. VEGF Production

Src-dependent FAK phosphorylation at Y861 is a novel VEGF-induced signaling pathway in endothelial cells, which implicates its involvement in VEGF-induced endothelial cell migration and anti-apoptosis [141]. There is also evidence that by amplifying the amount of secreted VEGF protein, Src plays a role in the expression of VEGF protein in response to tumor cell hypoxia [78, 142, 143]. Inhibiting Src PTK activity by an antisense strategy decreases both the inducible and constitutive VEGF expression from colon tumor cells, and decreases MVD in these cells when implanted in mice [78, 144]. Thus, logically, a pharmacologic Src inhibitor would be expected to exert an antiangiogenic effect.

4.7. PKC Function

PKC-delta may enact a tumor suppressor effect that can be amplified by the over-expression of Src [145]. Increased complex formation between PKC-delta and Src was shown to increase serine phosphorylation and activation of Src family kinases, whereas Src activity was dramatically inhibited in cells with an overexpression of PKC-delta [146]. Actin reorganization depends upon PKC-induced Src activation with a coincidentally decreased activity of the small G protein RhoA [29]. Inhibiting Src PTK activity would thus be expected to nullify some of the effects of PKC-delta on actin reorganization and tumor cell motility.

4.8. Protection Against Apoptosis

Inhibition of apoptosis is a frequent hallmark of cancer cells and its reversal is, therefore, viewed as a logical goal for anticancer therapies. As we learn more about the molecular (and genetic) basis of tumorigenicity, it is becoming increasingly apparent that multiple, and at times quite different, perturbations of signaling pathways occur in tumor cells. Predicting an apoptotic outcome for tumor cells is even more problematic when considered from an experimental therapy perspective in that some of the more recently recognized "specific" signaling pathway inhibitors are often less specific than originally thought and could activate unpredicted mechanisms leading to apoptosis. This possibility encourages investigations that seek to understand the impact on apoptosis of drugs that inhibit signaling pathways. Understanding the mechanisms of activity of inhibitors of Src activity is thus of particular interest.

In addition to association with the inner surface of the external limiting cell membrane, Src is associated with the inner membrane of mitochondria and has been shown to promote cytochrome c oxidase (Cyt-c) activity in osteoclasts [147]. Inhibition of cytochrome c oxidase can cause the release of Cyt-c from mitochondria into the cytoplasm [148], which leads to activation of caspase-3, a key event in apoptosis and a potential target in carcinomas and gliomas. Lastly, low levels of nitric oxide upregulate Bcl-2, leading to increased Bcl-2 tyrosine kinase activity, reduction in Cyt-c release from mitochondria, and caspase-3 inhibition in serum-deprived cells, ultimately abrogating apoptosis. Furthermore, these findings depend on Src and protein kinase G (PKG) activation [149] and such anti-apoptotic signaling is abrogated by PP1, an inhibitor of Src-family PTKs. Because a reduction in Src activity could lead to apoptosis as a consequence of caspase-3 activation, this putative mechanism warrants further investigation.

Another mechanism by which Src can influence apoptosis was inferred from studies with v-Src, a surrogate for Src without the inhibitory Y530 locus. Using v-Src it has been shown that a form of apoptosis develops that is not associated with increased p53 levels [150]. Complicating our understanding of how v-Src is implicated in apoptosis is the observation that whereas caspase-3 activation appears necessary for v-Src-dependent apoptosis, when caspase activation was blocked, cells still died [150]. That Src affects many pathways impacting apoptosis is further supported by the observation that Src positively regulates the transformed phenotype of cells overexpressing EGFR family kinases. Src also likely positively regulates Bcl-XL expression via Stat3 activation [151]. In this capacity, it acts not only as a potent mitogenic signaling element, but also as an anti-apoptotic signaling protein. Also, ERK activation, a response to cell stress (such as UV light) that helps to prevent apoptosis, is mediated through Src-dependent phosphorylation of EGFR [127].

A growing body of literature couples Src with Stat3 activity in different signal transduction pathways [152-154]. The function of Stat3 is intimately associated with oncogenesis, particularly through its action against apoptosis and its parallel activity with Src. Stat's upregulate genes encoding inhibitors of apoptosis, such as Bcl-XL and Mcl-1, and inhibition of constitutively active Stat signaling pathways has been shown to inhibit tumor cell growth [152, 155, 156]. Inhibition of Src kinase activity results in reduced Stat3 tyrosine phosphorylation, abrogated Bcl-XL expression, and induction of cell death [151]. It is also interesting that inactive Stat3, when pre-assembled with inactive PDGF receptors, is phosphorylated and activated by ligand binding and Src activity; and that PD180970, an inhibitor of Src activity, abolishes Stat3 phosphorylation at Y705 and activation in Balb/c-3T3 fibroblasts [157]. It would thus appear that as an important downstream substrate for Src, Stat3 phosphorylation at Y705 will reflect Src activity and should also parallel apoptosis [158-162].

Another observation related to a tendency to inhibit apoptosis is that of the phenomenon of anoikis: a property of anchorage-independent tumor cells that reflects resistance to apoptosis when such cells are detached from cell clusters. This has been shown to be influenced by Src and FAK, with Src activation leading to increased Akt phosphorylation and a commensurate decreased sensitivity of cells to anoikis [163] and by inference, a reduced tendency toward apoptosis.

5. WHICH THERAPEUTIC STRATEGIES HAVE BEEN DEVELOPED TO INHIBIT SRC ACTIVITY?

During the past 20 years, academic and pharmaceutical industry efforts have focused on drug approaches to [1] abrogate the catalytic activity of Src by reversible and irreversible inhibition of its catalytic site and/or the adjacent ATP-binding site, and [2] to interfere with a phosphotyrosine binding region unique to the SH2 domain of Src. The result of this interference is that the activated tertiary structure of Src is compromised, as are the functions of multiprotein complexes resulting from the association of the Src SH2 domain with adapter and effector proteins. Approaches to develop Src inhibitors have been structure-based and screening-based, with a combined approach found to be most productive in recent years. Many of the former approaches utilized three-dimensional homology models of Src-family PTKs as well as secondary information gleaned from crystallography of Src [164-166] and other PTKs [167, 168]. New advances in combinatorial chemistry and specific screening approaches have also helped pave the way for the development of Src-specific inhibitors.

Showalter and Kraker, in an excellent review, presented the status between 1995 and 1997 of small molecule inhibitors of PDGFR, FGFR, and Src-family PTKs culled from the patent and published literature [169]. Subsequent to that, reviews by Sawyer and colleagues [170, 171] appeared, which cover the patent literature to 1998 and drugs under development to 2000. While they discuss drugs directed toward the SH2 and SH3 domains, my goal in the remainder of the chapter will be to generalize about the progress made to date and ongoing opportunities that could impact our ability to inhibit Src PTK activity in the treatment of cancer. I will also introduce newer developments in small molecule Src PTK inhibitor research that occurred after those reviews.

A number of heterocyclic scaffolds have provided thousands of compounds for testing against purified Src enzyme, cultured tumor cells, and in animal model studies. The literature contains most of the enzyme and some of the cell culture work, but little to explicate the activity of Src inhibitors in murine tumor models. Relevant observations made using some of the more prominent drug scaffolds will be cited in the sections that follow.

5.1. Thio-reactive Agents

Some of their first compounds from Parke-Davis Pharmaceutical Research were N- and 3-substituted 2,2'-selenobis($1H$-indoles) such as 2,2'-diselenobis[N,1-dimethyl-$1H$-indole-3-carboxamide with 535 MW, Ki = 0.8 uM against v-src, and IC_{50} = 0.4 uM against 3T3 mouse fibroblasts and specificity compared to EGFR and PDGFR [172]. These were followed by extensive studies of pyrido[2,3-d]pyrimidines.

5.2. Ring-fused Pyrimidines

Kraker and colleagues reported on pyrido[2, 3-d]pyrimidines that were ATP-competitive inhibitors of Src kinase that had IC_{50} values <10 nM and specificity of 6 to >150-fold for bFGFR, PDGFR, and EGFR [173]. These Src inhibitors also blocked G_2/M progression in the cell cycle, produced growth delay, and reduced clonogenicity in soft agar (effect on transformed pheotype). In addition, phosphorylation of the Src cellular substrates paxillin, p130(cas), and Stat3 was also inhibited at concentrations <1 μM, whereas autophosphorylation of EGFR or PDGFR was not inhibited by these Src inhibitors. While seeking inhibitors of receptor and non-receptor PTK over the years the Parke-Davis Pharmaceutical Research group found (1) pyrido[2,3-d]pyrimidines with good inhibition of Src PTK but limited specificity for Src [174]; (2) found that 2-amino-6-(2,6-dichlorophenyl)-8-(3-dimethylaminopropyl)-8H-pyrido[2,3-d]pyrimidine-7-one was active against Src PTK with IC_{50} = 0.05 uM, 10 to 30-fold specificity over FGFR and PDGFR, and 394 MW [175]; (3) found that 3-[6-(2,6-dichlorophenyl)-8-methyl-7-oxo-7,8-dihydropyrido[2,3-d]pyrimidin-2-ylamino]benzoic acid with 441 MW had an IC_50 = 0.009 uM against Src PTK and greater than 10-fold specificity compared with PDGFR, FGFR, and EGFR [176]; (4) that another pyrido-[2,3-d]pyrimidine, PD173955, which inhibited both Src and c-Yes and produced a prophase block in mitosis [177]; (5) another pyrido[2,3-d]pyrimidine with 321 MW had an IC_{50} = 0.21 uM, and >10-fold specificity for FGFGR and PDGFR [178, 179]; and (6) 3-(2, 6- dichloro-phenyl)-1-methyl-7-[[4-(4-morpholinophenyl]amino]-1,6- naphthyridin-2(1H)-one with a 481 MW had an IC_{50} = 0.25 uM, >10-fold specificity for FGFGR and PDGFR, and IC_{50} against SW-620 and HT-29 colon cells of 0.17 to 0.20 uM [180].

In a structure-activity study of 2-substituted aminopyrido[2,3-d]pyrimidin-7-yl ureas, Schroeder and colleagues [181] found specificity for Src PTK at the expense of high MW drug candidates. For instance, 1-[6-(2,6-dichloro-phenyl)-2-(4-diethylamino-butylamino)-pyrido[2,3-d]pyrimidine-7-yl]-3-phenyl-urea had a 558 MW, an IC_{50} = 7.5 nM, specificity compared with FGFR, EGFR, and PDGFR of ≥15-fold, and IC_{50} against SW-620 and HT-29 cells of 2.6 to 0.8 uM. Based on molecular size, these compounds would be expected to have a pharmacokinetic disadvantage.

New drug entities used to inhibit Src PTK have also come from what is Novartis Pharma AG today. Like the Parke-Davis program, emphasis has been on [2,3-d]pyrimidines with considerable work with 5,7-diphenyl-pyrrolo[2,3-d]pyrimidines and N^7-alkyl-5-aryl-pyrrolo[2,3-d]pyrimidines [182-184]. The former group contained specific inhibitors of Src PTK (compounds 2g, 4e, 4f, and 4g) with IC_{50} = 0.3 nM to 27 nM, specificity compared to EGFR (5 to 267), Kdr (40 to 233), v-Abl (1 to 7), Cdc-2 (370-1167), Csk (76 to 2000), and Lck (6 to 100) and the latter group IC_{50} = 0.2 to 2.3 uM with much less specificity, especially for Lck. The series of 7-cycloalkyl-pyrrolo-pyrimidines yielded two Src PTK compounds (13d and 13p) with <400 MW with IC_{50} = 2 and 20 nM and specificity for EGFR (25-230) and v-Abl (21-85) [185].

5.5. Tyrphostins

The tyrphostin compounds have been proposed as Src PTK inhibitors by Levitski [23, 186, 187]. Ramdas and colleagues [188, 189] found that the unstable tyrphostin 23 yielded a biotransformation product, P3, that was a tyrphostin 23 dimer joined at the benzylidene carbon. P3 had a Src PTK Ki =

6 μM and specificity for Csk, EGFR, and FGFR ranging from 6- to 50-fold compared to Src. The growth and colony formation of HT-29 colon adenocarcinoma cells was inhibited by P3 with an IC_{50} value of 10 μM.

To improve on tryphostins as PTK inhibitors, Hori and colleagues synthesized a series of 2-hydroxyarylidene-4-cyclopentene-1,3-diones and found these compounds to be more potent antitumor agents than the tryphostin AG17 [190, 191]. Unfortunately, insufficient data are available to evaluate activity and specificity against Src at this time.

5.6. Quinazolines

Another chemical series is composed of the 4-anilinoquinazolines that were developed as dual site inhibitors competitive with both ATP-binding and peptide-binding sites. One such inhibitor, 4-(4'-phenoxyanilino)-6,7-dimethoxyquinazoline, is approximately 400 MW and has a Ki = 15 nM, and IC_{50} = 44 nM against Src PTK, but was not selective when compared to Lck and only 7-fold when compared to VEGFR2 [192].

5.7. SH2 Domain Inhibitors

Over the years, both GalaxoSmithKline Research and Development and Ariad Pharmaceuticals have focused discovery efforts on the development of drugs acting on the SH2 domain of Src. Of the two companies, Ariad has had the greatest longevity and its initial goal appeared to be the development of Src inhibitors directed at osteoclasts, envisioned as a treatment for osteoporosis. This has led to an understanding of peptidic and non-peptidic features necessary to interfere with SH2 function in Src [193, 194]. Later work led to the incorporation of the phosphotyrosine mimetic, 4'-carboxymethyloxy-3'-phosphonophenylalanine, into nonpeptide inhibitors of the Src SH2 domain [195]. Two compounds (1a and 1b) from this last study with >500 MW have IC_{50} = 3.2 to 3.6 μM with specificity for Yes and Zap70 of >140 and to Yes 28 to 30 compared with Src. Efforts to develop specific non-peptidic SH2 inhibitors of Src continue [196].

6. CONCLUSIONS

Reviewing Src as a target for cancer therapy is a daunting process, but not so much so as to become a hard sell for its potential role in oncogenesis, invasion, angiogenesis, and proliferation. Certainly, others also appreciate its importance as a cancer target [75]. The contribution of Src to the malignant phenotype is extensive and frequently protected by signaling redundancy within the cell. My hope and expectation, supported by the experimental evidence described in section 2.3 above, is that such redundancy will not be operant in all signaling processes in which Src participates. Thus, drugs targeting Src will provide opportunities that, while varying from tumor to tumor, should abound in tumor cells compared to normal cells. Hence, I believe that some of the drugs under preclinical study will find their way to the clinic for the treatment of human cancer. While many associated patents for different agents have been issued during the past decade, unfortunately, at the time of this review none has yet led to clinical trials in cancer patients, although phase I trials appear imminent.

6.1. How will Src PTK Inhibitors be Used?

As single agents for some tumors, my expectation is that Src PTK inhibitors will produce apoptosis to a degree that will have a major impact on tumor growth. In addition, it is quite likely that inhibiting Src PTK activity will compromise the ability of some tumor cells to invade and migrate. As a consequence, Src inhibitors should help to limit metastasis of breast and colon adenocarcinomas and invasion of glial neoplasms. These concepts may be difficult to assess, as inhibitors might not always be "successful" as determined by our standard criteria for remission of the *primary* tumor, but would be expected, by virtue of their ability to inhibit metastases, to lead to increased survival of patients with solid tumors. Src inhibition might also be especially efficacious for specific tumors such as breast and prostate cancer with a propensity to metastasize to bone. Based on the relationship of Src to osteoclast function, Src inhibitors would also be expected to not only inhibit bone resorption, but also to inhibit growth of tumor cells in bone. In these and other circumstances, Src PTK inhibitors could be combined with other signal inhibitors to compromise additional cell signaling pathways that will help guide the tumor cell toward an apoptotic end and, thereby, increase patient survival.

Furthermore, given the large number of and interaction between intracellular signaling molecules, it is really no wonder that drugs that inhibit one target in one signaling pathway are likely to be insufficient to stop and/or reverse the growth of cancers in patients. The one exception today is the use of Gleevec to inhibit Bcr-Abl in patients with chronic myelogenous leukemia (CML) and c-kit in gastrointestinal stromal tumors [197]. Even with Gleevec, there are an increasing number of CML cases where intersecting pathways are being discovered that appear to rescue some tumor cells from inhibition of Bcr-Abl [198-200], and importantly, one of these pathways involves activation of Src-family members. For all these reasons, I believe that Src PTK inhibitors may be successful alone against some cancers, but more likely this activity will be garnered from Src in combination with other drugs inhibiting other signaling pathways and cell proliferation functions. As an optimist, I believe that Src PTK inhibitors have the potential to become one of the most effective agents to use in future drug combinations and are likely to win the Academy Awards in the category of "best supporting anticancer drug".

7. ACKNOWLEDGEMENTS

I would like to thank Joann Aaron for outstanding editorial assistance and Melissa McLane for manuscript preparation. I also owe a great deal to Gary Gallick for his continued commitment to understand Src in cancer, his thoughtful comments and suggestions to improve this review, and for his friendship.

Victor A. Levin
Department of Neuro-Oncology, The University of Texas M. D. Anderson Cancer Center, Houston Texas

8. REFERENCES

1. Stehelin D, Varmus HE, Bishop JM, Vogt PK. DNA related to the transforming gene(s) of avian sarcoma viruses is present in normal avian DNA. Nature 1976;260(5547):170-3.
2. Rous P. A sarcoma of the fowl transmissible by an agent separable from the tumor cells. J Exp Med 1911;13:397-411.
3. Bishop JM, Baker B, Fujita D, McCombe P, Sheiness D, Smith K, et al. Genesis of a virus-transforming gene. Natl Cancer Inst Monogr 1978(48):219-23.
4. Parker RC, Varmus HE, Bishop JM. Cellular homologue (c-src) of the transforming gene of Rous sarcoma virus: isolation, mapping, and transcriptional analysis of c-src and flanking regions. Proc Natl Acad Sci U S A 1981;78(9):5842-6.
5. Purchio AF, Erikson E, Brugge JS, Erikson RL. Identification of a polypeptide encoded by the avian sarcoma virus src gene. Proc Natl Acad Sci U S A 1978;75(3):1567-71.
6. Collett MS, Erikson RL. Protein kinase activity associated with the avian sarcoma virus src gene product. Proc Natl Acad Sci U S A 1978;75(4):2021-4.
7. Brugge JS, Erikson RL. Identification of a transformation-specific antigen induced by an avian sarcoma virus. Nature 1977;269(5626):346-8.
8. Levinson AD, Oppermann H, Levintow L, Varmus HE, Bishop JM. Evidence that the transforming gene of avian sarcoma virus encodes a protein kinase associated with a phosphoprotein. Cell 1978;15(2):561-72.
9. Levinson AD, Oppermann H, Varmus HE, Bishop JM. The purified product of the transforming gene of avian sarcoma virus phosphorylates tyrosine. J Biol Chem 1980;255(24):11973-80.
10. Varmus H, Bishop JM. Biochemical mechanisms of oncogene activity: proteins encoded by oncogenes. Introduction. Cancer Surv 1986;5(2):153-8.
11. Varmus H, Hirai H, Morgan D, Kaplan J, Bishop JM. Function, location, and regulation of the src protein-tyrosine kinase. Princess Takamatsu Symp 1989;20:63-70.

12. Thomas SM, Brugge JS. Cellular functions regulated by Src family kinases. Annu Rev Cell Dev Biol 1997;13:513-609.

13. Boerner RJ, Kassel DB, Barker SC, Ellis B, DeLacy P, Knight WB. Correlation of the phosphorylation states of pp60c-src with tyrosine kinase activity: the intramolecular pY530-SH2 complex retains significant activity if Y419 is phosphorylated. Biochemistry 1996;35(29):9519-25.

14. Cartwright CA, Kamps MP, Meisler AI, Pipas JM, Eckhart W. pp60c-src activation in human colon carcinoma. J Clin Invest 1989;83(6):2025-33.

15. Jun HS, Yoon JW. The role of viruses in Type I diabetes: two distinct cellular and molecular pathogenic mechanisms of virus-induced diabetes in animals. Diabetologia 2001;44(3):271-285.

16. Bellagamba C, Hubaishy I, Bjorge JD, Fitzpatrick SL, Fujita DJ, Waisman DM. Tyrosine phosphorylation of annexin II tetramer is stimulated by membrane binding. J Biol Chem 1997;272(6):3195-9.

17. Brown MT, Andrade J, Radhakrishna H, Donaldson JG, Cooper JA, Randazzo PA. ASAP1, a phospholipid-dependent arf GTPase-activating protein that associates with and is phosphorylated by Src. Mol Cell Biol 1998;18(12):7038-51.

18. Randazzo PA, Andrade J, Miura K, Brown MT, Long YQ, Stauffer S, et al. The Arf GTPase-activating protein ASAP1 regulates the actin cytoskeleton [see comments]. Proc Natl Acad Sci U S A 2000;97(8):4011-6.

19. Li S, Couet J, Lisanti MP. Src tyrosine kinases, Galpha subunits, and H-Ras share a common membrane-anchored scaffolding protein, caveolin. Caveolin binding negatively regulates the auto-activation of Src tyrosine kinases. J Biol Chem 1996;271(46):29182-90.

20. Huang RP, Hossain MZ, Huang R, Gano J, Fan Y, Boynton AL. Connexin 43 (cx43) enhances chemotherapy-induced apoptosis in human glioblastoma cells. International Journal of Cancer 2001;92(1):130-138.

21. Lau AF, Kurata WE, Kanemitsu MY, Loo LW, Warn-Cramer BJ, Eckhart W, et al. Regulation of connexin43 function by activated tyrosine protein kinases. J Bioenerg Biomembr 1996;28(4):359-68.

22. Okamura H, Resh MD. p80/85 cortactin associates with the Src SH2 domain and colocalizes with v-Src in transformed cells. J Biol Chem 1995;270(44):26613-8.

23. Agbotounou WK, Levitzki A, Jacquemin-Sablon A, Pierre J. Effects of tyrphostins on the activated c-src protein in NIH/3T3 cells. Mol Pharmacol 1994;45(5):922-31.

24. Biscardi JS, Maa MC, Tice DA, Cox ME, Leu TH, Parsons SJ. c-Src-mediated phosphorylation of the epidermal growth factor receptor on Tyr(845) and Tyr(1101) is associated with modulation of receptor function. Journal of Biological Chemistry 1999;274(12):8335-8343.

25. Seidel-Dugan C, Meyer BE, Thomas SM, Brugge JS. Effects of SH2 and SH3 deletions on the functional activities of wild-type and transforming variants of c-Src. Mol Cell Biol 1992;12(4):1835-45.

26. Schieffer B, Bernstein KE, Marrero MB. The role of tyrosine phosphorylation in angiotensin II mediated intracellular signaling and cell growth. J Mol Med 1996;74(2):85-91.

27. Guan JL. Role of focal adhesion kinase in integrin signaling. Int J Biochem Cell Biol 1997;29(8-9):1085-96.

28. Alexandropoulos K, Baltimore D. Coordinate activation of c-Src by SH3- and SH2-binding sites on a novel p130Cas-related protein, Sin. Genes Dev 1996;10(11):1341-55.

29. Brandt D, Gimona M, Hillmann M, Haller H, Mischak H. Protein kinase C induces actin reorganization via a Src- and Rho-dependent pathway. J Biol Chem 2002;277(23):20903-10.

30. Arthur WT, Petch LA, Burridge K. Integrin engagement suppresses RhoA activity via a c-Src-dependent mechanism. Curr Biol 2000;10(12):719-22.

31. Zhang-Sun G, Yang C, Viallet J, Feng G, Bergeron JJ, Posner BI. A 60-kilodalton protein in rat hepatoma cells overexpressing insulin receptor was tyrosine phosphorylated and associated with Syp, phophatidylinositol 3-kinase, and Grb2 in an insulin-dependent manner [see comments]. Endocrinology 1996;137(7):2649-58.

32. Turner CE. Paxillin. Int J Biochem Cell Biol 1998;30(9):955-9.

33. Hildebrand JD, Schaller MD, Parsons JT. Paxillin, a tyrosine phosphorylated focal adhesion-associated protein binds to the carboxyl terminal domain of focal adhesion kinase. Mol Biol Cell 1995;6(6):637-47.

34. Jabado N, Jauliac S, Pallier A, Bernard F, Fischer A, Hivroz C. Sam68 association with p120GAP in CD4+ T cells is dependent on CD4 molecule expression. J Immunol 1998;161(6):2798-803.

35. Weng Z, Thomas SM, Rickles RJ, Taylor JA, Brauer AW, Seidel-Dugan C, et al. Identification of Src, Fyn, and Lyn SH3-binding proteins: implications for a function of SH3 domains. Mol Cell Biol 1994;14(7):4509-21.

36. van der Geer P, Wiley S, Gish GD, Pawson T. The Shc adaptor protein is highly phosphorylated at conserved, twin tyrosine residues (Y239/240) that mediate protein-protein interactions. Curr Biol 1996;6(11):1435-44.

37. Schreiner SJ, Schiavone AP, Smithgall TE. Activation of STAT3 by the Src family kinase Hck requires a functional SH3 domain. Journal of Biological Chemistry 2002;277(47):45680-7.

38. Turkson J, Jove R. STAT proteins: novel molecular targets for cancer drug discovery. Oncogene 2000;19(56):6613-6626.

39. Berchtold S, Moriggl R, Gouilleux F, Silvennoinen O, Beisenherz C, Pfitzner E, et al. Cytokine receptor-independent, constitutively active variants of STAT5. J Biol Chem 1997;272(48):30237-43.

40. Chin H, Nakamura N, Kamiyama R, Miyasaka N, Ihle JN, Miura O. Physical and functional interactions between Stat5 and the tyrosine- phosphorylated receptors for erythropoietin and interleukin-3. Blood 1996;88(12):4415-25.

41. Weng Z, Taylor JA, Turner CE, Brugge JS, Seidel-Dugan C. Detection of Src homology 3-binding proteins, including paxillin, in normal and v-Src-transformed Balb/c 3T3 cells. J Biol Chem 1993;268(20):14956-63.

42. Reynolds AB, Daniel JM, Mo YY, Wu J, Zhang Z. The novel catenin p120cas binds classical cadherins and induces an unusual morphological phenotype in NIH3T3 fibroblasts. Exp Cell Res 1996;225(2):328-37.

43. Lampugnani MG, Corada M, Andriopoulou P, Esser S, Risau W, Dejana E. Cell confluence regulates tyrosine phosphorylation of adherens junction components in endothelial cells. J Cell Sci 1997;110(Pt 17):2065-77.

44. 44. Cooper JA, Gould KL, Cartwright CA, Hunter T. Tyr527 is phosphorylated in pp60c-src: implications for regulation. Science 1986;231(4744):1431-4.

45. 45. Kmiecik TE, Shalloway D. Activation and suppression of pp60c-src transforming ability by mutation of its primary sites of tyrosine phosphorylation. Cell 1987;49(1):65-73.

46. 46. Piwnica-Worms H, Saunders KB, Roberts TM, Smith AE, Cheng SH. Tyrosine phosphorylation regulates the biochemical and biological properties of pp60c-src. Cell 1987;49(1):75-82.

47. Yamaguchi H, Hendrickson WA. Structural basis for activation of human lymphocyte kinase Lck upon tyrosine phosphorylation. Nature 1996;384(6608):484-9.

48. Sicheri F, Kuriyan J. Structures of Src-family tyrosine kinases. Current Opinion in Structural Biology 1997;7(6):777-85.

49. Blume-Jensen P, Hunter T. Oncogenic kinase signalling. Nature 2001;411(6835):355-65.

50. Smart JE, Oppermann H, Czernilofsky AP, Purchio AF, Erikson RL, Bishop JM. Characterization of sites for tyrosine phosphorylation in the transforming protein of Rous

sarcoma virus (pp60v-src) and its normal cellular homologue (pp60c-src). Proc Natl Acad Sci U S A 1981;78(10):6013-7.

51. Chiang GG, Sefton BM. Phosphorylation of a Src kinase at the autophosphorylation site in the absence of Src kinase activity. Journal of Biological Chemistry 2000;275(9):6055-6058.

52. Sun G, Sharma AK, Budde RJ. Autophosphorylation of Src and Yes blocks their inactivation by Csk phosphorylation. Oncogene 1998;17(12):1587-95.

53. Shenoy S, Choi JK, Bagrodia S, Copeland TD, Maller JL, Shalloway D. Purified maturation promoting factor phosphorylates pp60c-src at the sites phosphorylated during fibroblast mitosis. Cell 1989;57(5):763-74.

54. Bagrodia S, Chackalaparampil I, Kmiecik TE, Shalloway D. Altered tyrosine 527 phosphorylation and mitotic activation of p60c-src. Nature 1991;349(6305):172-5.

55. Bagrodia S, Laudano AP, Shalloway D. Accessibility of the c-Src SH2-domain for binding is increased during mitosis. J Biol Chem 1994;269(14):10247-51.

56. Chackalaparampil I, Bagrodia S, Shalloway D. Tyrosine dephosphorylation of pp60c-src is stimulated by a serine/threonine phosphatase inhibitor. Oncogene 1994;9(7):1947-55.

57. Stover DR, Liebetanz J, Lydon NB. Cdc2-mediated modulation of pp60c-src activity. J Biol Chem 1994;269(43):26885-9.

58. Bjorge JD, Jakymiw A, Fujita DJ. Selected glimpses into the activation and function of Src kinase. Oncogene 2000;19(49):5620-5635.

59. Superti-Furga G, Fumagalli,S.,Koegl,M.,Courtneidge,S.A.,and Draetta,G. Csk inhibition of c-Src activity requires both the SH2 and SH3 domains of Src. EMBO Journal 1993;12:2625-2634.

60. Fincham VJ, Unlu M, Brunton VG, Pitts JD, Wyke JA, Frame MC. Translocation of Src kinase to the cell periphery is mediated by the actin cytoskeleton under the control of the Rho family of small G proteins. J Cell Biol 1996;135(6 Pt 1):1551-64.

61. Fincham VJ, Frame MC. The catalytic activity of Src is dispensable for translocation to focal adhesions but controls the turnover of these structures during cell motility. Embo J 1998;17(1):81-92.

62. Okamura H, Resh MD. Differential binding of pp60c-src and pp60v-src to cytoskeleton is mediated by SH2 and catalytic domains. Oncogene 1994;9(8):2293-303.

63. Kypta RM, Goldberg Y, Ulug ET, Courtneidge SA. Association between the PDGF receptor and members of the src family of tyrosine kinases. Cell 1990;62(3):481-92.

64. Alonso G, Koegl M, Mazurenko N, Courtneidge SA. Sequence requirements for binding of Src family tyrosine kinases to activated growth factor receptors. J Biol Chem 1995;270(17):9840-8.

65. Cobb BS, Schaller MD, Leu TH, Parsons JT. Stable association of pp60src and pp59fyn with the focal adhesion-associated protein tyrosine kinase, pp125FAK. Mol Cell Biol 1994;14(1):147-55.

66. Schaller MD, Parsons JT. Focal adhesion kinase and associated proteins. Curr Opin Cell Biol 1994;6(5):705-10.

67. Burnham MR, Bruce-Staskal PJ, Harte MT, Weidow CL, Ma A, Weed SA, et al. Regulation of c-SRC activity and function by the adapter protein CAS. Mol Cell Biol 2000;20(16):5865-78.

68. Broome MA, Hunter T. The PDGF receptor phosphorylates Tyr 138 in the c-Src SH3 domain in vivo reducing peptide ligand binding. Oncogene 1997;14(1):17-34.

69. Stover DR, Furet P, Lydon NB. Modulation of the SH2 binding specificity and kinase activity of Src by tyrosine phosphorylation within its SH2 domain. J Biol Chem 1996;271(21):12481-7.

70. Schaller MD, Hildebrand JD, Shannon JD, Fox JW, Vines RR, Parsons JT. Autophosphorylation of the focal adhesion kinase, pp125FAK, directs SH2-dependent binding of pp60src. Mol Cell Biol 1994;14(3):1680-8.

71. Eide BL, Turck CW, Escobedo JA. Identification of Tyr-397 as the primary site of tyrosine phosphorylation and pp60src association in the focal adhesion kinase, pp125FAK. Mol Cell Biol 1995;15(5):2819-27.

72. Calalb MB, Polte TR, Hanks SK. Tyrosine phosphorylation of focal adhesion kinase at sites in the catalytic domain regulates kinase activity: a role for Src family kinases. Mol Cell Biol 1995;15(2):954-63.

73. Schlaepfer DD, Jones KC, Hunter T. Multiple Grb2-mediated integrin-stimulated signaling pathways to ERK2/mitogen-activated protein kinase: summation of both c-Src- and focal adhesion kinase-initiated tyrosine phosphorylation events. Mol Cell Biol 1998;18(5):2571-85.

74. Richardson A, Malik RK, Hildebrand JD, Parsons JT. Inhibition of cell spreading by expression of the C-terminal domain of focal adhesion kinase (FAK) is rescued by coexpression of Src or catalytically inactive FAK: a role for paxillin tyrosine phosphorylation. Mol Cell Biol 1997;17(12):6906-14.

75. Summy JM, Sudol M, Eck MJ, Monteiro AN, Gatesman A, Flynn DC. Title. Front Biosci 2003;8:S185-205.

76. Schlessinger J. New roles for Src kinases in control of cell survival and angiogenesis. Cell 2000;100(3):293-6.

77. Frame MC. Src in cancer: deregulation and consequences for cell behaviour. Biochim Biophys Acta 2002;1602(2):114-30.

78. Ellis LM, Staley CA, Liu W, Fleming RY, Parikh NU, Bucana CD, et al. Down-regulation of vascular endothelial growth factor in a human colon carcinoma cell line transfected with an antisense expression vector specific for c-src. J Biol Chem 1998;273(2):1052-7.

79. Garcia R, Parikh NU, Saya H, Gallick GE. Effect of herbimycin A on growth and pp60c-src activity in human colon tumor cell lines. Oncogene 1991;6(11):1983-9.

80. Biscardi JS, Belsches AP, Parsons SJ. Characterization of human epidermal growth factor receptor and c-Src interactions in human breast tumor cells. Molecular Carcinogenesis 1998;21(4):261-272.

81. Licato LL, Brenner DA. Analysis of signaling protein kinases in human colon or colorectal carcinomas. Digestive Diseases & Sciences 1998;43(7):1454-1464.

82. Coutinho P, Goodyear R, Legan PK, Richardson GP. Chick alpha-tectorin: molecular cloning and expression during embryogenesis. Hear Res 1999;130(1-2):62-74.

83. Lei S, Lu WY, Xiong ZG, Orser BA, Valenzuela CF, MacDonald JF. Platelet-derived growth factor receptor-induced feed-forward inhibition of excitatory transmission between hippocampal pyramidal neurons. J Biol Chem 1999;274(43):30617-23.

84. Schaller MD, Parsons JT. pp125FAK-dependent tyrosine phosphorylation of paxillin creates a high-affinity binding site for Crk. Mol Cell Biol 1995;15(5):2635-45.

85. Ottenhoff-Kalff AE, Rijksen G, van, Beurden EA, Hennipman A, Michels AA, et al. Characterization of protein tyrosine kinases from human breast cancer: involvement of the c-src oncogene product. Cancer Research 1992;52(17):4773-4778.

86. Verbeek BS, Vroom TM, Adriaansen-Slot SS, Ottenhoff-Kalff AE, Geertzema JG, Hennipman A, et al. c-Src protein expression is increased in human breast cancer. An immunohistochemical and biochemical analysis. Journal of Pathology 1996;180(4):383-388.

87. Takeshima E, Hamaguchi M, Watanabe T, Akiyama S, Kataoka M, Ohnishi Y, et al. Aberrant elevation of tyrosine-specific phosphorylation in human gastric cancer cells. Jpn J Cancer Res 1991;82(12):1428-35.

88. Rosen N, Bolen JB, Schwartz AM, Cohen P, DeSeau V, Israel MA. Analysis of pp60c-src protein kinase activity in human tumor cell lines and tissues. Journal of Biological Chemistry 1986;261(29):13754-13759.

89. Bolen JB, Rosen N, Israel MA. Increased pp60c-src tyrosyl kinase activity in human neuroblastomas is associated with amino-terminal tyrosine phosphorylation of the src gene product. Proc Natl Acad Sci U S A 1985;82(21):7275-9.

90. Bolen JB, Veillette A, Schwartz AM, DeSeau V, Rosen N. Activation of pp60c-src protein kinase activity in human colon carcinoma. Proceedings of the National Academy of Sciences of the United States of America 1987;84(8):2251-2255.

91. Bolen JB, Veillette A, Schwartz AM, DeSeau V, Rosen N. Analysis of pp60c-src in human colon carcinoma and normal human colon mucosal cells. Oncogene Research 1987;1(2):149-168.

92. Cartwright CA, Meisler AI, Eckhart W. Activation of the pp60c-src protein kinase is an early event in colonic carcinogenesis. Proc Natl Acad Sci U S A 1990;87(2):558-62.

93. Talamonti MS, Roh MS, Curley SA, Gallick GE. Increase in activity and level of pp60c-src in progressive stages of human colorectal cancer. Journal of Clinical Investigation 1993;91(1):53-60.

94. Termuhlen PM, Curley SA, Talamonti MS, Saboorian MH, Gallick GE. Site-specific differences in pp60c-src activity in human colorectal metastases. J Surg Res 1993;54(4):293-298.

95. Lutz MP, Esser IB, Flossmann-Kast BB, Vogelmann R, Luhrs H, Friess H, et al. Overexpression and activation of the tyrosine kinase Src in human pancreatic carcinoma. Biochem Biophys Res Commun 1998;243(2):503-8.

96. Lynch SA, Brugge JS, Fromowitz F, Glantz L, Wang P, Caruso R, et al. Increased expression of the src proto-oncogene in hairy cell leukemia and a subgroup of B-cell lymphomas. Leukemia 1993;7(9):1416-22.

97. Fanning P, Bulovas K, Saini KS, Libertino JA, Joyce AD, Summerhayes IC. Elevated expression of pp60c-src in low grade human bladder carcinoma. Cancer Research 1992;52(6):1457-1462.

98. O'Shaughnessy J, Deseau V, Amini S, Rosen N, Bolen JB. Analysis of the c-src gene product structure, abundance, and protein kinase activity in human neuroblastoma and glioblastoma cells. Oncogene Res 1987;2(1):1-18.

99. Bjelfman C, Hedborg F, Johansson I, Nordenskjold M, Pahlman S. Expression of the neuronal form of pp60c-src in neuroblastoma in relation to clinical stage and prognosis. Cancer Res 1990;50(21):6908-14.

100. Budde RJ, Ke S, Levin VA. Activity of pp60c-src in 60 different cell lines derived from human tumors. Cancer Biochemistry Biophysics 1994;14(3):171-175.

101. Takenaka N, Mikoshiba K, Takamatsu K, Tsukada Y, Ohtani M, Toya S. Immunohistochemical detection of the gene product of Rous sarcoma virus in human brain tumors. Brain Research 1985;337(2):201-7.

102. Bjelfman C, Meyerson G, Cartwright CA, Mellstrom K, Hammerling U, Pahlman S. Early activation of endogenous pp60src kinase activity during neuronal differentiation of cultured human neuroblastoma cells. Mol Cell Biol 1990;10(1):361-70.

103. Cartwright CA, Coad CA, Egbert BM. Elevated c-Src tyrosine kinase activity in premalignant epithelia of ulcerative colitis. Journal of Clinical Investigation 1994;93(2):509-515.

104. Honda H, Oda H, Nakamoto T, Honda Z, Sakai R, Suzuki T, et al. Cardiovascular anomaly, impaired actin bundling and resistance to Src- induced transformation in mice lacking p130Cas [see comments]. Nat Genet 1998;19(4):361-5.

105. Brumell JH, Burkhardt AL, Bolen JB, Grinstein S. Endogenous reactive oxygen intermediates activate tyrosine kinases in human neutrophils. J Biol Chem 1996;271(3):1455-61.

106. Iravani S, Mao W, Fu L, Karl R, Yeatman T, Jove R, et al. Elevated c-Src protein expression is an early event in colonic neoplasia. Laboratory Investigation 1998;78(3):365-371.

107. Aligayer H, Boyd DD, Heiss MM, Abdalla EK, Curley SA, Gallick GE. Activation of Src kinase in primary colorectal carcinoma: an indicator of poor clinical prognosis. Cancer 2002;94(2):344-51.

108. Parker RC, Mardon G, Lebo RV, Varmus HE, Bishop JM. Isolation of duplicated human c-src genes located on chromosomes 1 and 20. Mol Cell Biol 1985;5(4):831-8.

109. Sakaguchi AY, Mohandas T, Naylor SL. A human c-src gene resides on the proximal long arm of chromosome 20 (cen----q131). Cancer Genet Cytogenet 1985;18(2):123-9.

110. Le Beau MM, Westbrook CA, Diaz MO, Rowley JD. c-src is consistently conserved in the chromosomal deletion (20q) observed in myeloid disorders. Proc Natl Acad Sci U S A 1985;82(19):6692-6.

111. Le Beau MM, Westbrook CA, Diaz MO, Rowley JD. Evidence for two distinct c-src loci on human chromosomes 1 and 20. Nature 1984;312(5989):70-1.

112. Irby RB, Mao W, Coppola D, Kang J, Loubeau JM, Trudeau W, et al. Activating SRC mutation in a subset of advanced human colon cancers. Nat Genet 1999;21(2):187-90.

113. Wang NM, Yeh KT, Tsai CH, Chen SJ, Chang JG. No evidence of correlation between mutation at codon 531 of src and the risk of colon cancer in Chinese. Cancer Lett 2000;150(2):201-4.

114. Daigo Y, Furukawa Y, Kawasoe T, Ishiguro H, Fujita M, Sugai S, et al. Absence of genetic alteration at codon 531 of the human c-src gene in 479 advanced colorectal cancers from Japanese and Caucasian patients. Cancer Research 1999;59(17):4222-4224.

115. Nilbert M, Fernebro E. Lack of activating c-SRC mutations at codon 531 in rectal cancer. Cancer Genet Cytogenet 2000;121(1):94-5.

116. Soriano P, Montgomery C, Geske R, Bradley A. Targeted disruption of the c-src proto-oncogene leads to osteopetrosis in mice. Cell 1991;64(4):693-702.

117. Muthuswamy SK, Muller WJ. Activation of Src family kinases in Neu-induced mammary tumors correlates with their association with distinct sets of tyrosine phosphorylated proteins in vivo. Oncogene 1995;11(9):1801-10.

118. Staley CA, Parikh NU, Gallick GE. Decreased tumorigenicity of a human colon adenocarcinoma cell line by an antisense expression vector specific for c-Src. Cell Growth Differ 1997;8(3):269-74.

119. Wiener JR, Nakano K, Kruzelock RP, Bucana CD, Bast RC, Jr., Gallick GE. Decreased Src tyrosine kinase activity inhibits malignant human ovarian cancer tumor growth in a nude mouse model [published erratum appears in Clin Cancer Res 1999 Oct;5(10):2980]. Clin Cancer Res 1999;5(8):2164-70.

120. Guy CT, Muthuswamy SK, Cardiff RD, Soriano P, Muller WJ. Activation of the c-Src tyrosine kinase is required for the induction of mammary tumors in transgenic mice. Genes & Development 1994;8(1):23-32.

121. Pisters LL, Troncoso P, Zhau HE, Li W, von Eschenbach AC, Chung LW. c-met proto-oncogene expression in benign and malignant human prostate tissues. JOURNAL OF UROLOGY 1995;154(1):293-298.

122. Humphrey PA, Zhu X, Zarnegar R, Swanson PE, Ratliff TL, Vollmer RT, et al. Hepatocyte growth factor and its receptor (c-MET) in prostatic carcinoma. Am J Pathol 1995;147(2):386-96.

123. Mao W, Irby R, Coppola D, Fu L, Wloch M, Turner J, et al. Activation of c-Src by receptor tyrosine kinases in human colon cancer cells with high metastatic potential. Oncogene 1997;15(25):3083-90.

124. Rahimi N, Hung W, Tremblay E, Saulnier R, Elliott B. c-Src kinase activity is required for hepatocyte growth factor- induced motility and anchorage-independent growth of mammary carcinoma cells. Journal of Biological Chemistry 1998;273(50):33714-33721.

125. Elliott BE, Hung WL, Boag AH, Tuck AB. The role of hepatocyte growth factor (scatter factor) in epithelial-mesenchymal transition and breast cancer. Can J Physiol Pharmacol 2002;80(2):91-102.

126. Kim SJ, Johnson M, Koterba K, Herynk MH, Uehara H, Gallick GE. Reduced c-Met Expression by an Adenovirus Expressing a c-Met Ribozyme Inhibits Tumorigenic Growth and Lymph Node Metastases of PC3-LN4 Prostate Tumor Cells in an Orthotopic Nude Mouse Model. Cancer Res 2003;in press.

127. Kitagawa D, Tanemura S, Ohata S, Shimizu N, Seo J, Nishitai G, et al. Activation of extracellular signal-regulated kinase by ultraviolet is mediated through Src-dependent epidermal growth factor receptor phosphorylation. Its implication in an anti-apoptotic function. J Biol Chem 2002;277(1):366-71.

128. Sato K, Sato A, Aoto M, Fukami Y. c-Src phosphorylates epidermal growth factor receptor on tyrosine 845. Biochem Biophys Res Commun 1995;215(3):1078-87.

129. Kassenbrock CK, Hunter S, Garl P, Johnson GL, Anderson SM. Inhibition of Src family kinases blocks epidermal growth factor (EGF)-induced activation of Akt, phosphorylation of c-Cbl, and ubiquitination of the EGF receptor. J Biol Chem 2002;277(28):24967-75.

130. Olayioye MA, Badache A, Daly JM, Hynes NE. An essential role for Src kinase in ErbB receptor signaling through the MAPK pathway. Experimental Cell Research 2001;267(1).

131. Park S, Liu X, Pawson T, Jove R. Activated Src tyrosine kinase phosphorylates Tyr-457 of bovine GTPase-activating protein (GAP) in vitro and the corresponding residue of rat GAP in vivo. J Biol Chem 1992;267(24):17194-200.

132. Courtneidge SA. Role of Src in signal transduction pathways. Biochem Soc Trans 2002;30(2):11-7.

133. Conway AM, Rakhit S, Pyne S, Pyne NJ. Platelet-derived-growth-factor stimulation of the p42/p44 mitogen- activated protein kinase pathway in airway smooth muscle: role of pertussis-toxinsensitive G-proteins, c-Src tyrosine kinases and phosphoinositide 3-kinase. Biochemical Journal 1999;337(Pt 2):171-177.

134. Yamboliev IA, Chen J, Gerthoffer WT. PI 3-kinases and Src kinases regulate spreading and migration of cultured VSMCs. American Journal of Physiology-Cell Physiology 2001;281(2).

135. Newcomb LF, Mastick CC. Src family kinase-dependent phosphorylation of a 29-kDa caveolin- associated protein. Biochemical And Biophysical Research Communications 2002;290(5):1447-1453.

136. Furstoss O, Dorey K, Simon V, Barila D, Superti-Furga G, Roche S. c-Abl is an effector of Src for growth factor-induced c-myc expression and DNA synthesis. Embo J 2002;21(4):514-24.

137. Marx M, Warren SL, Madri JA. pp60(c-src) modulates microvascular endothelial phenotype and in vitro angiogenesis. Experimental and Molecular Pathology 2001;70(3).

138. Valentinis B, Morrione A, Peruzzi F, Prisco M, Reiss K, Baserga R. Anti-apoptotic signaling of the IGF-I receptor in fibroblasts following loss of matrix adhesion. Oncogene 1999;18(10):1827-1836.

139. Flossmann-Kast BB, Jehle PM, Hoeflich A, Adler G, Lutz, MP. Src stimulates insulin-like growth factor I (IGF-I)-dependent cell proliferation by increasing IGF-I receptor number in human pancreatic carcinoma cells. Cancer Research 1998;58(16):3551-3554.

140. Tanno S, Tanno S, Mitsuuchi Y, Altomare DA, Xiao GH, Testa JR. AKT activation up-regulates insulin-like growth factor I receptor expression and promotes invasiveness of human pancreatic cancer cells. CANCER RESEARCH 2001;61(2):589-593.

141. Abu-Ghazaleh R, Kabir J, Jia H, Lobo M, Zachary I. Src mediates stimulation by vascular endothelial growth factor of the phosphorylation of focal adhesion kinase at tyrosine 861, and migration and anti-apoptosis in endothelial cells. Biochemical Journal 2001;360(Pt 1):255-264.

142. Mukhopadhyay D, Tsiokas L, Zhou XM, Foster D, Brugge JS, Sukhatme VP. Hypoxic induction of human vascular endothelial growth factor expression through c-Src activation. Nature 1995;375(6532):577-81.

116

143. Theurillat JP, Hainfellner J, Maddalena A, Weissenberger J, Aguzzi A. Early induction of angiogenetic signals in gliomas of GFAP-v-src transgenic mice. American Journal of Pathology 1999;154(2):581-590.

144. Fleming RY, Ellis LM, Parikh NU, Liu W, Staley CA, Gallick GE. Regulation of vascular endothelial growth factor expression in human colon carcinoma cells by activity of src kinase. Surgery 1997;122(2):501-7.

145. Lu Z, Hornia A, Jiang YW, Zang Q, Ohno S, Foster DA. Tumor promotion by depleting cells of protein kinase C delta. Mol Cell Biol 1997;17(6):3418-28.

146. Song JS, Swann PG, Szallasi Z, Blank U, Blumberg PM, Rivera J. Tyrosine phosphorylation-dependent and -independent associations of protein kinase C-delta with Src family kinases in the RBL-2H3 mast cell line: regulation of Src family kinase activity by protein kinase C- delta. Oncogene 1998;16(26):3357-68.

147. Miyazaki T, Neff L, Tanaka S, Horne WC, Baron R. Regulation of cytochrome c oxidase activity by c-Src in osteoclasts. J Cell Biol 2003;160(5):709-18.

148. Yang WL, Iacono L, Tang WM, Chin KV. Novel function of the regulatory subunit of protein kinase A: regulation of cytochrome c oxidase activity and cytochrome c release. Biochemistry 1998;37(40):14175-80.

149. Tejedo JR, Ramirez R, Cahuana GM, Rincon P, Sobrino F, Bedoya FJ. Evidence for involvement of c-Src in the anti-apoptotic action of nitric oxide in serum-deprived RINm5F cells. Cell Signal 2001;13(11):809-17.

150. Webb BL, Jimenez E, Martin GS. v-Src generates a p53-independent apoptotic signal. Mol Cell Biol 2000;20(24):9271-80.

151. Karni R, Jove R, Levitzki A. Inhibition of pp60c-Src reduces Bcl-XL expression and reverses the transformed phenotype of cells overexpressing EGF and HER-2 receptors. Oncogene 1999;18(33):4654-62.

152. Buettner R, Mora LB, Jove R. Activated STAT signaling in human tumors provides novel molecular targets for therapeutic intervention. Clin Cancer Res 2002;8(4):945-54.

153. Zhang Y, Turkson J, Carter-Su C, Smithgall T, Levitzki A, Kraker A, et al. Activation of Stat3 in v-Src-transformed fibroblasts requires cooperation of Jak1 kinase activity. J Biol Chem 2000;275(32):24935-44.

154. Turkson J, Bowman T, Garcia R, Caldenhoven E, De Groot RP, Jove R. Stat3 activation by Src induces specific gene regulation and is required for cell transformation. Mol Cell Biol 1998;18(5):2545-52.

155. Rahaman SO, Harbor PC, Chernova O, Barnett GH, Vogelbaum MA, Haque SJ. Inhibition of constitutively active Stat3 suppresses proliferation and induces apoptosis in glioblastoma multiforme cells. Oncogene 2002;21(55):8404-13.

156. Niu G, Bowman T, Huang M, Shivers S, Reintgen D, Daud A, et al. Roles of activated Src and Stat3 signaling in melanoma tumor cell growth. Oncogene 2002;21(46):7001-10.

157. Wang YZ, Wharton W, Garcia R, Kraker A, Jove R, Pledger WU. Activation of Stat3 preassembled with platelet-derived growth factor beta receptors requires Src kinase activity. Oncogene 2000;19(17):2075-2085.

158. Kaptein A, Paillard V, Saunders M. Dominant negative stat3 mutant inhibits interleukin-6-induced Jak-STAT signal transduction. Journal of Biological Chemistry 1996;271(11):5961-4.

159. Schaefer TS, Sanders LK, Park OK, Nathans D. Functional differences between Stat3alpha and Stat3beta. Molecular & Cellular Biology 1997;17(9):5307-16.

160. Smith PD, Crompton MR. Expression of v-src in mammary epithelial cells induces transcription via STAT3. Biochem J 1998;331(Pt 2):381-5.

161. Schaefer LK, Wang S, Schaefer TS. c-Src activates the DNA binding and transcriptional activity of Stat3 molecules: serine 727 is not required for transcriptional activation under certain circumstances. Biochem Biophys Res Commun 1999;266(2):481-7.

162. Hung W, Elliott B. Co-operative effect of c-Src tyrosine kinase and Stat3 in activation of hepatocyte growth factor expression in mammary carcinoma cells. Journal of Biological Chemistry 2001;276(15):12395-403.

163. Windham TC, Parikh NU, Siwak DR, Summy JM, McConkey DJ, Kraker AJ, et al. Src activation regulates anoikis in human colon tumor cell lines. Oncogene 2002;21(51):7797-807.

164. Lamers MB, Antson AA, Hubbard RE, Scott RK, Williams DH. Structure of the protein tyrosine kinase domain of C-terminal Src kinase (CSK) in complex with staurosporine. J Mol Biol 1999;285(2):713-25.

165. Eck MJ, Shoelson SE, Harrison SC. Recognition of a high-affinity phosphotyrosyl peptide by the Src homology-2 domain of p56lck. Nature 1993;362(6415):87-91.

166. Xu W, Harrison SC, Eck MJ. Three-dimensional structure of the tyrosine kinase c-Src. Nature 1997;385(6617):595-602.

167. Eck MJ, Atwell SK, Shoelson SE, Harrison SC. Structure of the regulatory domains of the Src-family tyrosine kinase Lck. Nature 1994;368(6473):764-9.

168. Zhu X, Kim JL, Newcomb JR, Rose PE, Stover DR, Toledo LM, et al. Structural analysis of the lymphocyte-specific kinase Lck in complex with non-selective and Src family selective kinase inhibitors. Structure Fold Des 1999;7(6):651-61.

169. Showalter HD, Kraker AJ. Small molecule inhibitors of the platelet-derived growth factor receptor, the fibroblast growth factor receptor, and Src family tyrosine kinases. Pharmacol Ther 1997;76(1-3):55-71.

170. Dalgarno DC, Metcalf CAI, Shakespeare WC, Sawyer TK. Signal transduction drug-discovery: Targets, mechanisms and structure-based design. Current Opinion in Drug Discovery & Development. 2000;3:549-564.

171. Sawyer T, Boyce B, Dalgarno D, Iuliucci J. Src inhibitors: genomics to therapeutics. Expert Opin Investig Drugs 2001;10(7):1327-44.

172. Showalter HD, Sercel AD, Leja BM, Wolfangel CD, Ambroso LA, Elliott WL, et al. Tyrosine kinase inhibitors. 6. Structure-activity relationships among N- and 3-substituted 2,2'-diselenobis(1H-indoles) for inhibition of protein tyrosine kinases and comparative in vitro and in vivo studies against selected sulfur congeners. J Med Chem 1997;40(4):413-26.

173. Kraker AJ, Hartl BG, Amar AM, Barvian MR, Showalter HD, Moore CW. Biochemical and cellular effects of c-Src kinase-selective pyrido[2, 3-d]pyrimidine tyrosine kinase inhibitors. Biochem Pharmacol 2000;60(7):885-98.

174. Hamby JM, Connolly CJ, Schroeder MC, Winters RT, Showalter HD, Panek RL, et al. Structure-activity relationships for a novel series of pyrido[2,3- d]pyrimidine tyrosine kinase inhibitors. J Med Chem 1997;40(15):2296-303.

175. Boschelli DH, Wu Z, Klutchko SR, Showalter HD, Hamby JM, Lu GH, et al. Synthesis and tyrosine kinase inhibitory activity of a series of 2- amino-8H-pyrido[2,3-d]pyrimidines: identification of potent, selective platelet-derived growth factor receptor tyrosine kinase inhibitors. J Med Chem 1998;41(22):4365-77.

176. Klutchko SR, Hamby JM, Boschelli DH, Wu Z, Kraker AJ, Amar AM, et al. 2-Substituted aminopyrido[2,3-d]pyrimidin-7(8H)-ones. structure- activity relationships against selected tyrosine kinases and in vitro and in vivo anticancer activity. J Med Chem 1998;41(17):3276-92.

177. Moasser MM, Srethapakdi M, Sachar KS, Kraker AJ, Rosen N. Inhibition of Src kinases by a selective tyrosine kinase inhibitor causes mitotic arrest. Cancer Res 1999;59(24):6145-52.

178. Blankley CJ, Doherty AM, Hamby JM, Panek RL, Schroeder MC, Showalter HDH, et al. Preparation of 6-aryl pyrido[2,3-d]pyrimidines and naphthyridines for inhibiting protein tyrosine kinase-mediated cellular proliferation. Chem. Abst. 1996;125(WO 9615128):114688k.

179. Thompson AM, Connolly CJ, Hamby JM, Boushelle S, Hartl BG, Amar AM, et al. 3-(3,5-Dimethoxyphenyl)-1,6-naphthyridine-2,7-diamines and related 2-urea derivatives are potent and selective inhibitors of the FGF receptor-1 tyrosine kinase. J Med Chem 2000;43(22):4200-11.

180. Thompson AM, Rewcastle GW, Boushelle SL, Hartl BG, Kraker AJ, Lu GH, et al. Synthesis and structure-activity relationships of 7-substituted 3-(2, 6- dichlorophenyl)-1,6-naphthyridin-2(1H)-ones as selective inhibitors of pp60(c-src). J Med Chem 2000;43(16):3134-47.

181. Schroeder MC, Hamby JM, Connolly CJ, Grohar PJ, Winters RT, Barvian MR, et al. Soluble 2-substituted aminopyrido[2,3-d]pyrimidin-7-yl ureas. Structure-activity relationships against selected tyrosine kinases and exploration of in vitro and in vivo anticancer activity. Journal of Medicinal Chemistry 2001;44(12):1915-26.

182. Missbach M, Altmann E, Widler L, Susa M, Buchdunger E, Mett H, et al. Substituted 5,7-diphenyl-pyrrolo[2,3d]pyrimidines: potent inhibitors of the tyrosine kinase c-Src [In Process Citation]. Bioorg Med Chem Lett 2000;10(9):945-9.

183. Widler L, Green J, Missbach M, Susa M, Altmann E. 7-alkyl- and 7-cycloalkyl-5-aryl-pyrrolo[2,3-d]pyrimidines - potent inhibitors of the tyrosine kinase c-Src. Bioorganic & Med. Chem. Lett. 2001;11(6):849-852.

184. Altmann E, Widler L, Missbach M. N(7)-substituted-5-aryl-pyrrolo[2,3-d]pyrimidines represent a versatile class of potent inhibitors of the tyrosine kinase c-Src. Mini Rev Med Chem 2002;2(3):201-8.

185. Widler L, Green J, Missbach M, Susa M, Altmann E. 7-Alkyl- and 7-cycloalkyl-5-aryl-pyrrolo[2,3-d]pyrimidines--potent inhibitors of the tyrosine kinase c-Src. Bioorganic & Medicinal Chemistry Letters 2001;11(6):849-52.

186. Levitzki A. SRC as a target for anti-cancer drugs. Anticancer Drug Des 1996;11(3):175-82.

187. Levitzki A. Protein tyrosine kinase inhibitors as novel therapeutic agents. Pharmacol Therap 1999;82(2-3):231-239.

188. Ramdas L, McMurray JS, Budde RJ. The degree of inhibition of protein tyrosine kinase activity by tyrphostin 23 and 25 is related to their instability. Cancer Res 1994;54(4):867-9.

189. Ramdas L, Obeyesekere NU, McMurray JS, Gallick GE, Seifert WE, Jr., Budde RJ. A tyrphostin-derived inhibitor of protein tyrosine kinases: isolation and characterization. Arch Biochem Biophys 1995;323(2):237-42.

190. Hori H, Nagasawa H, Uto Y. Structure-based design of the antitumor 2-hydroxyarylidene-4-cyclopentene-1,3-dione TX-1123 as a protein tyrosine kinase inhibitor having low mitochondrial toxicity. Cell Mol Biol Lett 2003;8(2A):528-30.

191. Hori H, Nagasawa H, Ishibashi M, Uto Y, Hirata A, Saijo K, et al. TX-1123: an antitumor 2-hydroxyarylidene-4-cyclopentene-1,3-dione as a protein tyrosine kinase inhibitor having low mitochondrial toxicity. Bioorg Med Chem 2002;10(10):3257-65.

192. Tian G, Cory M, Smith AA, Knight WB. Structural determinants for potent, selective dual site inhibition of human pp60c-src by 4-anilinoquinazolines. Biochemistry 2001;40(24):7084-91.

193. Plummer MS, Holland DR, Shahripour A, Lunney EA, Fergus JH, Marks JS, et al. Design, synthesis, and cocrystal structure of a nonpeptide Src SH2 domain ligand. J Med Chem 1997;40(23):3719-25.

194. Plummer MS, Lunney EA, Para KS, Shahripour A, Stankovic CJ, Humblet C, et al. Design of peptidomimetic ligands for the pp60src SH2 domain. Bioorg Med Chem 1997;5(1):41-7.

195. Kawahata N, Yang MG, Luke GP, Shakespeare WC, Sundaramoorthi R, Wang Y, et al. A novel phosphotyrosine mimetic 4'-carboxymethyloxy-3'-phosphonophenylalanine (Cpp):

exploitation in the design of nonpeptide inhibitors of pp60(Src) SH2 domain. Bioorg Med Chem Lett 2001;11(17):2319-23.

196. Sawyer TK, Bohacek RS, Dalgarno DC, Eyermann CJ, Kawahata N, Metcalf CA, 3rd, et al. SRC homology-2 inhibitors: peptidomimetic and nonpeptide. Mini Rev Med Chem 2002;2(5):475-88.

197. Kovacs B, Liossis SNC, Gist ID, Tsokos GC. Crosslinking of Fas/CD95 suppresses the CD3-mediated signaling events in Jurkat T cells by inhibiting the association of the T-cell receptor zeta chain with src-protein tyrosine kinases and ZAP70. APOPTOSIS 1999;4(5):327-334.

198. Nimmanapalli R, O'Bryan E, Huang M, Bali P, Burnette PK, Loughran T, et al. Molecular characterization and sensitivity of STI-571 (imatinib mesylate, Gleevec)-resistant, Bcr-Abl-positive, human acute leukemia cells to SRC kinase inhibitor PD180970 and 17-allylamino-17-demethoxygeldanamycin. Cancer Res 2002;62(20):5761-9.

199. Nimmanapalli R, Bhalla K. Mechanisms of resistance to imatinib mesylate in Bcr-Abl-positive leukemias. Current Opinion in Oncology 2002;14(6):616-20.

200. Tsuruo T, Naito M, Tomida A, Fujita N, Mashima T, Sakamoto H, et al. Molecular targeting therapy of cancer: drug resistance, apoptosis and survival signal. Cancer Science 2003;94(1):15-21.

THERAPEUTIC TARGETING OF THE RECEPTOR TYROSINE KINASE MET

MARTIN SATTLER, PATRICK C. MA AND RAVI SALGIA

1. INTRODUCTION

The Met receptor tyrosine kinase was originally identified as the cellular homologue of the Tpr-Met oncoprotein [1-3]. The Tpr-Met translocation was created by treatment of the human osteogenic sarcoma (HOS) cell line with the chemical carcinogen N-methyl-N'-nitro-N-nitrosoguanidine resulting in a fusion of the *Tpr* (chromosome 1) gene to the *Met* kinase gene (chromosome 7). The Tpr-Met oncoprotein has constitutively elevated Met tyrosine kinase activity, which is required for its transforming properties [4, 5]. Met itself is mainly expressed on epithelial cells and is the receptor for HGF (hepatocyte growth factor). HGF was originally identified as a growth factor for hepatocytes and as a fibroblast-derived cell motility factor or scatter factor [6-8]. HGF is a member of the plasminogen-related growth factor family. HGF precursor pro-HGF is cleaved by a protease to a disulfide-linked heterodimeric molecule, predominantly produced by mesenchymal cells. The mature HGF protein binds to its high affinity receptor Met and leads to its activation. The HGF/Met pathway has gained considerable interest through its apparent deregulation by overexpression or gain-of-function mutations in Met in various cancers. In this review the structural determinants for Met as a receptor tyrosine kinase and target of oncogenic transformation as well as its regulation of biological functions will be described. Based on the current knowledge of the regulation of Met activities and protein expression, various targeted therapies have been designed to interfere with these processes. The potential and targeted approach of these therapies will be summarized.

2. STRUCTURAL DETERMINANTS FOR MET AS A RECEPTOR TYROSINE KINASE

Met belongs to a subfamily of receptor tyrosine kinases that also includes the structurally related Ron and Sea kinases, which have extracellular structures related to the semaphorin receptor (or plexin) family [9,10]. The gene for human Met is located on chromosome 7 (7q21–q31) [11,12]. The Met receptor is translated as a single 170 kDa precursor that is post-transcriptionally digested into an α-chain and a ß-chain. The mature receptor is glycosylated and contains a transmembrane 140 kDa ß-chain linked through disulfide bonds to a 50 kDa extracellular α-chain. Met is mainly expressed in epithelial cells but its expression can be significantly altered in malignant cells. The ß-chain contains structural features that are shared with some other transmembrane receptors, including a Sema domain, a PSI domain (identified in plexins, semaphorins and integrins), four IPT repeats (found within immunoglobulins, plexins and transcription factors), a transmembrane domain, a juxtamembrane (JM) domain, a tyrosine kinase domain and a carboxy-terminal tail region (Figure 1).

Ligation of the Met receptor by its ligand HGF leads to receptor dimerization and activation of its intrinsic tyrosine kinase. An early activation step of Met includes phosphorylation of Y1230, Y1234 and Y1235 in the activation loop of the tyrosine kinase domain [13,14]. Phosphorylation at these sites correlates with increased Met tyrosine kinase activity. Activated Met can lead to autophosphorylation and phosphorylation of downstream signaling molecules. An important phosphorylation site in Met includes Y1313, a binding site for phosphatidylinositol-3'kinase. Recruitment of phosphatidylinositol-3'kinase to the receptor leads to its activation and generation of bioactive phospholipids. Y1003 within the juxtamembrane domain is an important regulatory site and recruits c-CBL when phosphorylated. c-CBL acts as a ubiquitin E3-ligase and leads to ubiquitination and degradation of Met as well as internalization through complex formation with endophilin-CIN85 [15]. A unique multisubstrate signal transducer docking site at Y1349 and Y1356 leads to the recruitment of a variety of proteins when phosphorylated, including SH2 (Src homology-2) domains, PTB (phosphotyrosine binding) domains, and MBD (Met binding domain) containing signaling proteins [16]. There are additional phosphorylation sites in Met that are likely to be involved in its functional regulation.

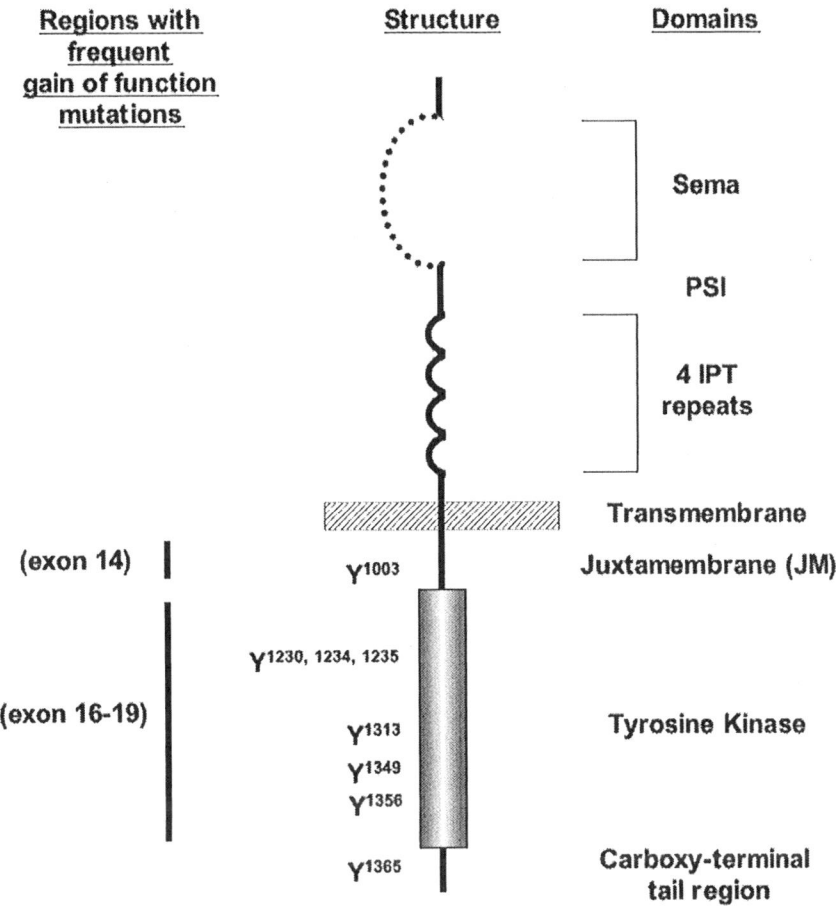

Figure 1: Schematic structure of the Met receptor (ß-chain). Represented are regions with activating mutations (exon location), important phospho-tyrosine sites and domains within the Kit receptor tyrosine kinase.

3. GAIN-OF-FUNCTION MUTATIONS IN MET

The Met receptor is typically expressed in epithelial cells [17, 18-20] but can be overexpressed in a variety of transformed cells [21-31]. Gain-of-function mutations within Met have been identified and are almost exclusively in the juxtamembrane domain (exon 14) or in the tyrosine kinase domain (exon 16-19). We have recently identified novel mutations of Met also in the N-terminus semaphorin (Sema) domain [32].

3.1. Mutations Within the Juxtamembrane (JM) Domain

JM domains of receptor tyrosine kinases have been identified as frequent targets of activating mutations. For example, there are gain-of-function mutations of the receptor tyrosine kinase Flt3 in about 20% of adult acute myeloid leukemias (AML) [33,34]. Mutations include insertion of duplicated sequences of variable positions and length that lead to ligand-independent constitutive kinase activity and transformation. It is believed that the JM domain regulates auto-inhibition, which is disrupted by mutations and consequently leads to deregulated tyrosine kinase activity and oncogenic properties. JM domain mutations have also been identified in Met-expressing cancers, however, with less transforming properties. A T1010I mutation has been found in hereditary papillary renal cell carcinoma and in a patient with breast cancer [35,36]. The T1010I mutation does not lead to apparent activation of the Met receptor, but athymic nude mice injected with mutant Met in NIH3T3 cells form tumors slightly faster than wild-type Met expressing cells [36]. Also, a P1009S mutation of Met was detected in a patient with gastric carcinoma [36]. Mutation at this site leads to a prolonged activation signal in HGF stimulated NIH3T3 cells compared to wild-type Met expressing cells. Interestingly, in mice there is an alternatively spliced form of Met with a 47 amino acid deletion in the JM domain [37]. Stimulation of this Met variant leads to increased tyrosine phosphorylation of cellular proteins and increased complex formation of Met with signaling proteins [38]. In tissue samples and cell lines from small cell lung cancer (SCLC), we have recently identified two specific JM domain mutations, namely R988C and T1010I, as well as potential deletion of the JM domain. The JM domain mutants in SCLC lead to enhanced tumorigenecity, increased cell migration, and increased phosphorylation of proteins [32].

3.2. Tyrosine Kinase Activating Mutations

Activating mutations of Met have been reported within the kinase domain resulting in an increase of the tyrosine kinase activity [39]. A majority of activating mutations in Met have been described in sporadic papillary renal carcinomas and hereditary papillary renal cell carcinomas [35,40,41]. There are a variety of mutations that can be located in exon 16 to exon 19. Identification of activating mutations of Met found in renal carcinomas provides direct evidence linking Met to oncogenic transformation. The unique Tpr-Met mutation led to the identification of Met itself [1-3]. The presence of Tpr-Met in human cancers is controversial however [21-23]. In the Tpr-Met fusion, the Tpr gene is fused to the intracellular tyrosine kinase domain of Met [1]. The Tpr sequence provides two leucine zipper domains, which facilitate oligomerization of the oncoprotein and substitute for HGF ligand stimulated activation, resulting in constitutive activation of its kinase activity. The Tpr-Met fusion also leads to altered localization of the Met kinase, which is membrane bound as a wild type protein but mainly cytoplasmic for the Tpr-Met oncoprotein. It is also interesting to note that by reintroducing the JM domain into Tpr-Met, the transformation ability of this oncogene is reverted back to normal [42].

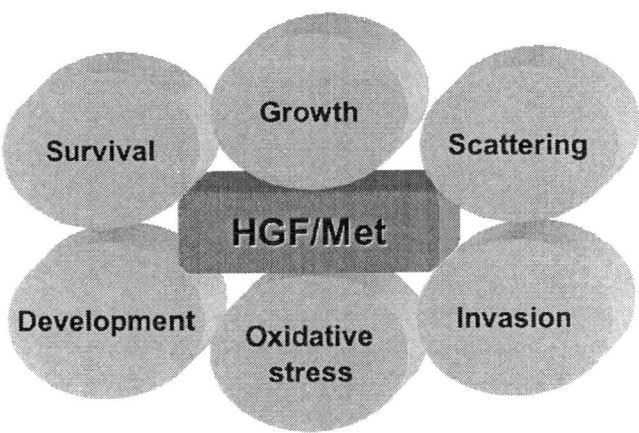

Figure 2: Biological activities of activated Met.

4. REGULATION OF BIOLOGICAL FUNCTIONS BY ACTIVATED MET

Despite the identification of many different signaling pathways activated by Met, it has been difficult to link any specific signaling event to a specific biologic effect. HGF-stimulated Met or gain-of-function Met mutants have been shown to trigger a variety of cellular responses (Figure 2) that may vary depending on the cellular context. In addition, there may be considerable redundancy in signaling pathways regulating certain biological activities. Below, we will summarize important biological functions (cell growth and survival, regulation of cytoskeletal functions, development and increased levels of reactive oxygen species) that have been shown to be regulated by activated Met.

4.1. Met Promotes Cell Growth and Survival

The role of Met as an anti-apoptotic factor is apparent through its role in cancer development. HGF/Met signaling has been shown to mediate a proliferative signal to tumor cells derived from cancers, including ovarian, gastric, glioma, lung adenocarcinoma and small cell cancer, pancreatic, colorectal, prostate, and breast cancer patients. The activated Met receptor leads to the downstream activation of several signaling molecules known to promote cell growth and survival. For example, we have shown that HGF ligation of the Met receptor induces activation of the focal adhesion kinase FAK [27]. Activated FAK has been shown to promote cell proliferation, cell survival and migration. Targets of FAK signaling implicated in the pro-survival pathways include RAS, RAC, phosphatidylinositol-3'-kinase or ERK [43]. Another example includes activation of NF-kB by HGF/Met, which has been shown to be crucial for proliferation and tubulogenesis in liver-derived MLP29 cells [44].

4.2. Met Regulates Embryonic Development

An interesting role for HGF/Met in embryonic development has been identified in mice with disruption of the Met gene [45-47]. These Met -/- mice displayed embryonic lethality with severe defects in liver and placenta development. Similar results were found in mice with a disruption of the HGF gene. In a developing organism Met may have a broader role in morphogenesis. The HGF/Met pathway may regulate invasive growth and morphogenesis in multiple embryonic tissues. Consistent with this idea, HGF/Met has been implicated in nervous system development [48].

4.3. Activated Met Leads to Cytoskeletal Abnormalities

The cytoskeleton is a complex network of actin filaments associated with regulatory proteins within the cytoplasm of eukaryotic cells. The activated cytoskeleton can regulate cell structure, cell division, growth, adhesion, motility and invasion. HGF was discovered as a secretory product of fibroblasts that induced scattering, a process that involves active regulation of cytoskeletal functions and as a mitogen for hepatocytes [6-8]. Cell scattering is a three-phase process that includes cell spreading, cell-cell dissociation and cell migration. Release of intercellular junctions is a process that has also been associated with tumor cell invasion and metastasis. The cytoskeleton can interact with extracellular signals from matrix proteins or cell contacts through transmembrane receptors, such as integrins. There is an accumulation of pivotal signaling proteins at these contact points or focal adhesions. The interactions of focal adhesion proteins with each other and other cytoplasmic proteins is often altered after cellular transformation [49]. A variety of studies have shown that HGF/Met can increase cell motility, a process regulated through the cytoskeleton and comprised of formation and retraction of filopodia/lamellipodia, uropod alteration and cell migration. For example, we have shown that cell motility of small cell lung cancer cells is increased after HGF stimulation [27]. Dysregulation of cell motility may be tightly linked to tumor invasion, a process distinct from tumor progression. During invasion cells degrade or remodel the surrounding extracellular matrix and migrate through the tissue boundary.

The molecular mechanisms behind HGF-dependent invasive growth are not fully understood and have just begun to be elucidated. It has been suggested that Met leads to the induction of genes that are actively involved in metastasis. Matrix glycoprotein osteopontin was identified as a major HGF transcriptional target in MLP-29 mouse embryo liver cells [50]. Also, activated Met has been associated with upregulation of urokinase-type plasminogen activator (uPA), plasminogen activator inhibitor-1 (PAI-1), and matrix metalloproteases (MMPs) in tumor cells [51-53]. There are likely additional factors that are involved in HGF-induced invasive growth.

Invasive growth through the Met pathway has been convincingly demonstrated in several animal models. Transgenic mice overexpressing Met in hepatocytes develop hepatocellular carcinoma, likely through a ligand-independent mechanism [54]. In another model transplantation of Met overexpressing tumor cells introduced into transgenic mice overexpressing HGF resulted in pulmonary metastasis [55]. The metastatic potential of this model system was equivalent to that of the HGF/Met autocrine signaling loops. HGF/Met autocrine loop have been found in human primary and metastatic tumors, including breast cancer, glioblastoma, osteosarcoma and melanoma [29,56-58]. Another mechanism of Met-induced metastasis may involve gain-of-function mutations in the receptor tyrosine kinase domain. Examples are the Y1230C and Y1235D Met mutations in head and neck squamous-cell carcinomas [59]. It has been suggested that these activating mutations deliver a positive selection signal during metastatic invasion of the cancer. There is also evidence that Met may be required for invasive growth mechanisms activated by other receptors. Plexin B1, a transmembrane tyrosine kinase related to Met, has been shown to require Met expression for invasive growth. Ligation of Plexin B1 leads to activation of Met, resulting in tyrosine phosphorylation of both receptors [60].

4.4. Met Increases Intracellular Levels of Reactive Oxygen Species (ROS)

We have shown that stimulation of the Met receptor with HGF results in an increase in reactive oxygen species (ROS) compared to quiescent cells [27]. Previous results suggest that increased production of ROS are sufficient to cause a transforming phenotype itself. Overexpression of the superoxide-generating NADPH oxidase Mox1 in NIH3T3 fibroblasts increases cell growth and induces tumors in athymic mice [61]. ROS are likely to act as second messengers and regulate activities of redox-sensitive enzymes, including protein kinases and protein phosphatases. Some enzymes are highly sensitive to oxidation because of a critical thiol group in the active site. For example, oxidation of Cys[118] in Ras is known to activate its GTPase activity [62]. Also, PTPases contain a redox sensitive cysteine residue in their active site that must be in the reduced state for full enzyme activity [63]. In addition to tyrosine phosphorylation, oxidative stress by ROS also has the potential to induce a specific response by activation or induction of transcription factors such as NF-κB [64] or forkhead [65]. Of particular relevance to cancer biology is the fact that ROS can react with DNA bases and a persistent increase in ROS could lead to accumulation of mutations that could contribute to progression of the disease [66].

5. INHIBITORS OF MET PATHWAYS

Aberrant activation of the Met pathway in oncogenic transformation points at multiple approaches for targeted therapies. It is expected that disruption of this pathway in Met-dependent tumors will lead to decreased cell growth, decreased viability or lower the metastatic potential of the cancer. A variety of Met pathway inhibitors have been identified that not only target Met itself but also disrupt the interaction of Met with its natural ligand HGF (Figure 3). It will now be necessary to determine the efficacy of these individual inhibitors and therapeutic interventions in clinical trials. With the success of Gleevec in chronic myelogenous leukemia and gastrointestinal tumors, it is predicted molecularly targeted therapy against Met will lead to dramatic inhibition of cancer growth and metastasis.

5.1. Inhibition of HGF-Mediated Activation of Met

The family of NK (N-terminal hairpin domain and kringle domain) inhibitors includes four variants of the HGF α-chain containing one (NK1) to four (NK4) kringle domains [67-70]. All variants can act as antagonists of HGF but retain some agonist activity towards Met. This activity is mediated through the receptor-binding region within the first kringle domain. NK inhibitors would therefore be very suitable to inhibit cancers with autocrine or paracrine activation of the Met receptor. The naturally occurring alternative splice variant NK2 and the HGF proteolytic fragment NK4 have been studied in more detail as potential Met inhibitors. Using an NK2-HGF double-

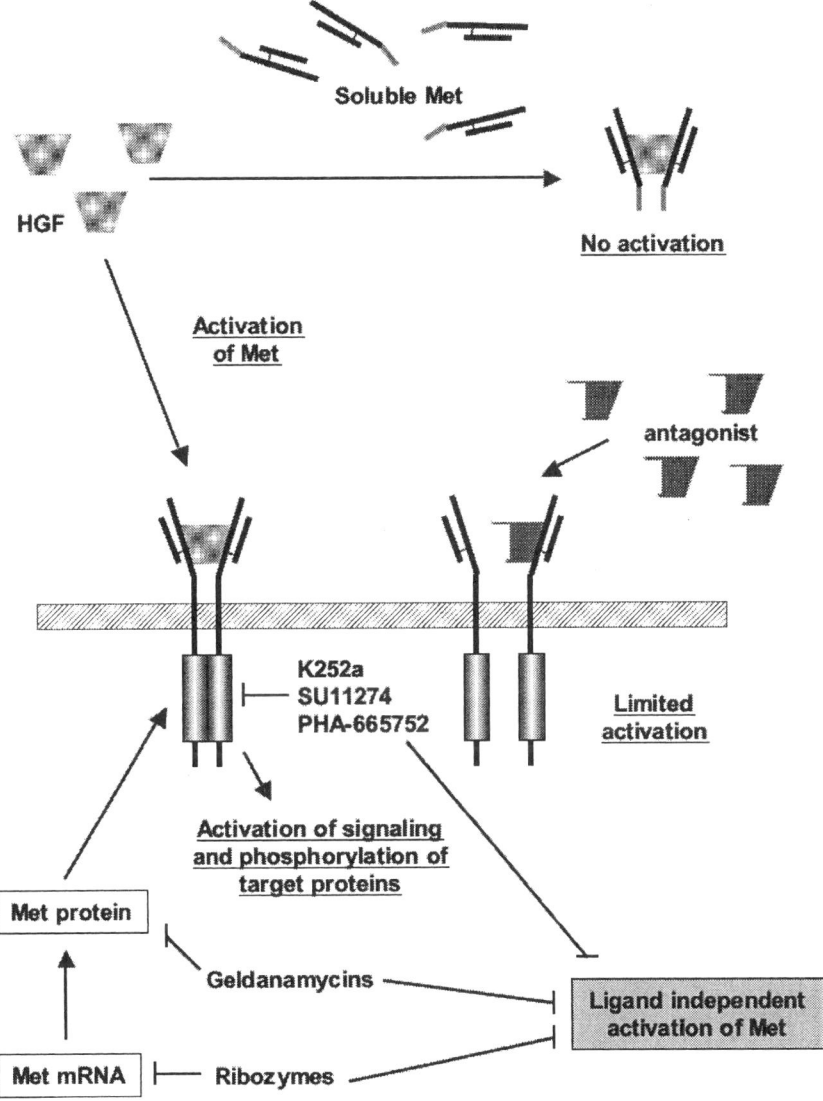

Figure 3: Inhibition of the HGF/Met pathway. The Met receptor can be activated through binding by its natural ligand HGF or by ligand-independent gain-of-function mutations. Inhibitors of HGF binding include receptor antagonists or soluble chimeric receptors. Receptor expression can be reduced by ribozymes or geldanamycin, and intracellular signaling can be blocked by small molecule drugs or inhibitory peptides.

transgenic mouse model, it has been reported that NK2 can inhibit growth, but it can also facilitate metastasis of transplanted Met-containing melanoma cells [71]. NK4 is of special interest since it not only antagonize HGF but also acts as an anti-angiogenesis factor [70,72,73]. Angiogenesis as well as motility and invasion of HT115 human colorectal cancer cells has been reported to be antagonized by NK4 [72]. Met inhibition also results in disruption of HGF-induced tyrosine phosphorylation. NK4 inhibited growth, invasion and disseminated metastasis of pancreatic cancer cells in mice, resulting in prolonged lifespan [73]. The efficacy of interference with extracellular signaling events in Met-dependent cancers was also demonstrated by using a decoy Met receptor. This soluble chimeric form of Met bound and therefore neutralized HGF activity as well as inhibited mutant Met-induced transformation of NIH3T3 cells [74, 75].

5.2. Targeting Met Expression

The expression of Met protein has been targeted at two distinct levels with either Met-specific ribozymes or inhibition of protein maturation. Ribozymes are RNA-based enzymes that bind to and cleave RNA molecules in a sequence-specific manner. The ribozyme inhibition approach has been shown to be quite challenging, and Met-positive breast cancer, colorectal carcinoma and glioblastoma have been targeted with ribozymes [76-78]. Interestingly, even though Met protein levels were reduced by only 25–35% in KM20 cell lines, the autophosphorylation and kinase activity of Met was reduced by 60–90% [78]. It still remains to be determined if ribozymes will sufficiently reduce Met levels in a clinical setting.

Inhibition of the heat shock protein (HSP90) by geldanamycin or related compounds of the anisomycin antibiotic family is of special interest since HSP90 appears to be in an activated state in cancer cells [79]. In these cancer cells HSP90 shows a high binding affinity for 17-allylaminogeldanamycin and increased ATPase activity, which regulates HSP90 chaperone function. Geldanamycins have previously been shown to inhibit maturation and functional expression of oncogenic proteins, including BCR-ABL [80] and mutant p53 [81]. We have shown that geldanamycin can inhibit growth and viability of small cell lung cancer cells [27], and 17-allylaminogeldanamycin is currently being developed for clinical cancer therapy [82,83]. We further showed that Met protein levels were downregulated in small cell lung cancer cells by treatment with geldanamycin [27]. Even though geldanamycins act by inhibiting HSP90 function, there was a large accumulation of HSP70. This might be related to the stress response of the small cell lung cancer cells. There are a variety of proteins that have been shown to be down-regulated by geldanamycins including serine/threonine and tyrosine kinases [84]. At low femtomolar concentrations geldanamycin inhibits HGF-mediated plasmin activation in MDCK-2 canine kidney cells, however, at nanomolar concentrations, the geldanamycins down-regulate Met protein expression and inhibit HGF-mediated cell motility and invasion in MDCK-2 cells. Geldanamycins also block transformation of NIH3T3 cells expressing activating mutations or coexpressing HGF and Met [85].

5.3. Small Molecule Met Kinase Inhibitors

Specific inhibition of tyrosine kinase activity by a small molecule drug and clinical efficacy had first been demonstrated for BCR-ABL in chronic myeloid leukemia and its inhibitor imatinib mesylate (STI571, Gleevec). However, thus far no tyrosine kinase inhibitor has been reported that is specific for only one single tyrosine kinase. Initial attempts to identify a Met kinase inhibitor led to the identification of the broad-spectrum inhibitor K252a [86], which can also inhibit serine/threonine kinases. K252a binds to the ATP binding site in the kinase domain and inhibits Trk family members and the PDGF receptor tyrosine kinase as well. *In vitro*, K252a may be a more potent inhibitor for Met with the activating M1268T mutation than for wild type Met [86]. We identified the novel small molecule Met inhibitor SU11274 using a TPR-MET transformed hematopoietic BaF3 cell line model and found it to be effective in lung cancer cells with activated Met as well [87]. The inhibitor did not affect other tyrosine kinase oncoproteins, including BCR-ABL, TEL-JAK2, TEL-PDGFßR or TEL-ABL. Inhibition of Met kinase activity by SU11274 led to time- and dose-dependent reduced cell growth. The Met inhibitor also induced G1 cell cycle arrest and apoptosis. The autophosphorylation of the Met kinase was reduced on sites that have been previously shown to be important for activation of pathways involved in cell growth and survival, especially the phosphatidylinositol-3'-kinase and the Ras pathway. The identification of SU11274 as an effective inhibitor of Met tyrosine kinase activity illustrates the potential therapeutic use of targeting for Met in cancers associated with activated forms of this kinase. There is also another specific Met inhibitor, PHA-665752 that is selective and inhibits Met dependent activity and function in tumor cells [88]. PHA-665752 causes regression of GTL-16 gastric carcinoma xenografts.

5.4. Met Peptide Inhibitors

Peptides derived from the Met receptor have also been shown to inhibit Met signaling, however, clinical efficacy of these potential large molecule drugs may be limited. Whereas peptides derived from the Met kinase activation loop are not biologically active, peptides derived form the carboxy-terminal tail region bind Met and inhibit its kinase activity *in vitro* [89]. These carboxy-terminal peptides inhibit Met kinase activity and HGF-mediated invasion, cell migration and branched morphogenesis by almost 50%. Likely due to protein homology, there is also inhibition of the Ron receptor tyrosine kinase by these peptides but no inhibition of the EGF, PDGF or VEGF receptors [89]. Peptides homologous to binding sites of crucial signaling proteins are also good candidates for interference and thus inhibition of Met signaling and Met-mediated transformation.

6. CONCLUSION

There are different options to inhibit Met activation and these are in part limited depending on the transforming phenotype. For example, inhibiting ligand/receptor interactions may be a useful approach when inhibiting a paracrine/autocrine mechanism but will show little effect on a gain-of-function mutation of Met. It still remains unknown as to how much Met inhibition is required to achieve clinically beneficial anti-tumor and anti-metastatic results. The answer may potentially be different depending upon whether the 'tumorigenic culprit' lies with Met expression or mutations of the RTK. It will be necessary to further study and optimize targeted approaches for clinical studies. Another intriguing area of research, with specific Met inhibitors available, would be to test the drug efficacies against different mutant forms of the Met oncoprotein. It is possible that some mutations of Met may cause resistance, or conversely a greater susceptibility, to the inhibitors. Among various strategies to inhibit HGF/Met signaling, developing specific small molecule inhibitors against Met may have the greatest potential in molecularly targeted anti-cancer therapy. However, lessons learned from Gleevec (imatinib mesylate) treatment in chronic myelogenous leukemia demonstrate that drug resistance is a real issue therapeutically. It will therefore be necessary to further characterize the signaling mechanisms behind Met-mediated proliferation and other biological effects to identify additional molecules for targeted therapies.

Martin Sattler
Department of Medical Oncology, Dana-Farber Cancer Institute, Brigham and Women's Hospital, and Harvard Medical School, Boston, MA

Patrick C. Ma and Ravi Salgia
Department of Medicine, Pritzker School of Medicine, University of Chicago, Chicago, IL

7. REFERENCES

1. Cooper CS, Park M, Blair DG, Tainsky MA, Huebner K, Croce CM, Vande Woude GF. Molecular cloning of a new transforming gene from a chemically transformed human cell line. Nature. 1984;311:29-33.
2. Park M, Dean M, Cooper CS, Schmidt M, O'Brien SJ, Blair DG, Vande Woude GF. Mechanism of met oncogene activation. Cell. 1986;45:895-904.
3. Tempest PR, Reeves BR, Spurr NK, Rance AJ, Chan AM, Brookes P. Activation of the met oncogene in the human MNNG-HOS cell line involves a chromosomal rearrangement. Carcinogenesis. 1986;7:2051-2057.
4. Rodrigues GA, Park M. Dimerization mediated through a leucine zipper activates the oncogenic potential of the met receptor tyrosine kinase. Mol Cell Biol. 1993;13:6711-6722.
5. Zhen Z, Giordano S, Longati P, Medico E, Campiglio M, Comoglio PM. Structural and functional domains critical for constitutive activation of the HGF-receptor (Met). Oncogene. 1994;9:1691-1697.
6. Gohda E, Tsubouchi H, Nakayama H, Hirono S, Takahashi K, Koura M, Hashimoto S, Daikuhara Y. Human hepatocyte growth factor in plasma from patients with fulminant hepatic failure. Exp Cell Res. 1986;166:139-150.
7. Nakamura T, Nawa K, Ichihara A, Kaise N, Nishino T. Purification and subunit structure of hepatocyte growth factor from rat platelets. FEBS Lett. 1987;224:311-316.
8. Stoker M, Gherardi E, Perryman M, Gray J. Scatter factor is a fibroblast-derived modulator of epithelial cell mobility. Nature. 1987;327:239-242.
9. Comoglio PM, Boccaccio C. The HGF receptor family: unconventional signal transducers for invasive cell growth. Genes Cells. 1996;1:347-354.
10. Maestrini E, Tamagnone L, Longati P, Cremona O, Gulisano M, Bione S, Tamanini F, Neel BG, Toniolo D, Comoglio PM. A family of transmembrane proteins with homology to the MET-hepatocyte growth factor receptor. Proc Natl Acad Sci U S A. 1996;93:674-678.
11. Duh FM, Scherer SW, Tsui LC, Lerman MI, Zbar B, Schmidt L. Gene structure of the human MET proto-oncogene. Oncogene. 1997;15:1583-1586.
12. Liu Y. The human hepatocyte growth factor receptor gene: complete structural organization and promoter characterization. Gene. 1998;215:159-169.
13. Kamada M, Komori A, Chiba S, Nakao T. A prospective study of congenital cytomegalovirus infection in Japan. Scand J Infect Dis. 1983;15:227-232.
14. Rodrigues GA, Park M. Autophosphorylation modulates the kinase activity and oncogenic potential of the Met receptor tyrosine kinase. Oncogene. 1994;9:2019-2027.
15. Petrelli A, Gilestro GF, Lanzardo S, Comoglio PM, Migone N, Giordano S. The endophilin-CIN85-Cbl complex mediates ligand-dependent downregulation of c-Met. Nature. 2002;416:187-190.
16. Furge KA, Zhang YW, Vande Woude GF. Met receptor tyrosine kinase: enhanced signaling through adapter proteins. Oncogene. 2000;19:5582-5589.
17. Iwazawa T, Shiozaki H, Doki Y, Inoue M, Tamura S, Matsui S, Monden T, Matsumoto K, Nakamura T, Monden M. Primary human fibroblasts induce diverse tumor invasiveness: involvement of HGF as an important paracrine factor. Jpn J Cancer Res. 1996;87:1134-1142.
18. Weimar IS, Miranda N, Muller EJ, Hekman A, Kerst JM, de Gast GC, Gerritsen WR. Hepatocyte growth factor/scatter factor (HGF/SF) is produced by human bone marrow stromal cells and promotes proliferation, adhesion and survival of human hematopoietic progenitor cells (CD34+). Exp Hematol. 1998;26:885-894.

19. Nakashiro K, Okamoto M, Hayashi Y, Oyasu R. Hepatocyte growth factor secreted by prostate-derived stromal cells stimulates growth of androgen-independent human prostatic carcinoma cells. Am J Pathol. 2000;157:795-803.
20. Takai K, Hara J, Matsumoto K, Hosoi G, Osugi Y, Tawa A, Okada S, Nakamura T. Hepatocyte growth factor is constitutively produced by human bone marrow stromal cells and indirectly promotes hematopoiesis. Blood. 1997;89:1560-1565.
21. Soman NR, Correa P, Ruiz BA, Wogan GN. The TPR-MET oncogenic rearrangement is present and expressed in human gastric carcinoma and precursor lesions. Proc Natl Acad Sci U S A. 1991;88:4892-4896.
22. Yu J, Miehlke S, Ebert MP, Hoffmann J, Breidert M, Alpen B, Starzynska T, Stolte Prof M, Malfertheiner P, Bayerdorffer E. Frequency of TPR-MET rearrangement in patients with gastric carcinoma and in first-degree relatives. Cancer. 2000;88:1801-1806.
23. Heideman DA, Snijders PJ, Bloemena E, Meijer CJ, Offerhaus GJ, Meuwissen SG, Gerritsen WR, Craanen ME. Absence of tpr-met and expression of c-met in human gastric mucosa and carcinoma. J Pathol. 2001;194:428-435.
24. Natali PG, Prat M, Nicotra MR, Bigotti A, Olivero M, Comoglio PM, Di Renzo MF. Overexpression of the met/HGF receptor in renal cell carcinomas. Int J Cancer. 1996;69:212-217.
25. Olivero M, Rizzo M, Madeddu R, Casadio C, Pennacchietti S, Nicotra MR, Prat M, Maggi G, Arena N, Natali PG, Comoglio PM, Di Renzo MF. Overexpression and activation of hepatocyte growth factor/scatter factor in human non-small-cell lung carcinomas. Br J Cancer. 1996;74:1862-1868.
26. Porte H, Triboulet JP, Kotelevets L, Carrat F, Prevot S, Nordlinger B, DiGioia Y, Wurtz A, Comoglio P, Gespach C, Chastre E. Overexpression of stromelysin-3, BM-40/SPARC, and MET genes in human esophageal carcinoma: implications for prognosis. Clin Cancer Res. 1998;4:1375-1382.
27. Maulik G, Kijima T, Ma PC, Ghosh SK, Lin J, Shapiro GI, Schaefer E, Tibaldi E, Johnson BE, Salgia R. Modulation of the c-Met/hepatocyte growth factor pathway in small cell lung cancer. Clin Cancer Res. 2002;8:620-627.
28. Ramirez R, Hsu D, Patel A, Fenton C, Dinauer C, Tuttle RM, Francis GL. Over-expression of hepatocyte growth factor/scatter factor (HGF/SF) and the HGF/SF receptor (cMET) are associated with a high risk of metastasis and recurrence for children and young adults with papillary thyroid carcinoma. Clin Endocrinol (Oxf). 2000;53:635-644.
29. Ferracini R, Di Renzo MF, Scotlandi K, Baldini N, Olivero M, Lollini P, Cremona O, Campanacci M, Comoglio PM. The Met/HGF receptor is over-expressed in human osteosarcomas and is activated by either a paracrine or an autocrine circuit. Oncogene. 1995;10:739-749.
30. Di Renzo MF, Olivero M, Giacomini A, Porte H, Chastre E, Mirossay L, Nordlinger B, Bretti S, Bottardi S, Giordano S, et al. Overexpression and amplification of the met/HGF receptor gene during the progression of colorectal cancer. Clin Cancer Res. 1995;1:147-154.
31. Di Renzo MF, Olivero M, Katsaros D, Crepaldi T, Gaglia P, Zola P, Sismondi P, Comoglio PM. Overexpression of the Met/HGF receptor in ovarian cancer. Int J Cancer. 1994;58:658-662.
32. Ma PC, Kijima T, Maulik G, Fox EA, Sattler M, Griffin JD, Johnson BE, Salgia R. c-MET mutaional anlysis in small cell lung cancer: Novel juxtamembrance domain mutations regulating cytoskeletal functions. Cancer Research. 2003;63:(in press).
33. Nakao M, Yokota S, Iwai T, Kaneko H, Horiike S, Kashima K, Sonoda Y, Fujimoto T, Misawa S. Internal tandem duplication of the flt3 gene found in acute myeloid leukemia. Leukemia. 1996;10:1911-1918.
34. Yokota S, Kiyoi H, Nakao M, Iwai T, Misawa S, Okuda T, Sonoda Y, Abe T, Kahsima K, Matsuo Y, Naoe T. Internal tandem duplication of the FLT3 gene is preferentially seen in

acute myeloid leukemia and myelodysplastic syndrome among various hematological malignancies. A study on a large series of patients and cell lines. Leukemia. 1997;11:1605-1609.

35. Schmidt L, Junker K, Nakaigawa N, Kinjerski T, Weirich G, Miller M, Lubensky I, Neumann HP, Brauch H, Decker J, Vocke C, Brown JA, Jenkins R, Richard S, Bergerheim U, Gerrard B, Dean M, Linehan WM, Zbar B. Novel mutations of the MET proto-oncogene in papillary renal carcinomas. Oncogene. 1999;18:2343-2350

36. Lee JH, Han SU, Cho H, Jennings B, Gerrard B, Dean M, Schmidt L, Zbar B, Vande Woude GF. A novel germ line juxtamembrane Met mutation in human gastric cancer. Oncogene. 2000;19:4947-4953.

37. Lee CC, Yamada KM. Identification of a novel type of alternative splicing of a tyrosine kinase receptor. Juxtamembrane deletion of the c-met protein kinase C serine phosphorylation regulatory site. J Biol Chem. 1994;269:19457-19461.

38. Lee CC, Yamada KM. Alternatively spliced juxtamembrane domain of a tyrosine kinase receptor is a multifunctional regulatory site. Deletion alters cellular tyrosine phosphorylation pattern and facilitates binding of phosphatidylinositol-3-OH kinase to the hepatocyte growth factor receptor. J Biol Chem. 1995;270:507-510.

39. effers M, Fiscella M, Webb CP, Anver M, Koochekpour S, Vande Woude GF. The mutationally activated Met receptor mediates motility and metastasis. Proc Natl Acad Sci U S A. 1998;95:14417-14422.

40. Schmidt L, Junker K, Weirich G, Glenn G, Choyke P, Lubensky I, Zhuang Z, Jeffers M, Vande Woude G, Neumann H, Walther M, Linehan WM, Zbar B. Two North American families with hereditary papillary renal carcinoma and identical novel mutations in the MET proto-oncogene. Cancer Res. 1998;58:1719-1722.

41. Schmidt L, Duh FM, Chen F, Kishida T, Glenn G, Choyke P, Scherer SW, Zhuang Z, Lubensky I, Dean M, Allikmets R, Chidambaram A, Bergerheim UR, Feltis JT, Casadevall C, Zamarron A, Bernues M, Richard S, Lips CJ, Walther MM, Tsui LC, Geil L, Orcutt ML, Stackhouse T, Zbar B, et al. Germline and somatic mutations in the tyrosine kinase domain of the MET proto-oncogene in papillary renal carcinomas. Nat Genet. 1997;16:68-73.

42. Vigna E, Gramaglia D, Longati P, Bardelli A, Comoglio PM. Loss of the exon encoding the juxtamembrane domain is essential for the oncogenic activation of TPR-MET. Oncogene. 1999;18:4275-4281.

43. Parsons JT, Martin KH, Slack JK, Taylor JM, Weed SA. Focal adhesion kinase: a regulator of focal adhesion dynamics and cell movement. Oncogene. 2000;19:5606-5613

44. Muller M, Morotti A, Ponzetto C. Activation of NF-kappaB is essential for hepatocyte growth factor-mediated proliferation and tubulogenesis. Mol Cell Biol. 2002;22:1060-1072.

45. Bladt F, Riethmacher D, Isenmann S, Aguzzi A, Birchmeier C. Essential role for the c-met receptor in the migration of myogenic precursor cells into the limb bud. Nature. 1995;376:768-771.

46. Schmidt C, Bladt F, Goedecke S, Brinkmann V, Zschiesche W, Sharpe M, Gherardi E, Birchmeier C. Scatter factor/hepatocyte growth factor is essential for liver development. Nature. 1995;373:699-702.

47. Uehara Y, Minowa O, Mori C, Shiota K, Kuno J, Noda T, Kitamura N. Placental defect and embryonic lethality in mice lacking hepatocyte growth factor/scatter factor. Nature. 1995;373:702-705.

48. Maina F, Klein R. Hepatocyte growth factor, a versatile signal for developing neurons. Nat Neurosci. 1999;2:213-217.

49. Weisberg E, Sattler M, Ewaniuk DS, Salgia R. Role of focal adhesion proteins in signal transduction and oncogenesis. Crit Rev Oncog. 1997;8:343-358.

136

50. Medico E, Gentile A, Lo Celso C, Williams TA, Gambarotta G, Trusolino L, Comoglio PM. Osteopontin is an autocrine mediator of hepatocyte growth factor-induced invasive growth. Cancer Res. 2001;61:5861-5868.
51. Besser D, Bardelli A, Didichenko S, Thelen M, Comoglio PM, Ponzetto C, Nagamine Y. Regulation of the urokinase-type plasminogen activator gene by the oncogene Tpr-Met involves GRB2. Oncogene. 1997;14:705-711.
52. Dunsmore SE, Rubin JS, Kovacs SO, Chedid M, Parks WC, Welgus HG. Mechanisms of hepatocyte growth factor stimulation of keratinocyte metalloproteinase production. J Biol Chem. 1996;271:24576-24582.
53. Wojta J, Nakamura T, Fabry A, Hufnagl P, Beckmann R, McGrath K, Binder BR. Hepatocyte growth factor stimulates expression of plasminogen activator inhibitor type 1 and tissue factor in HepG2 cells. Blood. 1994;84:151-157.
54. Wang R, Ferrell LD, Faouzi S, Maher JJ, Bishop JM. Activation of the Met receptor by cell attachment induces and sustains hepatocellular carcinomas in transgenic mice. J Cell Biol. 2001;153:1023-1034.
55. Yu Y, Merlino G. Constitutive c-Met signaling through a nonautocrine mechanism promotes metastasis in a transgenic transplantation model. Cancer Res. 2002;62:2951-2956.
56. Tuck AB, Park M, Sterns EE, Boag A, Elliott BE. Coexpression of hepatocyte growth factor and receptor (Met) in human breast carcinoma. Am J Pathol. 1996;148:225-232
57. Koochekpour S, Jeffers M, Rulong S, Taylor G, Klineberg E, Hudson EA, Resau JH, Vande Woude GF. Met and hepatocyte growth factor/scatter factor expression in human gliomas. Cancer Res. 1997;57:5391-5398.
58. Li G, Schaider H, Satyamoorthy K, Hanakawa Y, Hashimoto K, Herlyn M. Downregulation of E-cadherin and Desmoglein 1 by autocrine hepatocyte growth factor during melanoma development. Oncogene. 2001;20:8125-8135.
59. Di Renzo MF, Olivero M, Martone T, Maffe A, Maggiora P, Stefani AD, Valente G, Giordano S, Cortesina G, Comoglio PM. Somatic mutations of the MET oncogene are selected during metastatic spread of human HNSC carcinomas. Oncogene. 2000;19:1547-1555.
60. Giordano S, Corso S, Conrotto P, Artigiani S, Gilestro G, Barberis D, Tamagnone L, Comoglio PM. The semaphorin 4D receptor controls invasive growth by coupling with Met. Nat Cell Biol. 2002;4:720-724.
61. Suh YA, Arnold RS, Lassegue B, Shi J, Xu X, Sorescu D, Chung AB, Griendling KK, Lambeth JD. Cell transformation by the superoxide-generating oxidase Mox1. Nature. 1999;401:79-82.
62. Lander HM, Hajjar DP, Hempstead BL, Mirza UA, Chait BT, Campbell S, Quilliam LA. A molecular redox switch on p21^{ras}. Structural basis for the nitric oxide-p21^{ras} interaction. J Biol Chem. 1997;272:4323-4326.
63. Denu JM, Tanner KG. Specific and reversible inactivation of protein tyrosine phosphatases by hydrogen peroxide: evidence for a sulfenic acid intermediate and implications for redox regulation. Biochemistry. 1998;37:5633-5642.
64. Flohe L, Brigelius-Flohe R, Saliou C, Traber MG, Packer L. Redox regulation of NF-kappa B activation. Free Radic Biol Med. 1997;22:1115-1126.
65. Nemoto S, Finkel T. Redox regulation of forkhead proteins through a p66shc-dependent signaling pathway. Science. 2002;295:2450-2452.
66. Dreher D, Junod AF. Role of oxygen free radicals in cancer development. Eur J Cancer. 1996;32A:30-38.
67. Silvagno F, Follenzi A, Arese M, Prat M, Giraudo E, Gaudino G, Camussi G, Comoglio PM, Bussolino F. In vivo activation of met tyrosine kinase by heterodimeric hepatocyte growth factor molecule promotes angiogenesis. Arterioscler Thromb Vasc Biol. 1995;15:1857-1865.

68. Cioce V, Csaky KG, Chan AM, Bottaro DP, Taylor WG, Jensen R, Aaronson SA, Rubin JS. Hepatocyte growth factor (HGF)/NK1 is a naturally occurring HGF/scatter factor variant with partial agonist/antagonist activity. J Biol Chem. 1996;271:13110-13115.

69. Schwall RH, Chang LY, Godowski PJ, Kahn DW, Hillan KJ, Bauer KD, Zioncheck TF. Heparin induces dimerization and confers proliferative activity onto the hepatocyte growth factor antagonists NK1 and NK2. J Cell Biol. 1996;133:709-718.

70. Date K, Matsumoto K, Shimura H, Tanaka M, Nakamura T. HGF/NK4 is a specific antagonist for pleiotrophic actions of hepatocyte growth factor. FEBS Lett. 1997;420:1-6

71. Otsuka T, Jakubczak J, Vieira W, Bottaro DP, Breckenridge D, Larochelle WJ, Merlino G. Disassociation of met-mediated biological responses in vivo: the natural hepatocyte growth factor/scatter factor splice variant NK2 antagonizes growth but facilitates metastasis. Mol Cell Biol. 2000;20:2055-2065.

72. Parr C, Hiscox S, Nakamura T, Matsumoto K, Jiang WG. Nk4, a new HGF/SF variant, is an antagonist to the influence of HGF/SF on the motility and invasion of colon cancer cells. Int J Cancer. 2000;85:563-570.

73. Tomioka D, Maehara N, Kuba K, Mizumoto K, Tanaka M, Matsumoto K, Nakamura T. Inhibition of growth, invasion, and metastasis of human pancreatic carcinoma cells by NK4 in an orthotopic mouse model. Cancer Res. 2001;61:7518-7524.

74. Mark MR, Lokker NA, Zioncheck TF, Luis EA, Godowski PJ. Expression and characterization of hepatocyte growth factor receptor-IgG fusion proteins. Effects of mutations in the potential proteolytic cleavage site on processing and ligand binding. J Biol Chem. 1992;267:26166-26171.

75. Michieli P, Basilico C, Pennacchietti S, Maffe A, Tamagnone L, Giordano S, Bardelli A, Comoglio PM. Mutant Met-mediated transformation is ligand-dependent and can be inhibited by HGF antagonists. Oncogene. 1999;18:5221-5231.

76. Abounader R, Lal B, Luddy C, Koe G, Davidson B, Rosen EM, Laterra J. In vivo targeting of SF/HGF and c-met expression via U1snRNA/ribozymes inhibits glioma growth and angiogenesis and promotes apoptosis. Faseb J. 2002;16:108-110.

77. Jiang WG, Grimshaw D, Lane J, Martin TA, Abounader R, Laterra J, Mansel RE, Abounder R. A hammerhead ribozyme suppresses expression of hepatocyte growth factor/scatter factor receptor c-MET and reduces migration and invasiveness of breast cancer cells. Clin Cancer Res. 2001;7:2555-2562.

78. Herynk MH, Stoeltzing O, Reinmuth N, Parikh NU, Abounader R, Laterra J, Radinsky R, Ellis LM, Gallick GE. Down-regulation of c-Met inhibits growth in the liver of human colorectal carcinoma cells. Cancer Res. 2003;63:2990-2996.

79. Kamal A, Thao L, Sensintaffar J, Zhang L, Boehm MF, Fritz LC, Burrows FJ. A high-affinity conformation of Hsp90 confers tumour selectivity on Hsp90 inhibitors. Nature. 2003;425:407-410.

80. An WG, Schulte TW, Neckers LM. The heat shock protein 90 antagonist geldanamycin alters chaperone association with p210bcr-abl and v-src proteins before their degradation by the proteasome. Cell Growth Differ. 2000;11:355-360.

81. Blagosklonny MV, Toretsky J, Neckers L. Geldanamycin selectively destabilizes and conformationally alters mutated p53. Oncogene. 1995;11:933-939.

82. Isaacs JS, Xu W, Neckers L. Heat shock protein 90 as a molecular target for cancer therapeutics. Cancer Cell. 2003;3:213-217.

83. Workman P. Auditing the pharmacological accounts for Hsp90 molecular chaperone inhibitors: unfolding the relationship between pharmacokinetics and pharmacodynamics. Mol Cancer Ther. 2003;2:131-138.

84. Neckers L, Schulte TW, Mimnaugh E. Geldanamycin as a potential anti-cancer agent: its molecular target and biochemical activity. Invest New Drugs. 1999;17:361-373.

85. Webb CP, Hose CD, Koochekpour S, Jeffers M, Oskarsson M, Sausville E, Monks A, Vande Woude GF. The geldanamycins are potent inhibitors of the hepatocyte growth

138

factor/scatter factor-met-urokinase plasminogen activator-plasmin proteolytic network. Cancer Res. 2000;60:342-349.

86. Morotti A, Mila S, Accornero P, Tagliabue E, Ponzetto C. K252a inhibits the oncogenic properties of Met, the HGF receptor. Oncogene. 2002;21:4885-4893.

87. Sattler M, Pride YB, Ma P, Gramlich JL, Chu SC, Quinnan LA, Shirazian S, Liang C, Podar K, Christensen JG, Salgia R. A novel small molecule met inhibitor induces apoptosis in cells transformed by the oncogenic TPR-MET tyrosine kinase. Cancer Res. 2003;63:5462-5469.

88. Christensen JK, Schreck R, Burrows J, Kuruganti P, Chan E, Le P, Chen J, Wang X, Ruslim L, Blake R, Lipson KE, Ramphal J, Do S, Cui JJ, Cherrington JM, Mendel DB. A selective small molecule inhibitor of c-Met kinase inhibits c-Met-dependent phenotypes *in vitro* and exhibits cytoreductive anti-tumor activity *in vivo*. Cancer Research. 2003;63:(in press)

89. Bardelli A, Longati P, Williams TA, Benvenuti S, Comoglio PM. A peptide representing the carboxyl-terminal tail of the met receptor inhibits kinase activity and invasive growth. J Biol Chem. 1999;274:29274-29281.

NUCLEAR FACTOR-κB ACTIVATION MEDIATES CELLULAR TRANSFORMATION, PROLIFERATION, INVASION ANGIOGENESIS AND METASTASIS OF CANCER

SHISHIR SHISHODIA AND BHARAT B. AGGARWAL

1. INTRODUCTION

Transcription factors are proteins that act like genetic switches to regulate the expression of various genes. These factors regulate transcription by binding to specific sequences present within the promoter, enhancer, or other regulatory regions of DNA or even RNA. Transcription factors can interact directly with RNA polymerase or through other transcription factors. Eukaryotic transcription factors are nearly always positive regulators of transcription, although negative regulation is seen occasionally. Hundreds of transcription factors with functionally separable domains essential for DNA binding and activation have been identified and characterized in several organisms [1-3] Some of the most common transcription factors in mammalian cells include Sp1, AP-1/TRE, C/EBP, OCT-1, OCT-2, SRF, HSF, IRF1, STAT, myc, jun, fos, and NF-κB. The current review is exclusively concerned with NF-κB.

NF-κB, a nuclear transcription factor, was first identified in 1986 by Sen and Baltimore. It was so named because it was found to be a nuclear factor bound to an enhancer element of the immunoglobulin kappa light chain gene in B cells [4]. It was considered to be a B-cell transcription factor in the beginning, but now it is known to constitute a family of ubiquitous proteins. NF-κB proteins contain a Rel homology (RH) domain (DNA-binding domain/dimerization domain) with a nuclear localization sequence; the sequences are conserved from *Drosophila* to man. Class I proteins include p50, p52, p100, and p105. Multiple copies of ankyrin repeats are present in p100 and p105, and proteolytic cleavage of p100 forms p52 and that of p105 forms p50 (Figure 1). These in turn form dimers with class II proteins (c-Rel, RelB, and RelA/p65), which exclusively contain C-terminal activation

140

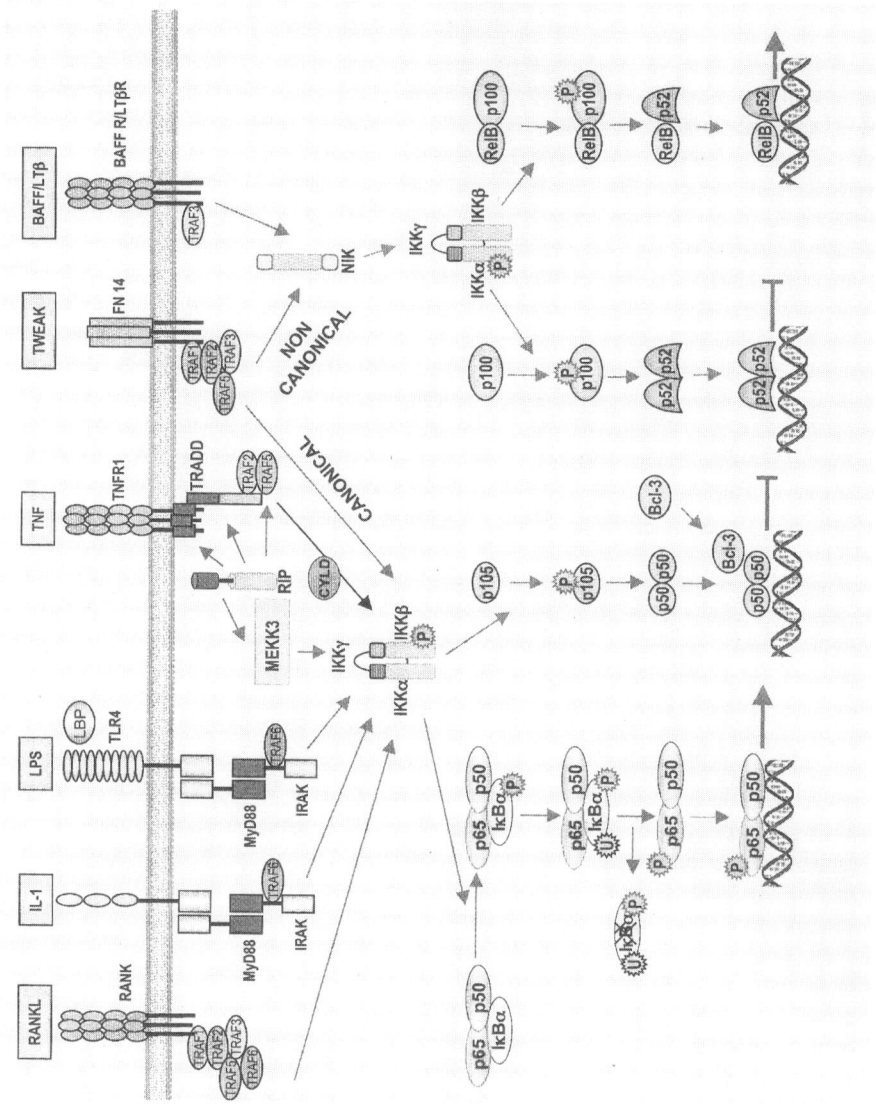

Figure 1: Diagrammatic representation of various NF-κB activation pathway induced by different cytokines and LPS.

domains. RelB forms only heterodimers, whereas all other members can form both homo and heterodimers. NF-κB is the most common heterodimer formed between Rel A and p50. Dimeric NF-κB transcription factors bind to the 10-base-pair consensus site GGGPuNNPyPyCC, where Pu is purine, Py is pyrimidine, and N is any base. The individual dimers have distinct DNA-binding specificities for a collection of related κB sites [5, 6].

Various inhibitors of NF-κB include IκBα, IκBβ, IκBγ (derived from the C-terminal of p100), IκB-ε, Bcl-3, pp40 (chicken homologue), cactus (*Drosophila* homologue), and avian swine fever virus protein p28.2. p105 and p100 can also function to retain NF-κB subunits in the cytoplasm. All of these proteins are characterized by the presence of multiple ankyrin repeats. Perhaps the most common and best understood form of NF-κB consists of p50, p65, and IκBα. IκBα mediates transient gene expression, whereas IκBβ mediates persistent response.

The IκB proteins are expressed in a tissue-specific manner and have distinct affinities for individual Rel/NF-κB complexes. IκBs contain six or more ankyrin repeats, an N-terminal regulatory domain, and a C-terminal domain that contains a PEST motif. IκBs bind to NF-κB dimers and sterically block the function of their nuclear localization sequences, thereby causing their cytoplasmic retention. Most agents that activate NF-κB mediate the phosphorylation-induced degradation of IκB. Upon receiving a signal, phosphorylation of IκBα takes place on two conserved serine residues (S32 and S36) in the N-terminal regulatory domain. Another member of the IκB family, Bcl-3, however, stimulates transcription after interacting with p50 and p52 subunits of NF-κB.

Several of the IκB kinases (IKKs), have now been characterized (IKKα, IKKβ and IKKγ). Mutation analysis revealed that IKKα and not IKKβ mediates proinflammatory signals. Once phosphorylated, the IκBs, still bound to NF-κB, almost immediately undergo a second post-translational modification, polyubiquitination. The major ubiquitin acceptor sites in human IκBα are lysines 21 and 22. Protein ubiquitination occurs through E1 ubiquitin activating enzyme, E2 ubiquitin-conjugating enzyme, and E3 ubiquitin protein ligases. After ubiquitination, IκBs are degraded in 26 S proteasomes, leading to the release of NF-κB dimers to translocate into the nucleus [6, 7]. This general mechanism of activation of NF-κB is depicted in Figure 1. In contrast, activation of NF-κB in response to UV radiation is accompanied by IκBα degradation but not phosphorylation on the N-terminus of IκBα [8]. Hypoxia or pervandate treatment stimulates the phosphorylation of IκBα at tyrosine 42, but other IκBs do not have a tyrosine at this position [9]. Phosphorylation on Ser-276 by the catalytic subunit of protein kinase A (PKAc) can contribute to the intrinsic transcriptional capacity of the p65 subunit of NF-κB. PKAc was also found associated with NF-κB and IκB in the cytoplasm and was able to phosphorylate p65 only after IκB degradation [10]. A site-directed mutant of p65 (Ser-276 to Ala) is still phosphorylated at Ser 529 in response to tumor necrosis factor (TNF), suggesting that multiple physiological stimuli modulate p65 through distinct phosphorylation sites to control transcriptional activity. RelA (C-terminus) has been shown to interact with basal transcriptional apparatus proteins like TBP, TFIIB and TAF 105 as well as coactivators such as CBP and p300, though the actual role of these interactions is not clear [10]. This pathway is well conserved, both in structure and function, from *Drosophila* to humans.

NF-κB is activated by many divergent stimuli including proinflammatory cytokines such as TNF-α, interleukin-1 (IL-1), T- and B-cell mitogens, bacteria and lipopolysaccharide (LPS), viruses, viral proteins, double-stranded RNA, and physical and chemical stresses [11]. Cellular stresses including ionizing radiation and chemotherapeutic agents, also activate NF-κB [12].

Although much has been learned since its discovery, the precise mechanism of NF-κB activation is still not fully understood. Depending on the stimulus, the mechanism of activation involves overlapping and nonoverlapping steps. For example TNF, one of the most potent activators of NF-κB, interacts with the receptor and then recruits a protein called TNF receptor-associated death domain (TRADD). This protein binds to TNF receptor-associated factor (TRAF)-2, which recruits NF-κB inducing kinase (NIK), which in turn activates IκBα kinase (IKK). IKK phosphorylates IκBα at serines 32 and 36, which leads to ubiquitination at lysines 21 and 22, and this leads to the degradation of IκBα by the 26S proteosome. This degradation results in translocation of NF-κB to the nucleus, where it binds to its consensus sequence (5'-GGGACTTTC-3') and activates gene expression. Thus NF-κB can be monitored by IκBα degradation (by western blotting), NF-κB binding to the DNA (by electrophoretic mobility shift assay), or NF-κB-dependent reporter gene expression (by transient transfection).

Besides the canonical NF-κB activation pathway decribed above, a noncanonical NF-κB activation pathway (Figure 1), is activated by CD40L, LT-β, RANKL, and BAFF, all members of the TNF family [13, 14]. This pathway does not involve IκBα but involves direct phosphorylation and ubiquitin-dependent degradation of p100. Current research indicates that NF-κB activation is highly complex and may involve dozens of different protein kinases. Besides NIK, IKK-α, and IKK-β, NF-κB activation may also require atypical protein kinase C, PKC-z, pp90rsk, dsRNA-dependent protein kinase (PKR), cot kinase (also called TPL2), mitogen activated protein kinase kinase kinase (MEKK1, 2 and 3), phosphatidylinositol 3 protein kinase, Akt, mixed linage kinase (MLK)-3, HPK-1, transforming growth factor (TGF)-β-activated kinase-1 (TAK)-1, and c-raf kinase. These kinases may form a cascade, and different cascades may be formed depending on the NF-κB activator. For instance IKK can be phosphorylated by either NIK, MEKK, or Akt. While IKK is required for NF-κB activation by most agents, some agents (such as HER2, H_2O_2, pervanadate, X-rays, γ-radiation) activate NF-κB through IKK-independent pathways [15]. Although several signaling proteins and protein kinases have been identified recently that mediate NF-κB activation, several more kinases and protein phosphatases remain to be identified. Degradation of IκBα involves the ubiquitin-dependent 26 S proteosome [16]. Besides this proteosome, other proteases have also been implicated in NF-κB activation [17].

The genetic deletion of different NF-κB proteins produces numerous phenotypic changes (Table I) [18]. For instance, deletion of the rel a gene induced embryonic lethality probably due to massive apoptosis in the liver [19]. The mouse embryonic fibroblasts from rel a-deletion mice were found to be hypersensitive to TNF-induced apoptosis. These results indicate a negative role for NF-κB in TNF-induced apoptosis. Mice lacking the RelA subunit are brought to term only by breeding onto a TNFR1-deficient background [20]. These mice lack lymph nodes, Peyer's patches, and an organized splenic microarchitecture, and they have a profound defect in T cell-dependent antigen responses. Analyses of TNFR1/RelA-deficient embryonic tissues and of radiation chimeras suggest that the dependence on RelA is manifest not in

Table 1. Genetic deletions of the NF-κB proteins

Gene	Phenotype
rela	Embryonic lethal; liver apoptosis; sensitivity to TNF
c-rel	Impaired T and B cell activation; systemic arthritis resistance
relb	Defects in acquired and innate immunity; T-cell infiltration of organs
nf-κb1	Immune response defects due to abnormal B-cell response; resistance to arthritis
nf-κb2	Abnormal spleen and lymph node architecture; defective T-cell response
c-rel delta C	Lymphoid cell hyperplasia
nf-κb1 delta C	Splenomegaly; enlarged lymph nodes; infiltrations of lung & liver; susceptibility to pathogens
nf-κb2 delta C	Multiple post-natal organ defects
nf-κb1/relB	Lethal 3-4 weeks post-natal; Increased organ inflammation
nf-κb1/ nf-κb2	Reduced growth; cranofacial abnormalities; bone defects; B-cell defects
nf-κb /c-rel	Decreased humoral immunity
c-rel/rela	Embryonic lethal early liver apoptosis
iκbα	Early neonatal lethal inflammatory dermatitis and granulocytosis
iκbε	Reduction in number of CDD44-CD25+ T cells
bcl-3	Defects in B- and T-cell responses to antigens
IKKα	Early neonatal lethal; Skin defects
IKKβ	Embryonic lethal; Liver apoptosis; Sensitivity to TNF
IKKα/IKKβ	Embryonic lethal; Liver apoptosis;
IKKγ (NEMO)	Embryonic lethal; Liver apoptosis; Sensitivity to TNF; Heterozygotes are model for incontinentia pigmenti

hematopoietic cells but rather in radioresistant stromal cells needed for the development of secondary lymphoid organs. In contrast to Rel A, the deletion of the iκbα gene leads to early neonatal lethality caused by inflammatory dermatitis and granulocytosis [21], most likely induced by constitutive activation of NF-κB leading to expression of granulocyte-colony-stimulating factor.

NF-κB activation has been implicated in a wide variety of diseases including including cancers, diabetes, cardiovascular diseases, autoimmune diseases, viral replication, septic shock, neurodegenerative disorders, ataxia telangiectasia(AT), arthritis, asthma, inflammatory bowel disease, and several other inflammatory conditions. Activation of NF-κB by LPS may have a role in the development of septic shock, since NF-κB activates transcription of the iNOS genes known to be involved in septic shock [22]. Autoimmune diseases such as systemic lupus erythromatosus (SLE) may also involve activation of NF-κB. In Alzheimer's disease, under chronic conditions, the amyloid β peptide causes production of reactive oxygen intermediates (ROIs) and indirectly activates gene expression through κB sites [23]. The influenza virus protein hemagglutinin activates NF-κB, and this activation may contribute to viral induction of cytokines and to some of the symptoms associated with flu [24]. The oxidized lipids from low density lipoproteins (LDL) associated with atherosclerosis activate NF-κB, which then activates other genes [25]. Mice that are susceptible to atherosclerosis exhibit NF-κB activation when fed an atherogenic diet [26]. Another important contributor to artherosclerosis is thrombin, which stimulates the proliferation of the vascular smooth muscle cells through the activation of NF-κB [27]. A truncated form of IκBα has been shown to protect AT cells, which express constitutive levels of an NF-κB-like activity, from ionizing radiation [28]. From these studies it is quite clear that abnormal activation or expression of NF-κB is associated with a wide variety of pathological conditions. How NF-κB activation mediates tumorigenesis is the focus of the current review.

2. NF-κB MEDIATES CARCINOGENESIS

NF-κB has been implicated in carcinogenesis because it plays a critical role in cell survival, cell adhesion, inflammation, differentiation, and cell growth. Cancer is a hyperproliferative disorder that results from tumor initiation and tumor promotion; it ultimately produces tumor metastasis, Several genes that are involved in cellular transformation, proliferation, invasion, and angiogenesis are regulated by NF-κB (Table II).

Table II. NF-κB regulated genes involved with carcinogenesis

Genes	Function
Cytokine	
CINC-1 Cytokine-induced neutrophil chemoattractant	
CXCL 11	Chemokine ligand for CXCR3
Gro a-g	Melanoma growth stimulating activity
Gro-1	Growth regulated oncogene; chemokine
IL-6	Interleukin-6, inflammatory cytokine
IL-8	Interleukin-8, α-chemokine
MIP	Macrophage inflammatory protein-1, β Chemokine
RANTES	Regulated upon Activation Normal T lymphocyte
Expressed and Secreted	
Cell adhesion molecules	
ELAM-1 (CD62E, E-selectin)	E-selectin, endothelial cell leukocyte adhesion
molecule	
ICAM-1	Intercellular adhesion molecule-1
VCAM-1	Vascular cell adhesion molecule
Acute phase proteins	
Tissue factor-1	Activates extrinsic pathway of complement activation
Urokinase-TPA	Activates fibrinogen for **fibrin clot lysis**
Cell surface receptors	
EGFR	Receptor for EGF
Lox-1	Receptor for oxidized low density lipoprotein
Mdr1	Multiple drug resistance mediator (P-glycoprotein)
Apoptosis regulators	
Bfl1/A1	Pro-survival Bcl-2 homologue
Bcl-xL	Pro-survival Bcl-2 homologue
Bcl-2	Pro-survival factor
c-FLIP	Pro-survival factor
IAPs	Inhibitors of Apoptosis
Growth factors	
BMP-2	Bone Morphogenic Protein-2, Colony Stimulating Factor
PDGF β chain	Platelet-Derived Growth Factor
VEGF C	Vascular Endothelial Growth Factor
Early response	
Egr-1	Mitogen-induced early response gene; zinc finger
Transcription factors	
A20	TNF-inducible zinc finger
c-myb	Proto-oncogene
c-myc	Proto-oncogene
c-rel	Proto-oncogene
IκBα	Inhibitor of Rel/NF-κB
junB	Proto-oncogene
p53	Tumor suppressor
Enzymes	
MMP-9,	matrix metalloproteinaase-9, secreted gelatinase involved in
metastasis	
iNOS	Inducible nitric oxide synthase
COX-2	Cyclooxygenase, prostaglandin endoperoxide synthase

Lipoxygenase synthesis	Arachidonic acid metabolic enzyme, leukotriene
Mn SOD	Superoxide dismutase
Others	
p21-CIP1	Cyclin-dependent kinase inhibitor
Cyclin D1	Cell-cycle regulation
Cyclin D3	Cell-cycle regulation
Prostate-specific antigen	Serum protein in prostate cancer
Wilm's Tumor Suppressor Gene	Tumor suppressor

2.1. NF-κB Genes are Protooncogenes

NF-κB genes are members of a protooncogene family, and many of the functions of the encoded proteins have obvious implications for the development of cancer and its therapy. Retroviruses encoding *v-rel* are oncogenic in avians and *rel* genes are prone to rearrangements and translocations [29]. Tumors with *rel* amplification had an increased frequency of chromosomal aberrations previously associated with tumor progression, suggesting an oncogenic effect of amplified *rel* in B-lymphoid cells that already contained a transforming genetic lesion. *rel* amplification is a frequent event in diffuse lymphoma with a large cell component and probably constitutes a progression-associated marker of primary extranodal lymphomas [30]. The *rel* proto-oncogene has been mapped to chromosome region 2p11.2-14, a site associated with nonrandom rearrangements in non-Hodgkin's lymphoma. Lu et al. characterized an abnormal rel mRNA from a cell line derived from a diffuse large cell lymphoma in which the evolutionarily conserved N-terminal half of the *rel* coding region was fused with the C-terminal coding region of an unrelated gene. In addition, rearrangement or amplification of the *rel* locus was found in the lymphomatous tissue of two follicular and one diffuse large cell lymphoma. The findings suggest involvement of rel in the pathogenesis of large cell lymphoma [31].

    Recent studies have implicated the p50 subunit of the NF-κB transcription factor complex in tumorigenesis. Mukhopadhyay et al. investigated the expression of the p50 subunit of the NF-κB transcription factor complex in paired normal and non-small cell lung carcinoma (NSCLC) tissues. They found out that 81.8% of fresh NSCLC tissues expressed from two- to 20-fold the levels of the p50 subunit as normal lung tissue. Thirteen NSCLC cell lines also exhibited high levels of p50. Such alterations in the normal NF-κB/rel pathway of regulation may play a role in the genesis of NSCLC [32]. Alterations in the Rel A protein have been implicated in squamous head and neck carcinoma [33], adenocarcinomas of the breast and stomach [34], thyroid carcinoma [35], and multiple myeloma [36]. Overexpression of the p50 subunit of NF-κB has been detected in cell lines derived from cancers of the lung, prostate, breast, bone, and brain γ[32].

In certain lymphomas, *bcl-3* and *nf-κb2* genes are known to be translocated [37]. Mutations in *iκbα* have been observed in Hodgkin's lymphoma [38], and blocking *iκbα* by antisense mechanisms appears to induce oncogenic transformation [39]. Antisense to *relA* blocks tumorigenesis induced by Tax, a protein derived from HTLV-1 [40]. A recent report shows that human breast tumors accumulate activated NF-κB complexes consisting of p50, p52 and Bcl-3 rather than p65 [10].

In 1992, Liptay et al. discovered that the NF-κB subunits p49/p100, as well as p105, map to regions associated with certain types of acute lymphoblastic leukemia [41]. The LYL1 gene was first identified during molecular characterization of translocations associated with some human T-cell acute lymphocytic leukemias (T-ALLs). In adult tissues, LYL1 expression was restricted to hematopoietic cells with the notable exception of the T cell lineage. LYL1 encodes a basic helix-loop-helix (bHLH) protein highly related to TAL-1 whose activation is also associated with a high proportion of human T-ALLs. They found that p105, the precursor of NF-κB1 p50, was the major LYL1-interacting protein in this system. Ectopic expression of LYL1 in a human T-cell line caused a significant decrease in NF-κB-dependent transcription, and a reduced level of NF-κB1 proteins [42]. It has been reported that p105/p50 is altered in cancers of the bone, colon, prostate, breast, and brain [43].

Alterations in p100/p52 have been implicated in chronic lymphocytic leukemia (CLL), multiple myeloma, and cutaneous T cell lymphomas [44]. NF-κB2/p52 (lyt-10) was shown to be involved in the breakpoint of a t(10;14)(q24;q32) chromosomal translocation in a case of B-cell lymphoma. Fracchiolla et al. have demonstrated that lyt-10 gene rearrangements are recurrent lesions that may be involved in the pathogenesis of both B- and T-cell malignancies and suggested that truncation of the ankyrin domain may be a common mechanism leading to abnormal lyt-10 activation in lymphoid neoplasia [45]. Also, it has been demonstrated in a MDA-MB-435 breast cancer model that most p65 protein is complexed with p100 in these cells, while it is complexed predominantly with IκBα in cell lines expressing less p100. Based on these observations Dejardin et al. hypothesized that NF-κB could be involved in carcinogenesis and suggested that the p100/p52 NF-κB subunit could play a role in the development of human breast cancers, possibly by sequestering other NF-κB-related proteins in the cytoplasm [46].

Activation of NF-κB also plays a key role in viral pathogenesis, resulting in inflammation and modulation of the immune response. Revilla et al. have shown that A238L, an open reading frame from African swine fever virus, encodes a protein with 40% homology to porcine IκBα and exerts a potent anti-inflammatory effect in host macrophages [47].

2.2. Constitutively Active NF-κB is Common in Tumor Cells

Constitutive expression of NF-κB has been reported in cell lines of breast, ovary, colon, pancreas, thyroid, prostate, lung, head and neck SCC, bladder and skin tumors [48]. Besides solid tumors, B-cell lymphoma, Hodgkin's disease, T-cell lymphoma, adult T-cell leukemia, acute lymphoblastic leukemia, multiple myeloma, chronic lymphocytic leukemia, and acute myelogenous leukemia also have constitutively active NF-κB.

Besides cell lines, activated NF-κB has also been noted in tissue samples derived from patients. For instance, NF-κB is constitutively activated in high-grade squamous intraepithelial lesions and squamous cell carcinomas of the human uterine cervix [49]. NF-κB has been implicated in an aggressive phenotype of renal cell carcinoma (RCC). Out of 45 cases of RCC, 33% showed >200% higher NF-κB activity than does normal renal tissue. In locally advanced cases, 64% showed an increased activity. Tissue from metastases showed an even greater increase in NF-κB activity. Serum CRP elevation correlated with the increase in NF-κB activation; therefore NF-κB may be a cause of the inflammatory paraneoplastic syndrome [50]. Yu et al. reported that increased expression of NF-κB in colorectal tumorigenesis plays an important role in the pathogenesis of colon cancer in humans by mediating the transition from colorectal adenoma with low-grade dysplasia to adenocarcinoma [51].

Why tumor cells express constitutively active NF-κB is not fully understood. An aberrant IKK activity and shorter half-life of IκBα as seen in B-cell lymphoma, mutated IκBα in Hodgkins lymphoma, IL-1β production in AML, and TNF production in cutaneous T-cell lymphoma and in Burkitt's cell lymphoma are some of the reasons that have been described for constitutive NF-κB activation [18].

2.3. Carcinogens and Tumor Promoters Activate NF-κB

During the last few years various carcinogens have been reported to activate NF-κB. These include DMBA [52], BPDE, NNK, and nicotine(for references see [53]). DMBA-induced NF-κB activation has been shown to occur not only in vitro but also when DMBA is administered to the animals. Our laboratory has shown that cigarette smoke condensate (CSC) is a potent activator of NF-κB in different cell types [54]. Unlike TNF-induced NF-κB activation, CSC-induced NF-κB activation was found to be persistent [53]. Besides tumor initiators, tumor promoters such as phorbol ester and okadaic acid have also been shown to activate NF-κB [55]. Additionally, NF-κB is activated by hypoxia [56] and acidic pH [57], both characterstic to the tumor microenvironment.

TNF is one of the most potent activators of NF-κB. Although initially identified as an anticancer agent [58], TNF has since been shown to be involved in cellular transformation [59], tumor promotion [60], and induction of metastasis [61]. In agreement with these observations, mice deficient in TNF have been shown to be resistant to skin carcinogenesis [62]. For several tumors, TNF has been shown to be a growth factor [63]. Like phorbol ester, TNF mediates these effects in part through activation of a protein kinase C pathway [64]. Other inflammatory cytokines have also been implicated in tumorigenesis [65].

2.4. Chemotherapeutic Agents and γ-Radiation Induce NF-κB Activation

Almost all chemotherapeutic agents, as well as inducing apoptosis, have also been shown to activate NF-κB. These include DNA-damaging agents such as doxorubicin [66], camptothecin [67], gemcitabine [68], and cisplatin, microtubule depolymerizing agents such as taxol [69], alkylating agents such as melphalan [70], and the glutathione reductase inhibitor 1,3-bis(2-chloroethyl)-1-nitrosourea [71]. Similarily, γ-radiation, x-rays, and UV radiation have also been shown to activate NF-κB [72]. Why these proapoptotic agents activate NF-κB is not clear. While in most instances, NF-κB mediates chemoresistance [73] or radioresistance [74], in other cases, it may mediate apoptosis [75]. Additionally, these agents activate NF-κB, by mechanisms that appear to differ from that of TNF.

2.5. Oncogenes Activate NF-κB

Several oncogene products that can activate NF-κB have been identified. These include ras [76], bcr-abl [77], and myc [78]. Oncogenic Ha-Ras-induced signaling also activates NF-κB transcriptional activity, which is required for cellular transformation [79]. How these genes induce NF-κB activation is poorly understood. Oncogenic Ras enhances NF-κB transcriptional activity through Raf-dependent and Raf-independent mitogen-activated protein kinase signaling pathways [80]. NF-κB activation was also found to be required for apoptosis induced by oncogenic Ras [81]. As many as 50% of all cancers (mostly solid tumors) express activated ras [82] which can lead to NF-κB activation. Bcr-abl, which has been shown to cause CML, can elicit survival signals through induction of NF-κB [83]. NF-κB activation is required in Bcr-Abl-mediated transformation.   The oncogene c-myc has been shown to be regulated by NF-κB [78] and can mediate tumor cell survival.

2.6. NF-κB Activation is Required for Cell Proliferation

Several genes that mediate cell proliferation are regulated by NF-κB. These include growth factors such as TNF, IL-1β, and IL-6 [84]. For instance TNF has been shown to be a growth factor for glioblastoma cells [85] and cutaneous T-cell lymphoma [63]; IL-1β is a growth factor for AML [86] and IL-6 for multiple myeloma [87], and head and neck squamous cell carcinoma [88]. Suppression of NF-κB in these tumors downregulates the cytokine expression and inhibit tumor cell proliferation. Besides growth factors, certain cell cycle regulatory proteins such as cyclin D1, required for transition of cells from G1 to S phase, are also regulated by NF-κB [89]. In some cells PGE2 has been shown to induce proliferation of tumor cells. The synthesis of cyclooxygenase (COX)-2 which controls PGE2 production, is also regulated by NF-kB activation [90].

Constitutive activation of NF-κB has been implicated in cell proliferation. Bargou et al. reported that proliferation of Hodgkin/Reed-Sternberg cells depended on activated NF-κB. Furthermore, constitutive NF-κB prevented Hodgkin's lymphoma cells from undergoing apoptosis under stress conditions [91].

It was shown that growth factors such as epithelial growth factor (EGF) and platelet-derived growth factor (PDGF) induce proliferation of tumor cells through activation of NF-κB [92]. The EGF receptor was found to engage receptor interacting protein and NF-kappa B-inducing kinase (NIK) to activate NF-κB, thus identifying a novel receptor-tyrosine kinase signalosome [93]. NF-κB signaling has also been shown to promote both cell survival and neurite process formation in nerve growth factor-stimulated PC12 cells [94].

2.7. Activation of NF-κB Inhibits Apoptosis

Both the pro-survival and antiapoptotic function of NF-κB was recently reviewed [95]. Several gene products that negatively regulate apoptosis in tumor cells are controlled by NF-κB activation. These include IAP-1, IAP-2, XIAP, cFLIP, TRAF1, TRAF2, A20, bcl-xl, A1, and survivin [95]. Bcl-xl suppresses cytochrome C release from mitochondria, the IAP inhibit caspase-3 and caspase-9 [96] and FLIP inhibits caspase-8 [97].

An anti-apoptotic role of NF-κB has been linked to T-cell lymphoma, osteoclasts, melanoma, pancreatic cancer, bladder cancer, and breast cancer. Cell types that display an anti-apoptotic role for NF-κB include B cells, T cells, granulocytes, macrophages, neuronal cells, and smooth muscle cells [18].

Although rare, there are systems in which NF-κB plays a pro-apoptotic role in addition to its more common anti-apoptotic role. Examples of its pro-apoptotic effects in cells include those found in B-cells, T-cells, neuronal cells, and endothelial cells. The opposing effects of NF-κB are thought to be cell type specific and/or dependent on the inducing signal (e.g. IL-1, TNF-, and UV radiation). Different activation pathways of NF-κB may cause the expression of proteins that promote apoptosis (eg Fas, c-myc, p53, IB) or inhibit apoptosis (eg TRAF2, IAP proteins, Bcl-2-like proteins). In addition, NF-κB activation variably controls the regulation of cell cycle proteins (e.g., cyclin D1 and CDK2 kinase) and the interaction with various cellular components (e.g., p300 and p53) that promote or induce apoptosis [95].

2.8. NF-κB Mediates the Invasion of Tumor Cells

When treatment is ineffective, the remaining tumor cells inevitably infiltrate the surrounding normal tissue, which leads to tumor recurrence. Recent studies indicate that the ability of tumor cells to digest the extracellular matrix by secreting proteolytic enzymes correlates well with their tissue invasiveness. For most primary human tumors, invasion is thought to be accomplished, at least in part, by proteases — serine, cysteine and metalloproteinases — that penetrate connective-tissue barriers, induce vascular remodeling and destroy normal tissue. Several proteases like matrix metalloproteinases and the serine protease urokinase-type plasminogen activator (uPA) that influence the invasive characteristics of tumors are regulated by NF-κB [98-100].

Matrix metalloproteinases (MMPs) are a pivotal family of zinc enzymes responsible for degradation of the extracellular matrix (ECM) components, including basement membrane collagen, interstitial collagen, fibronectin, and various proteoglycans. Matrix metalloproteinases (MMPs) promote growth of cancer cells through the interaction of extracellular matrix (ECM) molecules and integrins, cleaving insulin-like growth factors and shedding transmembrane precursors of growth factors, including transforming growth factor (TGF)β. MMPs promote angiogenesis by increasing the bioavailability of pro-angiogenic growth factors. MMPs also regulate invasion and migration by degrading structural ECM components — in particular, by cleaving laminin-5. It has been reported that MMP-9 expression is regulated transcriptionally through NF-κB elements within the MMP-9 gene [98]. Bond et al. reported, using an adenovirus that overexpresses the inhibitory subunit

IκBα, that NF-κB activation was an absolute requirement in upregulation of MMP-9.

Urokinase-type plasminogen activator (uPA) is another critical protease involved in tumor invasion and metastasis. Novak et al. reported that transcriptional activation of the uPA gene by PMA, IL-1, and TNFα requires the induction of NF-κB activity and the decay of its short-lived repressor protein, IκBα. Wang et al. reported that uPA was overexpressed in pancreatic tumor cells, and its overexpression was induced by constitutive RelA activity [101]. The uPA promoter contains an NF-κB binding site that directly mediates the induction of uPA expression by RelA. Treating the pancreatic tumor cell lines with the NF-κB inhibitors dexamethasone and n-tosylphenyalanine chloromethyl ketone (TPCK) abolished constitutive RelA activity and uPA overexpression. These results showed that uPA is one of the downstream target genes induced by constitutively activated RelA in human pancreatic tumor cells and suggested that constitutive RelA activity may play a critical role in tumor invasion and metastasis.

Recent studies have shown further that constitutively active PI 3-kinase controls cell motility by the regulation of expression of uPA through the activation of NF-κB [102]. In a very recent study, Mahabeleshwar et al. have demonstrated that activation of Syk, a protein-tyrosine kinase, suppressed cell motility and NF-κB-mediated secretion of uPA by inhibiting the phosphatidylinositol 3'-kinase activity in breast cancer cells [103]. Thus, one of the ways to block the invasion of tumors is to target NF-κB and thus its activation of genes involved in cancer progression.

2.9. NF-κB Activation is Needed for Angiogenesis

It is now well recognized that the induction of tumor vasculature (angiogenesis) is critical for the progression of tumors. Tumor vascularization has been shown to be dependent on chemokines (e.g; MCP-1, IL-8), and growth factors (e.g; TNF, VEGF) produced by macrophages, neutrophils, and other inflammatory cells [104]. The production of these angiogeneic factors has been shown to be regulated by NF-κB activation [105].

NF-κB has been shown to mediate the up-regulation of IL-8 and VEGF expression in bombesin-stimulated PC-3 cells [106]. Yu et al. demonstrated that NF-κB expression was associated with VEGF expression and microvessel density in human colorectal cancer [51]. They observed ten specimens from normal colorectal mucosa and 52 colorectal adenocarcinomas obtained by surgery or endoscopy. Immunohistochemical expression of NF-κB, VEGF, and CD34 was detected on paraffin-embedded tissue sections. NF-κB and VEGF were significantly overexpressed and associated with increased microvessel density in colorectal cancer. These results suggest that increased expression of NF-κB contributes to tumor angiogenesis in colorectal cancer and that VEGF may play an important role in mediating the NF-κB angiogenic pathway.

Highly metastatic melanoma cells expressed high levels of constitutive NF-κB activity that was suppressed by transfection with IκBαM. Suppression of constitutive NF-κB activity inhibited tumor growth, prevented lung metastasis, and a decreased microvessel density (angiogenesis), which correlated with a decrease in the level of interleukin-8 expression [107]. In another study Pollet et al. demonstrated that LPS directly stimulates endothelial sprouting in vitro, which is mediated through TRAF6. Inhibition of NF-κB activity, downstream of TRAF6, was sufficient to inhibit LPS-induced endothelial sprouting [108]. Also, inhibition of NF-κB activity blocked basic fibroblast growth factor (Bfgf)-induced angiogenesis.These studies further underscore the role of NF-κB activation in mediating angiogenesis.

2.10. NF-κB is Involved in Metastasis of Tumor Cells

The metastasis of cancer requires the migration of cancerous cells both into and out of the vessel walls that transport them to other parts of the body. The ability to penetrate through vessel walls is mediated by specific molecules that are expressed on the endothelial cells of the blood vessels in response to a number of signals from inflammatory cells, tumor cells, etc. Among those special molecules are ICAM-1, ELAM-1, and VCAM-1, all of which have been shown to be expressed in response to NF-κB activation [109].

Helbig et al. have demonstrated that NF-κB regulates the motility of breast cancer cells by directly up-regulating the expression of CXCR4 [110]. They showed that the NF-κB subunits p65 and p50 bind directly to sequences within the -66 to +7 region of the CXCR4 promoter and activate transcription. They also showed that the cell surface expression of CXCR4 and stromal derived factor-1α (SDF-1α)-mediated migration are enhanced in breast cancer cells isolated from mammary fat pad xenografts compared with parental cells grown in culture. A further increase in CXCR4 cell surface expression and stroma-derived factor-1α (SDF-1α) migration was observed with cancer cells that metastasized to the lungs. These results implicate NF-κB in the migration and organ-specific homing of metastatic breast cancer cells. Fujioka et al. showed that inhibiting constitutive NF-κB activity by expressing mutant IκBα (IκBαM) suppresses liver metastasis, but not tumorigenesis, from the metastatic human pancreatic tumor cell line AsPc-1 in an orthotopic nude mouse model [111]. These studies demonstrate the importance of suppressing NF-κB activation in reducing the metastasis of cancer cells to other sites.

2.11. NF-κB Activation Mediates Bone Loss

The adult skeleton is in a dynamic state, undergoing turnover by the coordinated actions of osteoclasts and osteoblasts. The focal net loss of bone at sites of inflammation in conditions such as arthritis and osteoporosis occur due to an imbalance in the favor of bone resorption. The loss of bone is common to certain type of cancers such as breast cancer, melanoma, and prostate cancer. Binding of RANKL, a member of the TNF superfamily, to its receptor, RANK, leads to the activation of NF-κB [112]. Osteoclast differentiation and activation is mediated by RANKL, which is induced by osteoblasts or bone-lining cells in response to stimuli by parathormone, IL-1, or PGE2. Double-knockout mice for NF-κB1 and NF-κB2 developed osteopetrosis because of a defect in osteoclast differentiation [113] and failed to generate mature osteoclasts and B cells [114]. These findings establish the criticality of NF-κB in bone loss as well as development.

NF-κB proteins transmit growth factor signals between the ectoderm and the underlying mesenchyme during embryonic limb formation. Interruption of NF-κB activity using virus-mediated delivery of an inhibitor resulted in a highly dysmorphic apical ectodermal ridge (AER), reduction in overall limb size, loss of distal elements, and reversal in the direction of limb outgrowth [115]. Yao et al showed that ethyl alcohol induced NF-κB nuclear translocation through p56lck in human osteoblast-like cell line (HOBIT) cells. This activation of NF-κB may contribute to bone loss through activation of signal transduction that results in production of an osteoclastogenic cytokine, IL-6, in osteoblasts [116].

Inhibition of NF-κB signaling activity in Saos-2 cells results in a marked decrease in cellular proliferation accompanied by the induction of bone morphogenic proteins (BMP) 4 and 7 and the osteoblast-specific transciption factor, Cbfa1, heralding osteoblast differentiation, induction of alkaline phosphatase, osteopontin, and osteocalcin production and the attendant increase in matrix deposition and mineralization in vitro. These results point to the negative regulation of osteoblast differentiation by NF-κB, with implications in the pathogenesis and progression of osteosarcomas [117]. These results suggest a critical involvement of NF-κB proteins in regulating the dynamics of bone development raising new directions in the treatment of bone disorders.

3. SUPPRESSION OF NF-κB INIHIBITS TUMORIGENESIS

Evidence presented above suggests that activation of NF-κB can lead to tumor cell proliferation, invasion, angiogenesis, and metastasis. Thus suppression of NF-κB in cancer cells may provide target for prevention of cancer.

3.1. Tumor Suppressor Genes Inhibit NF-κB Activation

Whereas oncogenes can stimulate NF-κB activation, several tumor suppressor genes inactivate NF-κB activation. These include PTEN [118], CYLD [119-121], and p53 [122]. Mutated in multiple advanced cancers/phosphatase and tensin homologue (MMAC/PTEN) is a natural antagonist of PI 3-kinase activity. PTEN is a lipid phosphatase responsible for down-regulating the phosphoinositide 3-kinase product phosphatidylinositol 3,4,5-triphosphate. In a glioma cell line that was stably transfected with MMAC/PTEN, IL-1-induced DNA binding and transcriptional activities of NF-κB were both inhibited [118]. The ability of IL-1 to induce IκBα degradation or nuclear translocation of NF-κB was, however, unaffected by MMAC/PTEN expression. Moreover, IL-1-induced phosphorylation of p50 NF-κB was potently inhibited in MMAC/PTEN-expressing cells. IL-1-induced interaction between the PI 3-kinase target Akt kinase and the IKK complex was antagonized by MMAC/PTEN.

Mayo et al. found that PTEN inhibits TNF-stimulated NF-κB transcriptional activity. PTEN failed to block TNF-induced IKK activation, IκBα degradation, p105 processing, p65 (RelA) nuclear translocation, and DNA binding of NF-κB [123]. However, PTEN inhibited NF-κB-dependent transcription by blocking the ability of TNF to stimulate the transactivation domain of the p65 subunit. Thus, maintenance of the PTEN tumor suppressor protein is required to modulate Akt activity and to concomitantly control the transcriptional activity of NF-κB as an anti-apoptotic transcription factor.

Familial cylindromatosis is caused by mutations in a gene encoding the CYLD, a tumor suppressor gene with deubiquitinating enzymatic activity that negatively regulates activation of NF-κB. Loss of the deubiquitinating activity of CYLD correlates with tumorigenesis. CYLD inhibits activation of NF-κB by the TNFR family members CD40, XEDAR, and EDAR in a manner that depends on the deubiquitinating activity of CYLD. Downregulation of CYLD by RNA-mediated interference augments both basal and CD40-mediated activation of NF-κB. The inhibition of NF-κB activation by CYLD is mediated, at least in part, by the deubiquitination and inactivation of TRAF2. CYLD also binds to the NEMO (also known as IKKγ) component of the IKK complex and appears to regulate its activity through deubiquitination of TRAF2, since TRAF2 ubiquitination can be modulated by CYLD [119-121]. These results indicate that CYLD is a negative regulator of the cytokine-mediated activation of NF-κB that is required for appropriate cellular homeostasis of skin appendages. Inhibition of CYLD increases resistance to apoptosis, suggesting a mechanism through which loss of CYLD contributes to oncogenesis. This effect can be relieved by aspirin derivatives that inhibit NF-κB activity, which suggests a therapeutic intervention strategy to restore growth control in patients suffering from familial cylindromatosis.

While NF-κB provides a proliferative signal, p53 is a mediator of antiproliferation. A mechanism by which p53 regulates NF-κB function and cell cycle progression has been proposed. P53 has been shown to downregulate the expression of cyclinD1 [124] through inhibition of the expression of Bcl-3 protein, a member of the IκB family that functions as a transcriptional coactivator for p52 NF-κB; p53 also reduced p52/Bcl-3 complex levels. Concomitant with this, p53 induced a significant increase in the association of p52 and histone deacetylase 1 (HDAC1). Importantly, p53-mediated suppression of the cyclin D1 promoter was reversed by coexpression of Bcl-3 and inhibition of p52 or deacetylase activity. p53 therefore induces a transcriptional switch in which p52/Bcl-3 activator complexes are replaced by p52/HDAC1 repressor complexes, resulting in active repression of cyclin D1 transcription.

3.2. Inhibition of NF-κB Inhibits Tumorigenesis

That NF-κB is critical for tumorigenesis is also indicated by reports that indicate that suppression of NF-κB blocks tumorigenesis. For instance, fibrosarcoma and colorectal tumors grown in nude mice were induced to undergo apoptosis when infected with an adenovirus expressing a modified form of IκBα along with systemic delivery of CPT-11 chemotherapy [10]. HT1080 fibrosarcoma cells exposed to ionizing radiation, TNF-α, or daunorubicin exhibited enhanced activation of NF-κB, and inhibition of NF-κB dramatically enhanced apoptosis in response to radiation or daunorubicin [12]. There are, however, reports which suggest that suppression of NF-κB has no effect on apoptosis. For instance, stable transfection of *ikbα* did not yield enhanced cytotoxicity in response to chemotherapy despite the ability of these agents to activate NF-κB [125].

3.3. Chempreventive Agents Inhibit NF-κB Activation

Because of the critical role of NF-κB in proliferation, invasion, angiogenesis, and metastasis of tumors, there has been great interest in the modulators of the NF-κB signaling pathway. Several agents which have been described as natural chemopreventive agents, have also been found to be potent inhibitors of NF-κB activation (see Table III). How these agents suppress NF-κB activation is becoming increasingly apparent. These inhibitors may block any one or more steps in the NF-κB signaling pathway such as the incoming signals that activate the NF-κB signaling cascade, translocation of NF-κB into the nucleus, DNA binding of the dimers, and interactions with the basal transcriptional machinery. For instance, while curcumin blocks IKK activation [87], resveratrol suppresses p65 translocation to the nucleus [52] and CAPE suppresses the binding of p50-p65 complex directly to the DNA [126].

Table 3. List of chemopreventive agents that block NF-κB or NF-κB regulated genes

Chemopreventive agent	Source	Chemical name	Mechanism
Curcumin	*Curcuma longa* Linn.	(E,E)-1,7-Bis(4-hydroxy-3-methoxy-phenyl)-1,6-heptadiene-3,5-dione; Diferuloylmethane	Blocks IKK
Retinoids	Vitamin A	N-(4-hydroxyphenyl) retinamide; All-trans retinoc acid	Blocks IKK
Capsaicin	Capsicum	N-(4-hydroxy-3-methoxybenzyl)-8-methylnon-trans-6-enamide	Blocks IκBα degradation
Vitamin E	Plant material	α-tocopherol, ; 5,7,8-trimethyltocol	
Quercetin	*Rhododendron cinnabarinum* Hook, Widely distributed in plant kindom in rinds and barks	3,3',4',5,7-pentahydroxyflavone, Meletin	Blocks IκBα degradation
Dihydroxy vitamin D3	Fish liver oils	1α,25-dihydroxycholecalciferol	

Resveratrol	Phytoalexin found in a variety of plants: abundan in *Polygonum cuspidatum*	Trans-3,5,4'-trihydroxystilbene	Blocks IκBα degradation
Silymarin	*Silybum marianum* L.	Silybin	Blocks IκBα degradation
Lapachone	Heartwood of bignoniaceous plants e.g. Lapach tree	β-Lapachone	Blocks IκBα degradation
Sulindac*	Synthetic	(Z)--Fluoro-2-methyl-1-[[4-(methyl-sulfinyl)phenyl]methylene]-1H-indene-3-acetic acid	Blocks IKK
Celecoxib*	synthetic	1,2-diarylpyrroles	
Flavonoids	Green or Black Tea	(-) Epigallocatechin 3-gallate; Theaflavin	Blocks IκBα degradation
Sulforaphane	Cruciferous vegitables eg. Broccoli	1-isothiocyanato-(4R)-(methylsulfinyl) butane	Blocks NF-κB-DNA bindiing
Aspirin	Synthetic	2-(Acetyloxy)benzoic acid	Blocks IκBα degradation
Caffeic acid phenethyl ester	Propolis from honey bee hives	3,4-Dihydroxycinnamic acid phenylethyl ester	Blocks trans-location of p65
PBIT	Synthetic	S,S'-1,4-Phenylene-bis(1,2-ethanediyl)bisisothiourea	
PDTC	Synthetic	Pyrrolidine dithocarbamate	Blocks IκBα degradation
Anethole	*Pimpinella anisum*, camphor, fennel	p-Propenylanisole	Blocks
Oleandrin	Nerium oleander	(3β,5β,16β)-16-(Acetyloxy)-3-[(2,6-di-deoxy-3-O-methyl-α-arabino-hexopyranosyl)oxy]-14-hydroxycard-20(22)-enolide	Blocks

Wortmannin	Antibiotic from *Penicillium wormanni*	[1S-(1α,6bα,9aβ,11α,11bβ)]-11-(Acetyloxy)-1,6b,7,8,9a,10,11,11b-octahydro-1-(methoxymethyl)-9a,11b-dimethyl-3H-furo[4,3,2-de]indeno[4,5-h]-2-benzopyran-3,6,9-trione	Blocks IκBα degradation
Emodin	Aloe barbandensis	1,3,8-Trihydroxy-6-methyl-9,10-anthracenedione	Blocks IκBα degradation
Selenium	Dietary trace element	selenium	
Betulinic acid	Bark of the white birch tree	3β-hydroxy-20(29)-lupaene-28-oic acid	Blocks
Ursolic acid	Prunes and plums	3b-Hydroxy-12-ursen-28-ic acid	Blocks
Oleanolic acid	Prunes and plums	3b-Hydroxy-12-olean-28-ic acid	

For references see [127]

Other chemopreventive agents which have been shown to suppress NF-κB activation include betulinic acid, ursolic acid, piceatannol, green tea polyphenols, oleandrin, and anethole for references see[127]].   Similarly leflunomide, a drug approved for the treatment of rheumatoid arthritis, has also been show to suppress NF-κB activation [128].

3.4. Some cytokines Block NF-κB Activation

Several cytokines have also been found to suppress NF-κB activation. These include cytokines produced by Th2 cells such as IL-4, IL-11, IL-13, and IL-10 [129] and hormones of the endocrine system such as human chorionic gonadotropin (HCG) [130], melanocyte stimulating hormone (MSH) [131] and growth hormone (GH) [132]. Besides these cytokines, IFNα was found to be a potent suppressor NF-κB activation [133]. How these agents suppress NF-κB activation appears to vary. For instance nuclear translocation of the RelA subunit and degradation of IκBα induced by TNF was prevented by IL-13 and human chorionic gonadotropin [130]. Both IL-10 and IL-13 suppress nuclear localization of NF-κB and increase IκBα mRNA expression [134].

4. NF-κB AS A POTENTIAL TARGET FOR DRUG DEVELOPMENT

NF-κB is an ideal target for anticancer drug development. Cancer is a hyperproliferative disorder that involves transformation, initiation, promotion, angiogenesis, invasion, and metastasis. The diversity of its clinical presentation, aggressiveness, and current treatment strategies imply an equally diverse number of potential targets in the molecular pathways leading to its formation. Several strategies have been employed to block the activation of NF-κB. A wide variety of compounds like IKK inhibitors, inhibitory peptides, antisense RNA, and proteasome inhibitors have been screened for their ability to suppress NF-κB. These compounds block various steps leading to NF-κB activation.

4.1. Inhibitors of Proteasomes Block NF-κB Activation

Proteasome inhibitors block the 26S proteasome necessary to degrade the IκBα inhibitory subunit after its phosphorylation and ubiquitination in the cytoplasm and thus its release from the NF-κB complex [16]. Some of the well known proteasome inhibitors are peptide aldehydes such as ALLnL, LLM, Z-LLnV, and Z-LLL, lactacystine, PS-341, N-cbz-Leu-Leu-leucinal (MG132), MG115 and ubiquitin ligase inhibitors (for references see [18]).

PS-341 blocks TNFα-induced NF-κB activation in multiple myeloma cells through inhibition of IκBα phosphorylation and degradation of IκBα [135]. Proteasome inhibitors inhibit the chymotrypsin-like activity of the proteasome complex. Fiedler et al. have demonstrated that pretreatment of cells with the proteasome inhibitor N-cbz-Leu-Leu-leucinal (MG-132) reverses TNFα-induced NF-κB activation by inhibiting proteasome-mediated IκBα degradation and NF-κB-mediated gene transcription [136].

Calpain inhibitor I is a cysteine protease inhibitor and is less potent than MG132 and MG115. Molecules like lactacystine irreversibly block proteasome activity by acylating a threonine residue in the active site of the subunit X of the mammalian proteasome. Some serine protease inhibitors also act as proteasome inhibitors and block the phosphorylation and degradation of IκBα [137]. Retaining the NF-κB dimers in the cytoplasm by preventing the phosphorylation and degradation of IκBα is an effective way to inhibit NF-κB, and this can be achieved by several signaling molecules such as NO, estrogen, oxidized LDL, 4-hydroxynonenal, and prostaglandin A.

4.2. Inhibitors of IKK Block NF-κB Activation

IκBα phosphorylation is a critical step for NF-κB activation, and compounds that block this phosphorylation prevent its ubiquitination and further degradation. Recently, 4-hydroxy-2-nonenal, a lipid peroxidation product, has been shown to block phosphorylation by direct inhibition of IKK [138]. Also, 5-bromo-6-methoxy-beta-carboline, a natural product derivative, is a nonspecific IKK inhibitor that inhibits the phosphorylation of IκBα and subsequent activation of NF-κB with an IC_{50} in the nanomolar range [139]. SC-514 is another novel inhibitor of IKK that is selective to IKK2 with an IC_{50} of about 10 μM; it does not inhibit other IKK isoforms or other serine-threonine and tyrosine kinases. SC-514 inhibits the native IKK complex or recombinant human IKK-1/IKK-2 heterodimer and IKK-2 homodimer in a similar manner [140].

BMS-345541(4('-aminoethyl)amino-1,8-dimethylimidazo(1,2-a]quinoxaline) was identified as a selective inhibitor of the catalytic subunits of IKK with an IC_{50} of 4 μM for IKKα and 0.3 μM for IKKβ. Burke et al. proposed a binding model in which BMS-345541 binds to similar allosteric sites on IKKα and IKKβ, which then affects the active sites of the subunits differently. BMS-345541 was also shown to have excellent pharmacokinetics in mice, and its administration showed the compound to dose-dependently inhibit the production of serum TNFα following intraperitoneal challenge with lipopolysaccharide in mice and represents an important tool for investigating the role of IKK in disease models [141].

4.3. Acetylation Inhibitors can Block NF-κB Activation

Histone acetylation modulates gene expression, cellular differentiation, and survival and is regulated by the opposing activities of histone acetyltransferases (HATs) and histone deacetylases (HDACs). HDAC3 is a histone deacetylase that acts directly upon nuclear Rel A enabling its association with IκBα and its subsequent export from the nucleus. HDAC3 is another potential target for gene transfer. Expression of HDAC3 in TNFα-stimulated HeLa cells repressed both NF-κB DNA-binding and levels of Rel A with a corresponding increase in inactive cytoplasmic IκBα-NF-κB complexes [142]. This mechanism was shown to control the duration of NF-κB activation and thus may be a potential weapon against constitutive NF-κB activation.

4.4. Gene Transfer of Inhibitory Proteins can Block NF-κB Activation

Transfer of genes that encode for inhibitory proteins is another strategy used to block the activation of NF-κB. The most direct target is the IκB gene. This entails the modification of IκB at the specific phosphorylation sites (ser 32 and 36 switched with ala) and ubiquitination sites (lys 21 and 22 switched with arg) to prevent its degradation. In several studies this super-repressor of NF-κB activity was a mutated nondegradable IκBα resistant to phosphorylation and degradation that could be delivered into intestinal epithelial cell (IEC) using an adenoviral vector (Ad5 IκB). There was a strong inhibition of NF-κB activity because the 'superrepressor' kept the NF-κB complex in the cytoplasm indefinitely [125]. These studies suggest an exciting approach for in vivo intestinal gene therapy and illustrate a key role for NF-κB in transcriptional regulation of the inflammatory phenotype of IEC. Recently, a nonphosphorylatable form of IκBα was shown to inhibit osteoclastogenesis and block bone resorption when injected into bone marrow macrophages [143].

4.5. Antisense & siRNA Can Block NF-κB Activation

Antisense agents that inhibit the expression of a target gene in a sequence-specific manner may be used for therapeutic purposes against NF-κB. Three types of anti-mRNA strategies can be distinguished. Firstly, the use of single-stranded antisense-oligonucleotides; secondly, the triggering of RNA cleavage through catalytically active oligonucleotides referred to as ribozymes; and thirdly, RNA interference induced by small interfering RNA (siRNA) molecules.

Phosphorothioate antisense oligonucleotides to p65 inhibited in vitro growth, reduced soft-agar colony formation, and eliminated the ability of cells to adhere to an extracellular matrix in diverse transformed cell lines. Stable transfectants of a fibrosarcoma cell line expressing dexamethasone-inducible antisense RNA to p65 showed inhibition of in vitro growth and in vivo tumor development. In response to inducible expression of antisense RNA, a pronounced tumor regression was seen in nude mice. Also, the administration of antisense but not sense p65 oligonucleotides caused a pronounced inhibition of tumorigenicity in nude mice injected with diverse tumor-derived cell lines [144].

RNAi is the mechanism of sequence-specific, post-transcriptional gene silencing initiated by double-stranded RNAs (dsRNA) homologous to the gene being suppressed and have the potential to be a new and powerful tool in cancer therapeutics [145]. Surabhi and Gaynor have demonstrated the potential of siRNAs in decreasing the level of expression of p65 protein [146]. Also, when siRNAs were directed against IKKα, IKKβ and the upstream regulatory kinase TAK1, both IKKα and IKKβ were found to be important in activating the NF-κB pathway [147]. Another report showed that blocking c-Rel in primary macrophages using siRNA, selectively blocked IL-12 production and normalized the minimal, residual IL-12 levels in lupus- and diabetes prone mouse strain with aberrant IL-12 production [148].

4.6. Peptides That Can Cross the Cell Membrane and Block NF-κB Activation

Another approach to inhibiting NF-κB activation is to use peptides that cross the cell membrane and block the nuclear localization of the NF-κB complex. For example, SN-50 and o,o'-bismyristoyl thiamine disulfide [149] works by mimicking the sequence of p50 responsible for transporting the NF-κB complex from the cytoplasm to the nucleus to block the normal import machinery. A dual NLS peptide has been shown to block the karyopherin-NF-κB interaction, which is required for the translocation of NF-κB into the nucleus [150]. Also, a novel peptide that selectively blocks the association of IKKγ (NEMO) with the rest of the IKK complex has been shown to inhibit NF-κB activation in response to pro-inflammatory cytokines in mice while preserving basal NF-κB activity [151].

Another novel hybrid peptide, P1, derived from cecropin-A and magainin-2, reduced osteoclast differentiation in various osteoclast culture systems through the inhibition of NF-κB activation induced by the RANKL [152].

4.7. Antiinflammatory Agents Block NF-κB Activation

Additionally, several anti-inflammatory agents have been identified to suppress NF-κB activation. Examples include aspirin, ibuprofen, indomethacin, tamoxifen dexamethasone, and sulindac [18]. However, the exact mechanism of their action is not fully understood. Ghosh and Kopp demonstrated that the anti-inflammatory drugs sodium salicylate and aspirin inhibited the activation of NF-κB by preventing the degradation of the NF-κB inhibitor, IκB, so that NF-κB was retained in the cytosol [145].

The inhibition of NF-κB with the methods illustrated above represents possible approaches to the more complicated issue of creating drug therapies that are effective in preventing or attenuating tumorigenesis. Indeed, aspirin has been shown to be an effective therapeutic agent against cylindromatosis, colorectal cancers, and human hepatoma. An understanding of their precise mechanisms of action, specificity, and even toxicity with respect to NF-κB is still incomplete. However, targeting NF-κB is central to designing an effective therapy for cancer.

5. CONCLUSIONS

Overall these studies clearly demonstrate that NF-κB plays a critical role in tumorigenesis. Similarily, suppression of NF-κB has the ability to inhibit various steps in the tumorigenic process. Design of inhibitors that are pharmacologically safe will be critical for the treatment of cancer. Because of the multiple agents that activate NF-κB through diverse pathways, it is unlikely that any single inhibitor of NF-κB would be effective against all tumors.

Shishir Shishodia and Bharat B. Aggarwal
Department of Bioimmunotherapy, The University of Texas M. D. Anderson Cancer Center, Houston, Texas

6. REFERENCES

1. Latchman D, Gene regulation: A eukaryotic perspective. *Chapman and Hall, London, UK.*, 1995.
2. Darnell JE, Jr., Transcription factors as targets for cancer therapy. *Nat Rev Cancer* 2(10): 740-9, 2002.
3. Karungaran D and Aggarwal BB, Transcription factors as targets for drug development. In molecular pathomechanisms and new trends in drug research (eds. Drs. Gyorgy Keri and Istvan Toth), Taylor & Francis Publisher: 16-91, 2002.
4. Sen R and Baltimore D, Inducibility of kappa immunoglobulin enhancer-binding protein Nf-kappa B by a posttranslational mechanism. *Cell* 47(6): 921-8, 1986.
5. Gilmore TD, The Rel/NF-kappaB signal transduction pathway: introduction. *Oncogene* 18(49): 6842-4, 1999.
6. Ghosh S, May MJ and Kopp EB, NF-kappa B and Rel proteins: evolutionarily conserved mediators of immune responses. *Annu Rev Immunol* 16: 225-60, 1998.
7. Karin M, How NF-kappaB is activated: the role of the IkappaB kinase (IKK) complex. *Oncogene* 18(49): 6867-74, 1999.
8. Li N and Karin M, Ionizing radiation and short wavelength UV activate NF-kappaB through two distinct mechanisms. *Proc Natl Acad Sci U S A* 95(22): 13012-7, 1998.
9. Singh S, Darnay BG and Aggarwal BB, Site-specific tyrosine phosphorylation of IkappaBalpha negatively regulates its inducible phosphorylation and degradation. *J Biol Chem* 271(49): 31049-54, 1996.
10. Mayo MW and Baldwin AS, The transcription factor NF-kappaB: control of oncogenesis and cancer therapy resistance. *Biochim Biophys Acta* 1470(2): M55-62, 2000.
11. Pahl HL, Activators and target genes of Rel/NF-kappaB transcription factors. *Oncogene* 18(49): 6853-66, 1999.
12. Wang CY, Mayo MW and Baldwin AS, Jr., TNF- and cancer therapy-induced apoptosis: potentiation by inhibition of NF-kappaB. *Science* 274(5288): 784-7, 1996.
13. Dejardin E, Droin NM, Delhase M, Haas E, Cao Y, Makris C, Li ZW, Karin M, Ware CF and Green DR, The lymphotoxin-beta receptor induces different patterns of gene expression via two NF-kappaB pathways. *Immunity* 17(4): 525-35, 2002.

14. Cao Y, Bonizzi G, Seagroves TN, Greten FR, Johnson R, Schmidt EV and Karin M, IKKalpha provides an essential link between RANK signaling and cyclin D1 expression during mammary gland development. *Cell* 107(6): 763-75, 2001.

15. Fan C, Li Q, Ross D and Engelhardt JF, Tyrosine phosphorylation of I kappa B alpha activates NF kappa B through a redox-regulated and c-Src-dependent mechanism following hypoxia/reoxygenation. *J Biol Chem* 278(3): 2072-80, 2003.

16. Palombella VJ, Rando OJ, Goldberg AL and Maniatis T, The ubiquitin-proteasome pathway is required for processing the NF-kappa B1 precursor protein and the activation of NF-kappa B. *Cell* 78(5): 773-85, 1994.

17. Parcellier A, Schmitt E, Gurbuxani S, Seigneurin-Berny D, Pance A, Chantome A, Plenchette S, Khochbin S, Solary E and Garrido C, HSP27 is a ubiquitin-binding protein involved in I-kappaBalpha proteasomal degradation. *Mol Cell Biol* 23(16): 5790-802, 2003.

18. Garg A and Aggarwal BB, Nuclear transcription factor-kappaB as a target for cancer drug development. *Leukemia* 16(6): 1053-68, 2002.

19. Rosenfeld ME, Prichard L, Shiojiri N and Fausto N, Prevention of hepatic apoptosis and embryonic lethality in RelA/TNFR-1 double knockout mice. *Am J Pathol* 156(3): 997-1007, 2000.

20. Alcamo E, Hacohen N, Schulte LC, Rennert PD, Hynes RO and Baltimore D, Requirement for the NF-kappaB family member RelA in the development of secondary lymphoid organs. *J Exp Med* 195(2): 233-44, 2002.

21. Beg AA, Sha WC, Bronson RT, Ghosh S and Baltimore D, Embryonic lethality and liver degeneration in mice lacking the RelA component of NF-kappa B. *Nature* 376(6536): 167-70, 1995.

22. MacMicking JD, Nathan C, Hom G, Chartrain N, Fletcher DS, Trumbauer M, Stevens K, Xie QW, Sokol K, Hutchinson N and et al., Altered responses to bacterial infection and endotoxic shock in mice lacking inducible nitric oxide synthase. *Cell* 81(4): 641-50, 1995.

23. Behl PN, Vitiligo: treatment by dermabrasion and epithelial sheet grafting. *J Am Acad Dermatol* 30(6): 1044-5, 1994.

24. Pahl HL and Baeuerle PA, A novel signal transduction pathway from the endoplasmic reticulum to the nucleus is mediated by transcription factor NF-kappa B. *Embo J* 14(11): 2580-8, 1995.

25. Berliner JA, Navab M, Fogelman AM, Frank JS, Demer LL, Edwards PA, Watson AD and Lusis AJ, Atherosclerosis: basic mechanisms. Oxidation, inflammation, and genetics. *Circulation* 91(9): 2488-96, 1995.

26. Liao F, Andalibi A, Qiao JH, Allayee H, Fogelman AM and Lusis AJ, Genetic evidence for a common pathway mediating oxidative stress, inflammatory gene induction, and aortic fatty streak formation in mice. *J Clin Invest* 94(2): 877-84, 1994.

27. Nakajima T, Kitajima I, Shin H, Takasaki I, Shigeta K, Abeyama K, Yamashita Y, Tokioka T, Soejima Y and Maruyama I, Involvement of NF-kappa B activation in thrombin-induced human vascular smooth muscle cell proliferation. *Biochem Biophys Res Commun* 204(2): 950-8, 1994.

28. Jung M, Zhang Y, Lee S and Dritschilo A, Correction of radiation sensitivity in ataxia telangiectasia cells by a truncated I kappa B-alpha. *Science* 268(5217): 1619-21, 1995.

29. Luque I and Gelinas C, Rel/NF-kappa B and I kappa B factors in oncogenesis. *Semin Cancer Biol* 8(2): 103-11, 1997.

30. Houldsworth J, Mathew S, Rao PH, Dyomina K, Louie DC, Parsa N, Offit K and Chaganti RS, REL proto-oncogene is frequently amplified in extranodal diffuse large cell lymphoma. *Blood* 87(1): 25-9, 1996.

31. Lu D, Thompson JD, Gorski GK, Rice NR, Mayer MG and Yunis JJ, Alterations at the rel locus in human lymphoma. *Oncogene* 6(7): 1235-41, 1991.

166

32. Mukhopadhyay T, Roth JA and Maxwell SA, Altered expression of the p50 subunit of the NF-kappa B transcription factor complex in non-small cell lung carcinoma. *Oncogene* 11(5): 999-1003, 1995.

33. Mathew S, Murty VV, Dalla-Favera R and Chaganti RS, Chromosomal localization of genes encoding the transcription factors, c-rel, NF-kappa Bp50, NF-kappa Bp65, and lyt-10 by fluorescence in situ hybridization. *Oncogene* 8(1): 191-3, 1993.

34. Wang CY, Guttridge DC, Mayo MW and Baldwin AS, Jr., NF-kappaB induces expression of the Bcl-2 homologue A1/Bfl-1 to preferentially suppress chemotherapy-induced apoptosis. *Mol Cell Biol* 19(9): 5923-9, 1999.

35. Visconti R, Cerutti J, Battista S, Fedele M, Trapasso F, Zeki K, Miano MP, de Nigris F, Casalino L, Curcio F, Santoro M and Fusco A, Expression of the neoplastic phenotype by human thyroid carcinoma cell lines requires NFkappaB p65 protein expression. *Oncogene* 15(16): 1987-94, 1997.

36. Trecca D, Guerrini L, Fracchiolla NS, Pomati M, Baldini L, Maiolo AT and Neri A, Identification of a tumor-associated mutant form of the NF-kappaB RelA gene with reduced DNA-binding and transactivating activities. *Oncogene* 14(7): 791-9, 1997.

37. Baldwin AS, Jr., The NF-kappa B and I kappa B proteins: new discoveries and insights. *Annu Rev Immunol* 14: 649-83, 1996.

38. Cabannes E, Khan G, Aillet F, Jarrett RF and Hay RT, Mutations in the IkBa gene in Hodgkin's disease suggest a tumour suppressor role for IkappaBalpha. *Oncogene* 18(20): 3063-70, 1999.

39. Beauparlant P, Kwan I, Bitar R, Chou P, Koromilas AE, Sonenberg N and Hiscott J, Disruption of I kappa B alpha regulation by antisense RNA expression leads to malignant transformation. *Oncogene* 9(11): 3189-97, 1994.

40. Kitajima I, Shinohara T, Bilakovics J, Brown DA, Xu X and Nerenberg M, Ablation of transplanted HTLV-I Tax-transformed tumors in mice by antisense inhibition of NF-kappa B. *Science* 258(5089): 1792-5, 1992.

41. Liptay S, Schmid RM, Perkins ND, Meltzer P, Altherr MR, McPherson JD, Wasmuth JJ and Nabel GJ, Related subunits of NF-kappa B map to two distinct loci associated with translocations in leukemia, NFKB1 and NFKB2. *Genomics* 13(2): 287-92, 1992.

42. Ferrier R, Nougarede R, Doucet S, Kahn-Perles B, Imbert J and Mathieu-Mahul D, Physical interaction of the bHLH LYL1 protein and NF-kappaB1 p105. *Oncogene* 18(4): 995-1005, 1999.

43. Motokura T and Arnold A, PRAD1/cyclin D1 proto-oncogene: genomic organization, 5' DNA sequence, and sequence of a tumor-specific rearrangement breakpoint. *Genes Chromosomes Cancer* 7(2): 89-95, 1993.

44. Nakshatri H, Bhat-Nakshatri P, Martin DA, Goulet RJ, Jr. and Sledge GW, Jr., Constitutive activation of NF-kappaB during progression of breast cancer to hormone-independent growth. *Mol Cell Biol* 17(7): 3629-39, 1997.

45. Fracchiolla NS, Lombardi L, Salina M, Migliazza A, Baldini L, Berti E, Cro L, Polli E, Maiolo AT and Neri A, Structural alterations of the NF-kappa B transcription factor lyt-10 in lymphoid malignancies. *Oncogene* 8(10): 2839-45, 1993.

46. Dejardin E, Bonizzi G, Bellahcene A, Castronovo V, Merville MP and Bours V, Highly-expressed p100/p52 (NFKB2) sequesters other NF-kappa B-related proteins in the cytoplasm of human breast cancer cells. *Oncogene* 11(9): 1835-41, 1995.

47. Revilla Y, Callejo M, Rodriguez JM, Culebras E, Nogal ML, Salas ML, Vinuela E and Fresno M, Inhibition of nuclear factor kappaB activation by a virus-encoded IkappaB-like protein. *J Biol Chem* 273(9): 5405-11, 1998.

48. Rayet B and Gelinas C, Aberrant rel/nfkb genes and activity in human cancer. *Oncogene* 18(49): 6938-47, 1999.

49. Nair A, Venkatraman M, Maliekal TT, Nair B and Karunagaran D, NF-kappaB is constitutively activated in high-grade squamous intraepithelial lesions and squamous cell carcinomas of the human uterine cervix. *Oncogene* 22(1): 50-8, 2003.

50. Oya M, Takayanagi A, Horiguchi A, Mizuno R, Ohtsubo M, Marumo K, Shimizu N and Murai M, Increased nuclear factor-kappa B activation is related to the tumor development of renal cell carcinoma. *Carcinogenesis* 24(3): 377-84, 2003.

51. Yu HG, Yu LL, Yang Y, Luo HS, Yu JP, Meier JJ, Schrader H, Bastian A, Schmidt WE and Schmitz F, Increased expression of RelA/nuclear factor-kappa B protein correlates with colorectal tumorigenesis. *Oncology* 65(1): 37-45, 2003.

52. Banerjee S, Bueso-Ramos C and Aggarwal BB, Suppression of 7,12-dimethylbenz(a)anthracene-induced mammary carcinogenesis in rats by resveratrol: role of nuclear factor-kappaB, cyclooxygenase 2, and matrix metalloprotease 9. *Cancer Res* 62(17): 4945-54, 2002.

53. Shishodia S, Potdar P, Gairola CG and Aggarwal BB, Curcumin (diferuloylmethane) down-regulates cigarette smoke-induced NF-kappaB activation through inhibition of IkappaBalpha kinase in human lung epithelial cells: correlation with suppression of COX-2, MMP-9 and cyclin D1. *Carcinogenesis* 24(7): 1269-79, 2003.

54. Anto RJ, Mukhopadhyay A, Shishodia S, Gairola CG and Aggarwal BB, Cigarette smoke condensate activates nuclear transcription factor-kappaB through phosphorylation and degradation of IkappaB(alpha): correlation with induction of cyclooxygenase-2. *Carcinogenesis* 23(9): 1511-8, 2002.

55. Shishodia S, Majumdar S, Banerjee S and Aggarwal BB, Ursolic acid inhibits nuclear factor-kappaB activation induced by carcinogenic agents through suppression of IkappaBalpha kinase and p65 phosphorylation: correlation with down-regulation of cyclooxygenase 2, matrix metalloproteinase 9, and cyclin D1. *Cancer Res* 63(15): 4375-83, 2003.

56. Koong AC, Chen EY and Giaccia AJ, Hypoxia causes the activation of nuclear factor kappa B through the phosphorylation of I kappa B alpha on tyrosine residues. *Cancer Res* 54(6): 1425-30, 1994.

57. Schutze S, Potthoff K, Machleidt T, Berkovic D, Wiegmann K and Kronke M, TNF activates NF-kappa B by phosphatidylcholine-specific phospholipase C-induced "acidic" sphingomyelin breakdown. *Cell* 71(5): 765-76, 1992.

58. Aggarwal BB, Tumour necrosis factors receptor associated signalling molecules and their role in activation of apoptosis, JNK and NF-kappaB. *Ann Rheum Dis* 59 Suppl 1: i6-16, 2000.

59. Komori A, Yatsunami J, Suganuma M, Okabe S, Abe S, Sakai A, Sasaki K and Fujiki H, Tumor necrosis factor acts as a tumor promoter in BALB/3T3 cell transformation. *Cancer Res* 53(9): 1982-5, 1993.

60. Suganuma M, Okabe S, Marino MW, Sakai A, Sueoka E and Fujiki H, Essential role of tumor necrosis factor alpha (TNF-alpha) in tumor promotion as revealed by TNF-alpha-deficient mice. *Cancer Res* 59(18): 4516-8, 1999.

61. Hafner M, Orosz P, Kruger A and Mannel DN, TNF promotes metastasis by impairing natural killer cell activity. *Int J Cancer* 66(3): 388-92, 1996.

62. Moore RJ, Owens DM, Stamp G, Arnott C, Burke F, East N, Holdsworth H, Turner L, Rollins B, Pasparakis M, Kollias G and Balkwill F, Mice deficient in tumor necrosis factor-alpha are resistant to skin carcinogenesis. *Nat Med* 5(7): 828-31, 1999.

63. Giri DK and Aggarwal BB, Constitutive activation of NF-kappaB causes resistance to apoptosis in human cutaneous T cell lymphoma HuT-78 cells. Autocrine role of tumor necrosis factor and reactive oxygen intermediates. *J Biol Chem* 273(22): 14008-14, 1998.

64. Arnott CH, Scott KA, Moore RJ, Hewer A, Phillips DH, Parker P, Balkwill FR and Owens DM, Tumour necrosis factor-alpha mediates tumour promotion via a PKC alpha- and AP-1-dependent pathway. *Oncogene* 21(31): 4728-38, 2002.

168

65. Hehlgans T, Stoelcker B, Stopfer P, Muller P, Cernaianu G, Guba M, Steinbauer M, Nedospasov SA, Pfeffer K and Mannel DN, Lymphotoxin-beta receptor immune interaction promotes tumor growth by inducing angiogenesis. *Cancer Res* 62(14): 4034-40, 2002.

66. Ashikawa K, Shishodia S, Fokt I, Priebe W and Aggarwal BB, Evidence That Activation of Nuclear Factor-kappa B Is Essential for Doxorubicin-Induced Cell Death in Myeloid and Lymphoid Cells. *Bochemical Pharmacology* In press, 2003.

67. Singh S, Raju U, Mendoza J, Pantazis P and Aggarwal BB, Acquisition of cellular resistance to 9-nitro-camptothecin correlates with suppression of transcription factor NF-kappa B activation and potentiation of cytotoxicity by tumor necrosis factor in human histiocytic lymphoma U-937 cells. *Anticancer Drugs* 9(8): 703-14, 1998.

68. Arlt A, Gehrz A, Muerkoster S, Vorndamm J, Kruse ML, Folsch UR and Schafer H, Role of NF-kappaB and Akt/PI3K in the resistance of pancreatic carcinoma cell lines against gemcitabine-induced cell death. *Oncogene* 22(21): 3243-51, 2003.

69. Hwang S and Ding A, Activation of NF-kappa B in murine macrophages by taxol. *Cancer Biochem Biophys* 14(4): 265-72, 1995.

70. Donepudi M, Raychaudhuri P, Bluestone JA and Mokyr MB, Mechanism of melphalan-induced B7-1 gene expression in P815 tumor cells. *J Immunol* 166(11): 6491-9, 2001.

71. Galter D, Mihm S and Droge W, Distinct effects of glutathione disulphide on the nuclear transcription factor kappa B and the activator protein-1. *Eur J Biochem* 221(2): 639-48, 1994.

72. Brach MA, Hass R, Sherman ML, Gunji H, Weichselbaum R and Kufe D, Ionizing radiation induces expression and binding activity of the nuclear factor kappa B. *J Clin Invest* 88(2): 691-5, 1991.

73. Wang CY, Cusack JC, Jr., Liu R and Baldwin AS, Jr., Control of inducible chemoresistance: enhanced anti-tumor therapy through increased apoptosis by inhibition of NF-kappaB. *Nat Med* 5(4): 412-7, 1999.

74. Chen X, Shen B, Xia L, Khaletzkiy A, Chu D, Wong JY and Li JJ, Activation of nuclear factor kappaB in radioresistance of TP53-inactive human keratinocytes. *Cancer Res* 62(4): 1213-21, 2002.

75. Zhang Q, Siebert R, Yan M, Hinzmann B, Cui X, Xue L, Rakestraw KM, Naeve CW, Beckmann G, Weisenburger DD, Sanger WG, Nowotny H, Vesely M, Callet-Bauchu E, Salles G, Dixit VM, Rosenthal A, Schlegelberger B and Morris SW, Inactivating mutations and overexpression of BCL10, a caspase recruitment domain-containing gene, in MALT lymphoma with t(1;14)(p22;q32). *Nat Genet* 22(1): 63-8, 1999.

76. Finco TS and Baldwin AS, Jr., Kappa B site-dependent induction of gene expression by diverse inducers of nuclear factor kappa B requires Raf-1. *J Biol Chem* 268(24): 17676-9, 1993.

77. Hamdane M, David-Cordonnier MH and D'Halluin JC, Activation of p65 NF-kappaB protein by p210BCR-ABL in a myeloid cell line (P210BCR-ABL activates p65 NF-kappaB). *Oncogene* 15(19): 2267-75, 1997.

78. Duyao MP, Kessler DJ, Spicer DB, Bartholomew C, Cleveland JL, Siekevitz M and Sonenshein GE, Transactivation of the c-myc promoter by human T cell leukemia virus type 1 tax is mediated by NF kappa B. *J Biol Chem* 267(23): 16288-91, 1992.

79. Finco TS, Westwick JK, Norris JL, Beg AA, Der CJ and Baldwin AS, Jr., Oncogenic Ha-Ras-induced signaling activates NF-kappaB transcriptional activity, which is required for cellular transformation. *J Biol Chem* 272(39): 24113-6, 1997.

80. Norris JL and Baldwin AS, Jr., Oncogenic Ras enhances NF-kappaB transcriptional activity through Raf-dependent and Raf-independent mitogen-activated protein kinase signaling pathways. *J Biol Chem* 274(20): 13841-6, 1999.

81. Mayo MW, Norris JL and Baldwin AS, Ras regulation of NF-kappa B and apoptosis. *Methods Enzymol* 333: 73-87, 2001.

82. Balmain A and Pragnell IB, Mouse skin carcinomas induced in vivo by chemical carcinogens have a transforming Harvey-ras oncogene. *Nature* 303(5912): 72-4, 1983.

83. Lu Y, Jamieson L, Brasier AR and Fields AP, NF-kappaB/RelA transactivation is required for atypical protein kinase C iota-mediated cell survival. *Oncogene* 20(35): 4777-92, 2001.

84. Libermann TA and Baltimore D, Activation of interleukin-6 gene expression through the NF-kappa B transcription factor. *Mol Cell Biol* 10(5): 2327-34, 1990.

85. Aggarwal BB, Schwarz L, Hogan ME and Rando RF, Triple helix-forming oligodeoxyribonucleotides targeted to the human tumor necrosis factor (TNF) gene inhibit TNF production and block the TNF-dependent growth of human glioblastoma tumor cells. *Cancer Res* 56(22): 5156-64, 1996.

86. Estrov Z, Thall PF, Talpaz M, Estey EH, Kantarjian HM, Andreeff M, Harris D, Van Q, Walterscheid M and Kornblau SM, Caspase 2 and caspase 3 protein levels as predictors of survival in acute myelogenous leukemia. *Blood* 92(9): 3090-7, 1998.

87. Bharti AC, Donato N, Singh S and Aggarwal BB, Curcumin (diferuloylmethane) down-regulates the constitutive activation of nuclear factor-kappa B and IkappaBalpha kinase in human multiple myeloma cells, leading to suppression of proliferation and induction of apoptosis. *Blood* 101(3): 1053-62, 2003.

88. Kato T, Duffey DC, Ondrey FG, Dong G, Chen Z, Cook JA, Mitchell JB and Van Waes C, Cisplatin and radiation sensitivity in human head and neck squamous carcinomas are independently modulated by glutathione and transcription factor NF-kappaB. *Head Neck* 22(8): 748-59, 2000.

89. Mukhopadhyay A, Banerjee S, Stafford LJ, Xia C, Liu M and Aggarwal BB, Curcumin-induced suppression of cell proliferation correlates with down-regulation of cyclin D1 expression and CDK4-mediated retinoblastoma protein phosphorylation. *Oncogene* 21(57): 8852-61, 2002.

90. Yamamoto K, Arakawa T, Ueda N and Yamamoto S, Transcriptional roles of nuclear factor kappa B and nuclear factor-interleukin-6 in the tumor necrosis factor alpha-dependent induction of cyclooxygenase-2 in MC3T3-E1 cells. *J Biol Chem* 270(52): 31315-20, 1995.

91. Bargou RC, Emmerich F, Krappmann D, Bommert K, Mapara MY, Arnold W, Royer HD, Grinstein E, Greiner A, Scheidereit C and Dorken B, Constitutive nuclear factor-kappaB-RelA activation is required for proliferation and survival of Hodgkin's disease tumor cells. *J Clin Invest* 100(12): 2961-9, 1997.

92. Romashkova JA and Makarov SS, NF-kappaB is a target of AKT in anti-apoptotic PDGF signalling. *Nature* 401(6748): 86-90, 1999.

93. Habib AA, Chatterjee S, Park SK, Ratan RR, Lefebvre S and Vartanian T, The epidermal growth factor receptor engages receptor interacting protein and nuclear factor-kappa B (NF-kappa B)-inducing kinase to activate NF-kappa B. Identification of a novel receptor-tyrosine kinase signalosome. *J Biol Chem* 276(12): 8865-74, 2001.

94. Foehr ED, Lin X, O'Mahony A, Geleziunas R, Bradshaw RA and Greene WC, NF-kappa B signaling promotes both cell survival and neurite process formation in nerve growth factor-stimulated PC12 cells. *J Neurosci* 20(20): 7556-63, 2000.

95. Shishodia S and Aggarwal BB, Nuclear factor-kappa B activation: A question of life and death. *J. Biochem. Mol. Biol.* 35(1): 28-40, 2002.

96. Kawamura K, Sato N, Fukuda J, Kodama H, Kumagai J, Tanikawa H, Shimizu Y and Tanaka T, Survivin acts as an antiapoptotic factor during the development of mouse preimplantation embryos. *Dev Biol* 256(2): 331-41, 2003.

97. Matta H, Eby MT, Gazdar AF and Chaudhary PM, Role of MRIT/cFLIP in protection against chemotherapy-induced apoptosis. *Cancer Biol Ther* 1(6): 652-60, 2002.

98. Farina AR, Tacconelli A, Vacca A, Maroder M, Gulino A and Mackay AR, Transcriptional up-regulation of matrix metalloproteinase-9 expression during spontaneous epithelial to neuroblast phenotype conversion by SK-N-SH neuroblastoma

cells, involved in enhanced invasivity, depends upon GT-box and nuclear factor kappaB elements. *Cell Growth Differ* 10(5): 353-67, 1999.

99. Bond M, Fabunmi RP, Baker AH and Newby AC, Synergistic upregulation of metalloproteinase-9 by growth factors and inflammatory cytokines: an absolute requirement for transcription factor NF-kappa B. *FEBS Lett* 435(1): 29-34, 1998.

100. Novak U, Cocks BG and Hamilton JA, A labile repressor acts through the NFkB-like binding sites of the human urokinase gene. *Nucleic Acids Res* 19(12): 3389-93, 1991.

101. Wang W, Abbruzzese JL, Evans DB and Chiao PJ, Overexpression of urokinase-type plasminogen activator in pancreatic adenocarcinoma is regulated by constitutively activated RelA. *Oncogene* 18(32): 4554-63, 1999.

102. Sliva D, English D, Lyons D and Lloyd FP, Jr., Protein kinase C induces motility of breast cancers by upregulating secretion of urokinase-type plasminogen activator through activation of AP-1 and NF-kappaB. *Biochem Biophys Res Commun* 290(1): 552-7, 2002.

103. Mahabeleshwar GH and Kundu GC, Syk, a protein-tyrosine kinase, suppresses the cell motility and nuclear factor kappa B-mediated secretion of urokinase type plasminogen activator by inhibiting the phosphatidylinositol 3'-kinase activity in breast cancer cells. *J Biol Chem* 278(8): 6209-21, 2003.

104. Loch T, Michalski B, Mazurek U and Graniczka M, [Vascular endothelial growth factor (VEGF) and its role in neoplastic processes]. *Postepy Hig Med Dosw* 55(2): 257-74, 2001.

105. Chilov D, Kukk E, Taira S, Jeltsch M, Kaukonen J, Palotie A, Joukov V and Alitalo K, Genomic organization of human and mouse genes for vascular endothelial growth factor C. *J Biol Chem* 272(40): 25176-83, 1997.

106. Levine L, Lucci JA, 3rd, Pazdrak B, Cheng JZ, Guo YS, Townsend CM, Jr. and Hellmich MR, Bombesin stimulates nuclear factor kappa B activation and expression of proangiogenic factors in prostate cancer cells. *Cancer Res* 63(13): 3495-502, 2003.

107. Huang S, DeGuzman A, Bucana CD and Fidler IJ, Nuclear factor-kappaB activity correlates with growth, angiogenesis, and metastasis of human melanoma cells in nude mice. *Clin Cancer Res* 6(6): 2573-81, 2000.

108. Pollet I, Opina CJ, Zimmerman C, Leong KG, Wong F and Karsan A, Bacterial lipopolysaccharide directly induces angiogenesis through TRAF6-mediated activation of NF-kappaB and c-Jun N-terminal kinase. *Blood* 102(5): 1740-2, 2003.

109. van de Stolpe A, Caldenhoven E, Stade BG, Koenderman L, Raaijmakers JA, Johnson JP and van der Saag PT, 12-O-tetradecanoylphorbol-13-acetate- and tumor necrosis factor alpha-mediated induction of intercellular adhesion molecule-1 is inhibited by dexamethasone. Functional analysis of the human intercellular adhesion molecular-1 promoter. *J Biol Chem* 269(8): 6185-92, 1994.

110. Helbig G, Christopherson KW, 2nd, Bhat-Nakshatri P, Kumar S, Kishimoto H, Miller KD, Broxmeyer HE and Nakshatri H, NF-kappaB promotes breast cancer cell migration and metastasis by inducing the expression of the chemokine receptor CXCR4. *J Biol Chem* 278(24): 21631-8, 2003.

111. Fujioka S, Sclabas GM, Schmidt C, Frederick WA, Dong QG, Abbruzzese JL, Evans DB, Baker C and Chiao PJ, Function of nuclear factor kappaB in pancreatic cancer metastasis. *Clin Cancer Res* 9(1): 346-54, 2003.

112. Darnay BG, Haridas V, Ni J, Moore PA and Aggarwal BB, Characterization of the intracellular domain of receptor activator of NF-kappaB (RANK). Interaction with tumor necrosis factor receptor-associated factors and activation of NF-kappab and c-Jun N-terminal kinase. *J Biol Chem* 273(32): 20551-5, 1998.

113. Iotsova V, Caamano J, Loy J, Yang Y, Lewin A and Bravo R, Osteopetrosis in mice lacking NF-kappaB1 and NF-kappaB2. *Nat Med* 3(11): 1285-9, 1997.

114. Franzoso G, Carlson L, Xing L, Poljak L, Shores EW, Brown KD, Leonardi A, Tran T, Boyce BF and Siebenlist U, Requirement for NF-kappaB in osteoclast and B-cell development. *Genes Dev* 11(24): 3482-96, 1997.

115. Bushdid PB, Brantley DM, Yull FE, Blaeuer GL, Hoffman LH, Niswander L and Kerr LD, Inhibition of NF-kappaB activity results in disruption of the apical ectodermal ridge and aberrant limb morphogenesis. *Nature* 392(6676): 615-8, 1998.

116. Yao Z, Zhang J, Dai J and Keller ET, Ethanol activates NFkappaB DNA binding and p56lck protein tyrosine kinase in human osteoblast-like cells. *Bone* 28(2): 167-73, 2001.

117. Andela VB, Sheu TJ, Puzas EJ, Schwarz EM, O'Keefe RJ and Rosier RN, Malignant reversion of a human osteosarcoma cell line, Saos-2, by inhibition of NFkappaB. *Biochem Biophys Res Commun* 297(2): 237-41, 2002.

118. Koul D, Yao Y, Abbruzzese JL, Yung WK and Reddy SA, Tumor suppressor MMAC/PTEN inhibits cytokine-induced NFkappaB activation without interfering with the IkappaB degradation pathway. *J Biol Chem* 276(14): 11402-8, 2001.

119. Trompouki E, Hatzivassiliou E, Tsichritzis T, Farmer H, Ashworth A and Mosialos G, CYLD is a deubiquitinating enzyme that negatively regulates NF-kappaB activation by TNFR family members. *Nature* 424(6950): 793-6, 2003.

120. Brummelkamp TR, Nijman SM, Dirac AM and Bernards R, Loss of the cylindromatosis tumour suppressor inhibits apoptosis by activating NF-kappaB. *Nature* 424(6950): 797-801, 2003.

121. Kovalenko A, Chable-Bessia C, Cantarella G, Israel A, Wallach D and Courtois G, The tumour suppressor CYLD negatively regulates NF-kappaB signalling by deubiquitination. *Nature* 424(6950): 801-5, 2003.

122. Rocha S, Martin AM, Meek DW and Perkins ND, p53 represses cyclin D1 transcription through down regulation of Bcl-3 and inducing increased association of the p52 NF-kappaB subunit with histone deacetylase 1. *Mol Cell Biol* 23(13): 4713-27, 2003.

123. Mayo MW, Madrid LV, Westerheide SD, Jones DR, Yuan XJ, Baldwin AS, Jr. and Whang YE, PTEN blocks tumor necrosis factor-induced NF-kappa B-dependent transcription by inhibiting the transactivation potential of the p65 subunit. *J Biol Chem* 277(13): 11116-25, 2002.

124. Martin AM, Kanetsky PA, Amirimani B, Colligon TA, Athanasiadis G, Shih HA, Gerrero MR, Calzone K, Rebbeck TR and Weber BL, Germline TP53 mutations in breast cancer families with multiple primary cancers: is TP53 a modifier of BRCA1? *J Med Genet* 40(4): e34, 2003.

125. Bentires-Alj M, Hellin AC, Ameyar M, Chouaib S, Merville MP and Bours V, Stable inhibition of nuclear factor kappaB in cancer cells does not increase sensitivity to cytotoxic drugs. *Cancer Res* 59(4): 811-5, 1999.

126. Natarajan K, Singh S, Burke TR, Jr., Grunberger D and Aggarwal BB, Caffeic acid phenethyl ester is a potent and specific inhibitor of activation of nuclear transcription factor NF-kappa B. *Proc Natl Acad Sci U S A* 93(17): 9090-5, 1996.

127. Bharti AC and Aggarwal BB, Nuclear factor-kappa B and cancer: its role in prevention and therapy. *Biochem Pharmacol* 64(5-6): 883-8, 2002.

128. Manna SK, Mukhopadhyay A and Aggarwal BB, Leflunomide suppresses TNF-induced cellular responses: effects on NF-kappa B, activator protein-1, c-Jun N-terminal protein kinase, and apoptosis. *J Immunol* 165(10): 5962-9, 2000.

129. Wang P, Wu P, Siegel MI, Egan RW and Billah MM, Interleukin (IL)-10 inhibits nuclear factor kappa B (NF kappa B) activation in human monocytes. IL-10 and IL-4 suppress cytokine synthesis by different mechanisms. *J Biol Chem* 270(16): 9558-63, 1995.

130. Manna SK, Mukhopadhyay A and Aggarwal BB, Human chorionic gonadotropin suppresses activation of nuclear transcription factor-kappa B and activator protein-1 induced by tumor necrosis factor. *J Biol Chem* 275(18): 13307-14, 2000.

131. Manna SK and Aggarwal BB, Alpha-melanocyte-stimulating hormone inhibits the nuclear transcription factor NF-kappa B activation induced by various inflammatory agents. *J Immunol* 161(6): 2873-80, 1998.

132. Haeffner A, Thieblemont N, Deas O, Marelli O, Charpentier B, Senik A, Wright SD, Haeffner-Cavaillon N and Hirsch F, Inhibitory effect of growth hormone on TNF-alpha secretion and nuclear factor-kappaB translocation in lipopolysaccharide-stimulated human monocytes. *J Immunol* 158(3): 1310-4, 1997.

133. Manna SK, Mukhopadhyay A and Aggarwal BB, IFN-alpha suppresses activation of nuclear transcription factors NF-kappa B and activator protein 1 and potentiates TNF-induced apoptosis. *J Immunol* 165(9): 4927-34, 2000.

134. Ehrlich LC, Hu S, Sheng WS, Sutton RL, Rockswold GL, Peterson PK and Chao CC, Cytokine regulation of human microglial cell IL-8 production. *J Immunol* 160(4): 1944-8, 1998.

135. Hideshima T, Chauhan D, Schlossman R, Richardson P and Anderson KC, The role of tumor necrosis factor alpha in the pathophysiology of human multiple myeloma: therapeutic applications. *Oncogene* 20(33): 4519-27, 2001.

136. Fiedler MA, Wernke-Dollries K and Stark JM, Inhibition of TNF-alpha-induced NF-kappaB activation and IL-8 release in A549 cells with the proteasome inhibitor MG-132. *Am J Respir Cell Mol Biol* 19(2): 259-68, 1998.

137. Page S, Fischer C, Baumgartner B, Haas M, Kreusel U, Loidl G, Hayn M, Ziegler-Heitbrock HW, Neumeier D and Brand K, 4-Hydroxynonenal prevents NF-kappaB activation and tumor necrosis factor expression by inhibiting IkappaB phosphorylation and subsequent proteolysis. *J Biol Chem* 274(17): 11611-8, 1999.

138. Ji C, Kozak KR and Marnett LJ, IkappaB kinase, a molecular target for inhibition by 4-hydroxy-2-nonenal. *J Biol Chem* 276(21): 18223-8, 2001.

139. Castro AC, Dang LC, Soucy F, Grenier L, Mazdiyasni H, Hottelet M, Parent L, Pien C, Palombella V and Adams J, Novel IKK inhibitors: beta-carbolines. *Bioorg Med Chem Lett* 13(14): 2419-22, 2003.

140. Kishore N, Sommers C, Mathialagan S, Guzova J, Yao M, Hauser S, Huynh K, Bonar S, Mielke C, Albee L, Weier R, Graneto M, Hanau C, Perry T and Tripp CS, A selective IKK-2 inhibitor blocks NF-kappa B-dependent gene expression in interleukin-1 beta-stimulated synovial fibroblasts. *J Biol Chem* 278(35): 32861-71, 2003.

141. Burke JR, Pattoli MA, Gregor KR, Brassil PJ, MacMaster JF, McIntyre KW, Yang X, Iotzova VS, Clarke W, Strnad J, Qiu Y and Zusi FC, BMS-345541 is a highly selective inhibitor of I kappa B kinase that binds at an allosteric site of the enzyme and blocks NF-kappa B-dependent transcription in mice. *J Biol Chem* 278(3): 1450-6, 2003.

142. Chen L, Fischle W, Verdin E and Greene WC, Duration of nuclear NF-kappaB action regulated by reversible acetylation. *Science* 293(5535): 1653-7, 2001.

143. Abu-Amer Y, Dowdy SF, Ross FP, Clohisy JC and Teitelbaum SL, TAT fusion proteins containing tyrosine 42-deleted IkappaBalpha arrest osteoclastogenesis. *J Biol Chem* 276(32): 30499-503, 2001.

144. Higgins KA, Perez JR, Coleman TA, Dorshkind K, McComas WA, Sarmiento UM, Rosen CA and Narayanan R, Antisense inhibition of the p65 subunit of NF-kappa B blocks tumorigenicity and causes tumor regression. *Proc Natl Acad Sci U S A* 90(21): 9901-5, 1993.

145. Kopp E and Ghosh S, Inhibition of NF-kappa B by sodium salicylate and aspirin. *Science* 265(5174): 956-9, 1994.

146. Surabhi RM and Gaynor RB, RNA interference directed against viral and cellular targets inhibits human immunodeficiency Virus Type 1 replication. *J Virol* 76(24): 12963-73, 2002.

147. Takaesu G, Surabhi RM, Park KJ, Ninomiya-Tsuji J, Matsumoto K and Gaynor RB, TAK1 is critical for IkappaB kinase-mediated activation of the NF-kappaB pathway. *J Mol Biol* 326(1): 105-15, 2003.

148. Liu J and Beller DI, Distinct pathways for NF-kappa B regulation are associated with aberrant macrophage IL-12 production in lupus- and diabetes-prone mouse strains. *J Immunol* 170(9): 4489-96, 2003.

149. Pieper GM and Riaz ul H, Activation of nuclear factor-kappaB in cultured endothelial cells by increased glucose concentration: prevention by calphostin C. *J Cardiovasc Pharmacol* 30(4): 528-32, 1997.

150. Cunningham MD, Cleaveland J and Nadler SG, An intracellular targeted NLS peptide inhibitor of karyopherin alpha:NF-kappa B interactions. *Biochem Biophys Res Commun* 300(2): 403-7, 2003.

151. May MJ, D'Acquisto F, Madge LA, Glockner J, Pober JS and Ghosh S, Selective inhibition of NF-kappaB activation by a peptide that blocks the interaction of NEMO with the IkappaB kinase complex. *Science* 289(5484): 1550-4, 2000.

152. Kwak HB, Lee SW, Lee DG, Hahm KS, Kim KK, Kim HH and Lee ZH, A hybrid peptide derived from cecropin-A and magainin-2 inhibits osteoclast differentiation. *Life Sci* 73(8): 993-1005, 2003.

THE P53 PATHWAY: TARGETS FOR THE DEVELOPMENT OF NOVEL CANCER THERAPEUTICS

SHULIN WANG AND WAFIK S. EL-DEIRY

1. INTRODUCTION

In the last two decades since the first identification of p53 as a host protein that is bound by the large T antigen of SV40 virus [1, 2], the investigation of the p53 gene has become a major focus of cancer research. In the late 1980's, p53 was recognized as a tumor suppressor and it was established that the oncogenic complex with large T antigen causes p53 inactivation [3]. The tumor suppressor protein p53 has been called "Cellular Gatekeeper" and " a Guardian of the Genome" for its well-documented activities in causing cell cycle arrest, apoptosis, and senescence. The p53 gene is the most commonly mutated gene in human cancer. Over 50% of the human tumors bear mutations in p53 that inactivate its function and there are additional mechanisms for p53 inactivation that do not involve intragenic mutation. Germline mutations in p53 are responsible for the majority of cases of the inherited Li-Fraumeni cancer family syndrome [4, 5], and mice in which the p53 locus has been inactivated are cancer prone [79, 80, 81]. The loss of p53 not only potentiates the oncogenic action of many oncogenes and renders animals susceptible to radiation and chemically induced tumors, but also leads to therapeutic resistance [6-8].

The p53 protein acts as a sequence-specific transcription factor that binds to at a minimum several hundred different promoters elements in the genome, broadly altering patterns of specific gene expression [9]. The activity of the p53 protein as a transcription factor is low in normal and unstressed cells, and appears to be tightly regulated. However, its basal activity is greatly enhanced in cells exposed to a wide variety of stress signals. Different stimuli activate signal-transduction pathways that culminate in post-translational modification and stabilization of p53. This accumulation of p53 activates the transcription of genes that are involved in various activities, including cell cycle arrest and apoptosis [10, 11].

In recent years, the genetic and biochemical analysis of the p53 pathway that lead from cellular stress to growth arrest and apoptosis, has identified many targets for therapeutic development [10, 12]. Many of the genes that are induced by p53 can be divided into groups that might mediate a specific p53 function, such as inhibition of cell growth, DNA repair, activation of apoptosis or regulation of angiogenesis [8]. The cell cycle inhibitory capacity of p53 appears to be primarily mediated by p21^{Waf1} [13], and the death-promoting arm of p53 may have tens of genes employed in its service [12]. p53 seems to hold shares in more than one apoptotic operation, being capable of transactivating genes encoding death receptors for example Fas and Killer/DR5, as well as those encoding a multitude of proteins that act through mitochondrial pathway such as Bax, Bak, Noxa, Puma, Bid and p53AIP1, along with others having known or speculated roles in different steps of the apoptotic program [14, 12]. The remarkable tissue specificity of p53 target gene activation response has been uncovered *in vivo* [15, 16]. Meanwhile, it has become clear that the toxicity and efficacy of many of the current cancer treatments are also profoundly affected by the activity of the p53 pathway.

2. P53 FAMILY MEMBERS AND HUMAN CANCERS

The p53 tumor suppressor gene encodes a sequence-specific transcription factor whose activity is either disabled or attenuated in the vast majority of human cancers [17, 8, 11]. The discovery in 1997 of the p73 gene at the lp36 locus, a location subject to recurrent loss of heterozygosity in various human cancers [18], led to the critical issue of whether this first identified p53 homologue functions as the archetypal tumor suppressor gene. One year later, the p53 gene family expanded with the cloning of a second homologue, the p63 gene [19], located at 3p27-28. This chromosome region is not subject to loss but to gain in cancers, and therefore suggests some oncogenic functions for altered p63 [20, 21, 19]. Both p73 and p63 are highly similar to p53 in the regions corresponding to the p53 N-terminal transactivation, central DNA-binding, and C-terminal oligomerization domains (Figure 1). In addition, the residues that bind to DNA are highly conserved among these proteins. Remarkably, all three genes share a similar exon/intro organization [18, 19]. Unlike p53, however, both p73 and p63 undergo complex alternative splicing [18, 19]. Some p73 and p63 proteins expressed in cells appear to act in a dominant negative manner with respect to p53 [19].

In stark contrast to p53, which is mutated in about 50% of human cancers (Figure 2), p73 and p63 genes were not inactivated by mutations in the vast majority of tumors from various tissues [22, 23]. Furthermore, mouse knockout studies indicated that loss of p73 and p63 did not predispose to cancer [24]. Thus, it appeared that p53, p63 and p73 were not acting simply as redundant tumor suppressors. However, recent studies reunited the p53 family [25] by studying the role of p73 and p63 in p53-dependent apoptosis. It has been found that p53 target genes are differentially affected by the loss of p63 and p73 and that apoptosis-related targets may be specifically regulated within the entire p53 family [25]. The coordinated activity of p73 and p63 has been found to serve as a determinant of whether of not apoptosis occurs [26, 25, 27, 28, 29]. p73 was recently identified as a determinant of chemotherapeutic efficacy in p53-defective tumor cells [30, 31].

So far, two tumor models highlight a direct role for p73 and p63 in

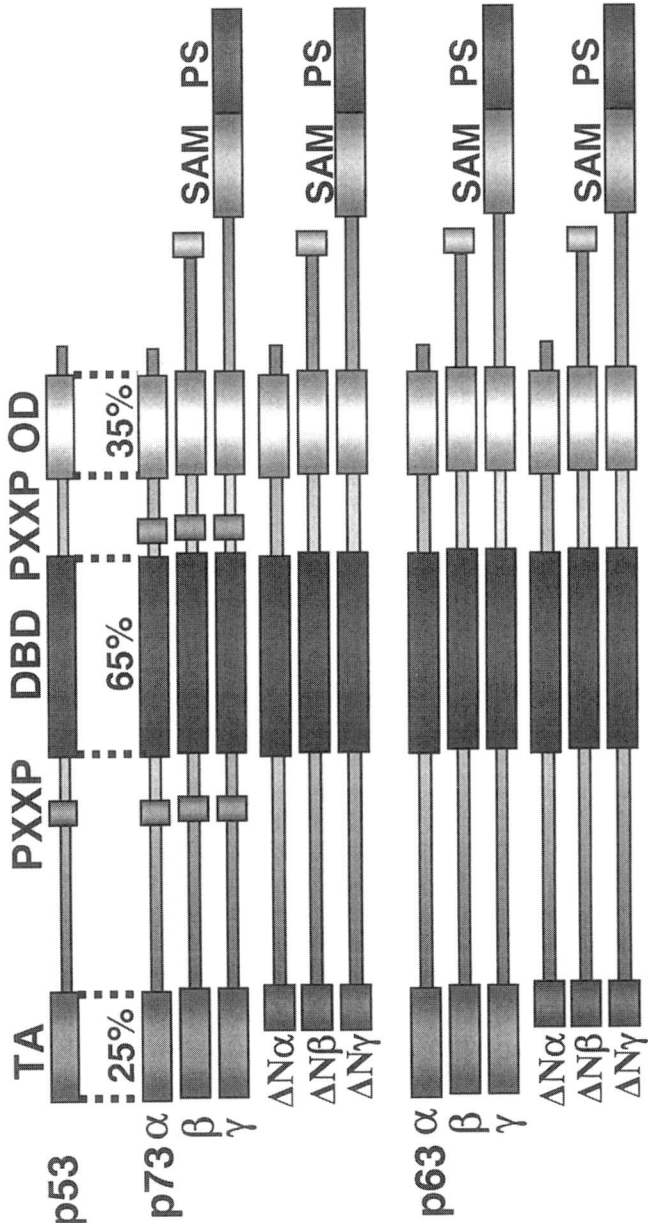

Figure 1. Functional domains of p53 family proteins. Comparison of the domain structure of p53 protein and the six major isoforms encoded by p63 and p73. The transactivation domain (TA), DNA binding domain (DBD), Oligomerization domain (OD), and a proline rich region (PXXP) are shown for each of the family members. SAM, Sterile alpha motif domain implicated in protein-protein interaction; PS, Post-SAM domain implicated in transcription suppression. ΔN, truncated amino-terminal domain of isoforms derived from an intronic promoter in the p63 and p73 genes.

tumorigenesis, neuroblastoma (NB) and squamous carcinoma cells. Remarkably, NB constitutes a first tumor model in which p73 status has been thoroughly studied and it provides a significant clinical impact [32]. In NB, p53 is not mutated but inactivated by cytoplasmic retention [33], making p73 a good candidate for NB tumorigenesis, given the recurrent loss of heterozygosity at p73 locus in these tumors. In direct correlation with its role in epithelia development and homeostatic maintenance, p63 levels significantly increased in squamous cell carcinoma [34].

3. THE REGULATION OF P53 FUNCTION

p53 is an efficient inhibitor of cell growth, including cell cycle arrest and/or apoptosis, depending on cell type and environment [35]. Regulation of p53 activity is therefore critical to allow both normal cell growth and tumor suppression. p53 levels remain low in most normally proliferating cells due to its rapid turnover. One of the key components regulating p53 stability is MDM2, a protein that functions as an ubiquitin ligase for p53, mediating ubiquitination of p53 and allowing it to be recognized and degraded by the proteasome [36]. MDM2 is a transcriptional target of p53 and establishes a feedback loop in which p53 drives expression of its negative regulator [37]. MDM2 interacts with the N-terminal region of p53, which contains the transcriptional trnasactivation domain, and inhibits the transcriptional activity by interfering with p53's ability to contact the transcriptional coactivators p300/CBP [38, 39](Figure 2). In addition to binding to p53, MDM2 plays another important role in controlling p53 protein stability [40, 36]. MDM2 is a RING-finger protein that functions as an ubiquitin ligase for both p53 and MDM2. The ability of MDM2 to target p53 for degradation of p53 is also regulated by other factors such as p300. p53 is stabilized by the binding of p300 to the oncoprotein E1A. In vitro, p300 with MDM2 catalyzed p53 polyubiquitination, whereas MDM2 catalyzed p53 monoubiquitination. E1A expression caused a decrease in polyubiquitinated but not monoubiquitinated p53 in cells [41]. MDM2 has also been found to aid the nuclear export of p53 and subsequent degradation of p53 [82]. MDM2 was identified as a target for phosphorylation by the serine/threonine kinase Akt . [42, 43]. Akt is activated by growth factor stimulation, and phosphorylation of MDM2 is associated with nuclear localization of MDM2 and enhanced degradation of p53 [42, 43]. This activity of Akt could therefore prevent activation of p53 during a normal proliferative response, and may also contribute to the ability of Akt to protect against p53-induced cell death [44]. Akt has also been shown to affect p73 transcription and apoptosis through a novel target YAP [45]. Apart from Akt, other components of signaling pathways associated with cell proliferation, such as Src [46] or C-Jun [47], can also function to block p53 activity. Recently, Pirh2, a gene

179

regulated by p53 that encodes a RING-H2 domain-containing protein with intrinsic ubiquitin-protein ligase activity, interacts with p53 and promotes ubiquitination of p53 independently of MDM2 [48]. Induction of p53 in response to stress occurs through several mechanisms, including

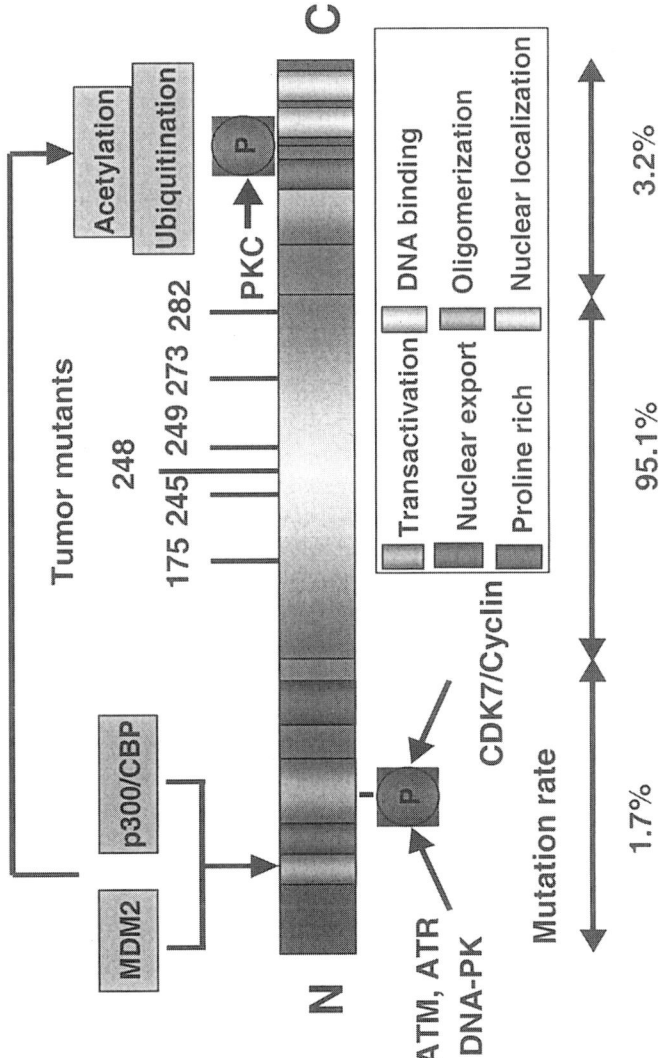

Figure 2. p53 structure, the post-translational modification and the location of tumor associated mutations. P53 contains several well-defined domains, including the amino-terminal transactivation domain, a central sequence-specific DNA-binding domain and a carboxy-terminal region that contains oligomerization sequences and nuclear-localization signal. Nuclear export of p53 is regulated by signals in the amino terminus and carboxyl termini. Interaction of proteins such as MDM2 or p300/CBP with the amino-terminal of p53 can lead to modification such as acetylation or ubiquitylation in the carboxyl terminus. Almost all of the point mutations that are found in cancers occur within the central DNA-binding domain. The percentage of mutations with each region detected in cancers to date is indicated at the bottom

180

stabilization of the protein and various modifications that produce a fully active p53 [35, 49]. Several stress-responsive kinases such as ATM, ATR, Chk1 and Chk2 have been shown to

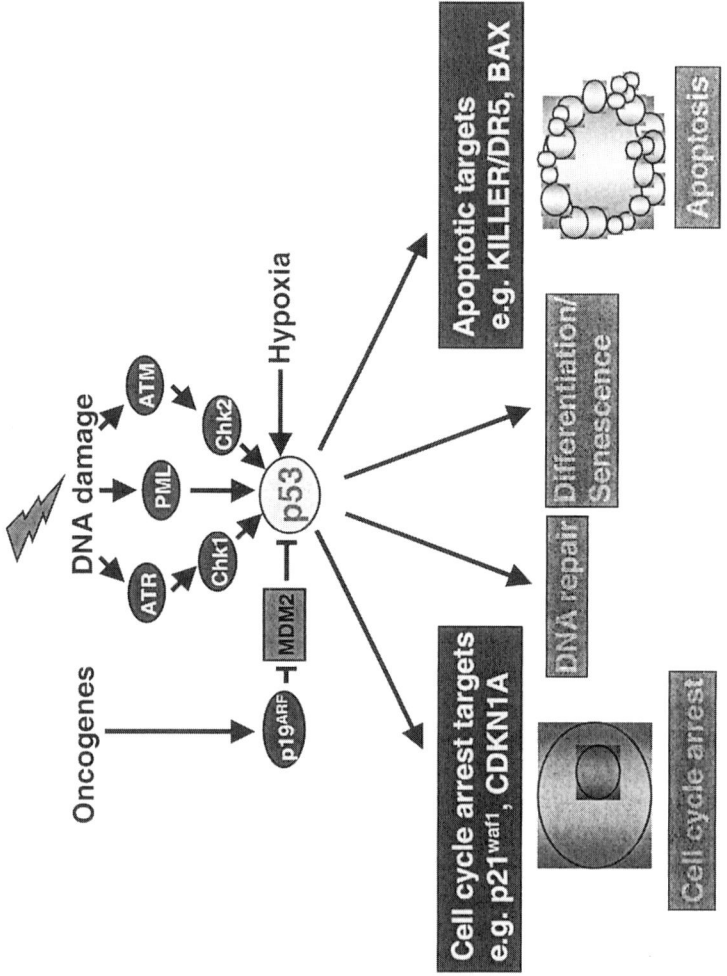

phosphorylate p53 (Figure 2), and inhibition or loss of these

Figure 3. The p53 pathway in cellular homeostasis and cancer. Several types of stress can activate p53, including DNA damage, oncogene activation, and hypoxia. Each type of stress is communicated to p53 by distinct mechanisms. P53 is the master switch that integrates signals from these pathways and transformed them into a second series of signals that trigger a cellular response. Several cellular responses to p53 activation have been described, and the choice of response depends on the factors such as cell type, cell environment and other oncogenic alterations that are sustained by the cell. In general, the effect of p53 activation is to inhibit cell growth, either through cell cycle arrest or induction of apoptosis, thereby preventing tumor development.

kinases renders cells deficient in mounting a p53 response to some signals. These kinase pathways appear to be particularly critical in the activation of p53 following exposure of cells to DNA-damaging events such as ionizing or UV radiation [50]. ATM-dependent events leading to p53 stabilization represent the initial major mechanisms by which cells utilize the p53 pathway for their response to stress leading to growth inhibition and tumor suppression (Figure 3). In addition to the phosphorylation of p53, ubiquitination, deubiquitination, acetylation, and sumoylation have also been observed to modulate p53 transcriptional activity [51, 52, 41, 48, 35] (Figure 2).

4. CHOICE OF RESPONSES TO P53

The stabilization and activation of p53 in response to cellular stress leads to
a number of cellular effects including cell cycle arrest or apoptosis [10]. The decision between apoptosis or growth arrest seems to be largely determined by p53, but it is by no means random [10](Figure 3). Several models have been proposed to explain how cells might choose between death and growth arrest in response to activation of p53. The first model indicates that the decision depends on the amount of activated p53 and the duration of its activation: the stronger and the longer activation of p53, the higher are the chance for apoptosis versus growth arrest [53, 54, 75]. It is possible that the differential binding affinity of p53 in different promoters could explain the observation that low levels of p53 protein result in cell cycle arrest, whereas higher levels of p53 induce apoptosis, which masks the concomitant cell cycle arrest response [56-58]. The second model indicates that the outcome is predetermined by the spectrum of p53-responsive genes that are available for modulation: different cell lineages might keep different sets of p53-responsive genes in active chromatin, thereby predetermining a specific pattern of transactivation and transrepression that dictates the final outcome. So, cells expressing more apoptotic genes would be more likely to undergo apoptosis than growth arrest. The third model claims that the choice between growth arrest and apoptosis depends on the availability of p53 co-factors that differentially regulate the ability of p53 to bind and interact with specific target genes [59]. In addition, the choice between cell death and survival is also dependent on the activity of survival signals that can be mediated by soluble ligand binding to cell-surface receptors, or direct interactions with neighboring cells. Rescue of p53-induced apoptosis by survival factors has been associated with the activation of the Akt kinase [44].

Comparison of p53-deficient mice with wild type mice shows that p53 is important for rapid apoptosis that occurs in sensitive tissues after treatment with irradiation. The most striking differences in radiosensitivity between p53-deficient and wild type mice are found in hematopoietic, lymphoid tissues and small intestine [60, 61]. The reasons why tissues of adults and embryos of different stages of development differ so much in their apoptotic response to irradiation might be due to the tissue specific activation of p53 regulated effector genes [16]. One could invoke tissue specific factors or modulations in p53 that might affect its ability to cause cell death or growth arrest.

5. THERAPEUTIC STRATEGIES BASED ON TARGETING P53

Due to the critical involvement of p53 in cell cycle control and apoptotic signaling (Figure 3), p53 provides an appealing target for gene therapy. The functional status of p53 in cells is an important determinant not only for the tumor development, but also for the therapeutic response [7]. P53-based gene therapeutic in vitro and in vivo approaches, which aimed at a direct correction of the specific molecular defect in p53-mutated malignant cells, showed efficacy in numerous cancer types, including head and neck, lung, prostate, cervical, and ovarian cancer [62]. The restitution of normal p53 function in malignant cells increases their sensitivity to undergo apoptosis, which may be paralleled by an increased sensitivity to chemotherapeutic drugs, whose cytotoxic effects are presumed to be dependent on the induction of apoptotic cell death. This hypothesis was taken into consideration in a number of clinical studies combining p53-based gene therapy with cytostatic agents [63]. One strategy already in clinical trials exploits the absence of p53 to enable selective viral replication (Bischoff et al., 1996). Onyx-015 (E1B deletion), a replication deficient E1B deleted adenovirus, has been found to selectively replicate in p53-deficient cells and early signs of efficacy are encouraging [64, 65].

Gudkov and his colleagues [66] have argued that conventional chemotherapy and radiation of p53 mutant tumors is limited by the p53-dependent toxicity induced in normal tissues. They therefore screened for the inhibitors of p53-induced transcription. They identified a compound, called Pifithrin that enabled normal cell growth but blocked the induction of the reporter gene by the chemotherapeutic agents. This approach was suggested as a way to mitigate the side effect of chemotherapy.

The search for small molecules that rescue mutant p53 function is attracting widespread interest. Recently, Rastinejad and colleagues reported the first results of these studies [67]. Two such compounds have been identified: CP-31398 [67-69] and Prima [70]. These compounds have been found to induce cell death of tumors that carry mutated p53 and to restore a wild type epitope in mutated p53. These compounds could suppress the growth of colon carcinoma cells by 75% and completely inhibit the growth of a melanoma cell line without obvious toxic side effects [67]. Recently, it has been demonstrated that Ellipticine can restore the transcription function of mutant p53 [71]. A screen of reported p53-binding peptides, using a sensitive physical analysis of the DNA binding core of the protein, has identified and characterized peptides derived from a known p53-binding protein that can stabilize the core folding of mutant p53 proteins [72]. The measurement of the p53 response to anticancer drugs and random compounds using a stably integrated p53-responsive luciferase reporter may have utility in the identification of novel chemotherapeutic agents, as an adjunct in the pharmaceutical optimization of lead compounds, in the exploration of environmental exposures, and in chemical probing of the p53 pathway [73].

Another area of therapeutic development attempts to find molecules that mimic the function of gene products whose synthesis is induced by p53. Two particular genes have attracted most attention. The first is the gene encoding p21^{Waf1} [13]. Small molecule mimics of p21 that act either at the ATP binding pocket of the CDK enzyme or at the cyclin substrate interface are being developed. A second target is the proapoptotic gene encoding Bax. Small peptides derived from Bax have shown activity in a variety of in vitro assays and drug-like small molecules that mimic the interaction of Bax with Bcl-2 have recently been characterized. One of these, Antimycin A, clearly shows selective and specific toxicity towards cells that over-express Bcl-2.

The identification of a subgroup of pro-apoptotic p53 targets has also recently shed some light on the long observed ability of wild type p53 to sensitize cells to killing by anticancer drugs or radiation [28, 74]. This group of genes has been referred to as "chemosensitivity genes" which include APAF1, Bid and Caspase 6 [12, 74]. The model for the action of chemosensitivity genes is that the increased levels of the chemosensitivity proteins in cells can lower the threshold for death in response to chemotherapy or radiation [12]. Therefore, agents that could specifically up-regulate these chemosensitivity genes might aid cancer treatment.

The growth or the death of tumor cells depends to a large extent on the fine-tuning of the p53-MDM2 interaction. As MDM2 controls p53 activity at the post-translational level, inhibiting the p53-MDM2 interaction is another promising approach for activating p53, because this association is well characterized at both structural and biological levels [75]. MDM2 inhibits p53 transcriptional activity, favors its nuclear export and stimulates its degradation. Therefore, inhibiting the p53-MDM2 interaction with synthetic molecules should lead to p53-mediated cell-cycle arrest or apoptosis in cells carrying wild type p53 [76-78]. Compounds that prevent the p53:MDM2 interaction might be useful as new anticancer agents that activate the p53 pathway in tumors.

6. CONCLUSIONS

P53 is a key component of the pathways regulating cellular responses to stress, and increasingly complex networks of mechanisms that contribute to the positive and negative control of p53 function are being described. Both wild type and mutant p53 remain as important targets for therapeutic modulation in cancers. There is clear consensus that reactivation of p53 function in cancer cells would be of therapeutic benefit, and many recent studies that define small molecules or peptides that restore function to mutant p53 proteins are beginning to illustrate the tremendous potential of this approach. The growing family of cofactors that are required for p53 to induce apoptosis is providing some auspicious new targets for the development of therapies to restore the apoptotic function of p53. The emergence of novel tools such as bioluminescence imaging as well as progress in imaging correlates of therapeutic response such as apoptosis, should accelerate the pace of drug development. Finally, exploiting siRNA to target p53 downstream target genes and subsequently modulate the apoptotic response continues to hold promise for improving therapeutic outcome.

7. ACKNOWLEDGEMENTS

W. S. El-Deiry is an Assistant Investigator of the Howard Hughes Medical Institute.

Shulin Wang and Wafik S. El-Deiry
Howard Hughes Medical Institute, Departments of Medicine, Genetics,
Pharmacology, and Cancer Center, University of Pennsylvania,
Philadelphia, Pennsylvania

8. REFERENCES

1. Lane, D.P. & Crawford, L.V. (1979). *Nature*, 278, 261-3.
2. Linzer, D.I. & Levine, A.J. (1979). *Cell*, 17, 43-52.
3. Lane, D.P. & Benchimol, S. (1990). *Genes Dev*, 4, 1-8.
4. Santibanez-Koref, M.F., Birch, J.M., Hartley, A.L., Jones, P.H., Craft, A.W., Eden, T., Crowther, D., Kelsey, A.M. & Harris, M. (1991). *Lancet*, 338, 1490-1.
5. Varley, J.M. (2003). *Hum Mutat*, 21, 313-20.
6. Vogelstein, B. & Kinzler, K.W. (1992). *Cell*, 70, 523-6.
7. Velculescu, V.E. & El-Deiry, W.S. (1996). *Clin Chem*, 42, 858-68.
8. Vogelstein, B., Lane, D. & Levine, A.J. (2000). *Nature*, 408, 307-10.
9. Zhao, R., Gish, K., Murphy, M., Yin, Y., Notterman, D., Hoffman, W.H., Tom, E., Mack, D.H. & Levine, A.J. (2000). *Genes Dev*, 14, 981-93.
10. El-Deiry, W.S. (2001). *Cell Death Differ*, 8, 1066-75.
11. Vousden, K.H. (2000). *Cell*, 103, 691-4.
12. Sax, J.K. & El-Deiry, W.S. (2003). *Cell Death Differ*, 10, 413-7.
13. El-Deiry, W.S., Tokino, T., Velculescu, V.E., Levy, D.B., Parsons, R., Trent, J.M., Lin, D., Mercer, W.E., Kinzler, K.W. & Vogelstein, B. (1993). *Cell*, 75, 817-25.
14. Burns, T.F. & el-Deiry, W.S. (2003). *Cancer Treat Res*, 115, 319-43.
15. Burns, T.F., Bernhard, E.J. & El-Deiry, W.S. (2001). *Oncogene*, 20, 4601-12.
16. Fei, P., Bernhard, E.J. & El-Deiry, W.S. (2002). *Cancer Res*, 62, 7316-27.
17. Oren, M., Damalas, A., Gottlieb, T., Michael, D., Taplick, J., Leal, J.F., Maya, R., Moas, M., Seger, R., Taya, Y. & Ben-Ze'ev, A. (2002). *Biochem Pharmacol*, 64, 865-71
18. Kaghad, M., Bonnet, H., Yang, A., Creancier, L., Biscan, J.C., Valent, A., Minty, A., Chalon, P., Lelias, J.M., Dumont, X., Ferrara, P., McKeon, F. & Caput, D. (1997). *Cell*, 90, 809-19.
19. Yang, A., Kaghad, M., Wang, Y., Gillett, E., Fleming, M.D., Dotsch, V.,

Andrews, N.C., Caput, D. & McKeon, F. (1998). *Mol Cell*, 2, 305-16.

20. Senoo, M., Seki, N., Ohira, M., Sugano, S., Watanabe, M., Inuzuka, S., Okamoto, T., Tachibana, M., Tanaka, T., Shinkai, Y. & Kato, H. (1998). *Biochem Biophys Res Commun*, 248, 603-7.

21. Trink, B., Okami, K., Wu, L., Sriuranpong, V., Jen, J. & Sidransky, D. (1998). *Nat Med*, 4, 747-8.

22. Benard, J., Douc-Rasy, S. & Ahomadegbe, J.C. (2003). *Hum Mutat*, 21, 182-91.

23. Kaelin, W.G., Jr. (1999). *Oncogene*, 18, 7701-5.

24. Yang, A. & McKeon, F. (2000). *Nat Rev Mol Cell Biol*, 1, 199-207.

25. Flores, E.R., Tsai, K.Y., Crowley, D., Sengupta, S., Yang, A., McKeon, F. & Jacks, T. (2002). *Nature*, 416, 560-4.

26. Bergamaschi, D., Samuels, Y., O'Neil, N.J., Trigiante, G., Crook, T., Hsieh, J.K., O'Connor, D.J., Zhong, S., Campargue, I., Tomlinson, M.L., Kuwabara, P.E. & Lu, X. (2003). *Nat Genet*, 33, 162-7.

27. Freeman, D.J., Li, A.G., Wei, G., Li, H.H., Kertesz, N., Lesche, R., Whale, A.D., Martinez-Diaz, H., Rozengurt, N., Cardiff, R.D., Liu, X. & Wu, H. (2003). *Cancer Cell*, 3, 117-30.

28. MacLachlan, T.K. & El-Deiry, W.S. (2002). *Proc Natl Acad Sci U S A*, 99, 9492-7.

29. Slee, E.A. & Lu, X. (2003). *Toxicol Lett*, 139, 81-7.

30. Bergamaschi, D., Gasco, M., Hiller, L., Sullivan, A., Syed, N., Trigiante, G., Yulug, I., Merlano, M., Numico, G., Comino, A., Attard, M., Reelfs, O., Gusterson, B., Bell, A.K., Heath, V., Tavassoli, M., Farrell, P.J., Smith, P., Lu, X. & Crook, T. (2003). *Cancer Cell*, 3, 387-402.

31. Irwin, M.S., Kondo, K., Marin, M.C., Cheng, L.S., Hahn, W.C. & Kaelin, W.G. (2003). *Cancer Cell*, 3, 403-10.

32. Melino, G., De Laurenzi, V. & Vousden, K.H. (2002). *Nat Rev Cancer*, 2, 605-15.

33. Moll, U.M., LaQuaglia, M., Benard, J. & Riou, G. (1995). *Proc Natl Acad Sci U S A*, 92, 4407-11.

34. Hibi, K., Trink, B., Patturajan, M., Westra, W.H., Caballero, O.L., Hill, D.E., Ratovitski, E.A., Jen, J. & Sidransky, D. (2000). *Proc Natl Acad Sci U S A*, 97, 5462-7.

35. Vousden, K.H. & Lu, X. (2002). *Nat Rev Cancer*, 2, 594-604.

36. Kubbutat, M.H., Jones, S.N. & Vousden, K.H. (1997). *Nature*, 387, 299-303.

37. Wu, X., Bayle, J.H., Olson, D. & Levine, A.J. (1993). *Genes Dev*, 7, 1126-32.

38. Grossman, S.R., Perez, M., Kung, A.L., Joseph, M., Mansur, C., Xiao, Z.X., Kumar, S., Howley, P.M. & Livingston, D.M. (1998). *Mol Cell*, 2, 405-15.

39. Oliner, J.D., Pietenpol, J.A., Thiagalingam, S., Gyuris, J., Kinzler, K.W. & Vogelstein, B. (1993). *Nature*, 362, 857-60.

40. Haupt, Y., Maya, R., Kazaz, A. & Oren, M. (1997). *Nature*, 387, 296-9.

41. Grossman, S.R., Deato, M.E., Brignone, C., Chan, H.M., Kung, A.L., Tagami, H., Nakatani, Y. & Livingston, D.M. (2003). *Science*, 300, 342-4.

42. Mayo, L.D. & Donner, D.B. (2001). *Proc Natl Acad Sci U S A*, 98, 11598-603.

43. Zhou, B.P., Liao, Y., Xia, W., Zou, Y., Spohn, B. & Hung, M.C. (2001). *Nat Cell Biol*, 3, 973-82.

44. Sabbatini, P. & McCormick, F. (1999). *J Biol Chem*, 274, 24263-9.

45.   Basu, S., Totty, N.F., Irwin, M.S., Sudol, M. & Downward, J. (2003). *Mol Cell*, 11, 11-23.

46.   Broome, M.A. & Courtneidge, S.A. (2000). *Oncogene*, 19, 2867-9.

47.   Schreiber, M., Kolbus, A., Piu, F., Szabowski, A., Mohle-Steinlein, U., Tian, J., Karin, M., Angel, P. & Wagner, E.F. (1999). *Genes Dev*, 13, 607-19.

48.   Leng, R.P., Lin, Y., Ma, W., Wu, H., Lemmers, B., Chung, S., Parant, J.M., Lozano, G., Hakem, R. & Benchimol, S. (2003). *Cell*, 112, 779-91.

49.   Woods, D.B. & Vousden, K.H. (2001). *Exp Cell Res*, 264, 56-66.

50.   Bakkenist, C.J. & Kastan, M.B. (2003). *Nature*, 421, 499-506.

51.   Ashcroft, M., Taya, Y. & Vousden, K.H. (2000) *Mol Cell Biol*, 20, 3224-33.

52.   Brooks, C.L. & Gu, W. (2003). *Curr Opin Cell Biol*, 15, 164-71.

53.   Blagosklonny, M.V. & el-Deiry, W.S. (1996). *Int J Cancer*, 67, 386-92.

54.   Blagosklonny, M.V. & El-Deiry, W.S. (1998). *Int J Cancer*, 75, 933-40.

55.   Wu, G.S. & El-Deiry, W.S. (1996). *Clin Cancer Res*, 2, 623-33.

56.   Chen, J., Willingham, T., Shuford, M., Bruce, D., Rushing, E., Smith, Y. & Nisen, P.D. (1996). *Oncogene*, 13, 1395-403.

57.   Kaeser, M.D. & Iggo, R.D. (2002). *Proc Natl Acad Sci U S A*, 99, 95-100.

58.   Szak, S.T., Mays, D. & Pietenpol, J.A. (2001). *Mol Cell Biol*, 21, 3375-86.

59.   Samuels-Lev, Y., O'Connor, D.J., Bergamaschi, D., Trigiante, G., Hsieh, J.K., Zhong, S., Campargue, I., Naumovski, L., Crook, T. & Lu, X. (2001). *Mol Cell*, 8, 781-94.

60.   Komarova, E.A., Christov, K., Faerman, A.I. & Gudkov, A.V. (2000). *Oncogene*, 19, 3791-8.

61.   Lowe, S.W., Schmitt, E.M., Smith, S.W., Osborne, B.A. & Jacks, T. (1993). *Nature*, 362, 847-9.

62.   Clayman, G.L., el-Naggar, A.K., Roth, J.A., Zhang, W.W., Goepfert, H., Taylor, D.L. & Liu, T.J. (1995). *Cancer Res*, 55, 1-6.

63.   Nishizaki, M., Meyn, R.E., Levy, L.B., Atkinson, E.N., White, R.A., Roth, J.A. & Ji, L. (2001) *Clin Cancer Res*, 7, 2887-97.

64.   Hamid, O., Varterasian, M.L., Wadler, S., Hecht, J.R., Benson, A., 3rd, Galanis, E., Uprichard, M., Omer, C., Bycott, P., Hackman, R.C. & Shields, A.F. (2003) *J Clin Oncol*, 21, 1498-504.

65.   Hecht, J.R., Bedford, R., Abbruzzese, J.L., Lahoti, S., Reid, T.R., Soetikno, R.M., Kirn, D.H. & Freeman, S.M. (2003) *Clin Cancer Res*, 9, 555-61.

66.   Komarov, P.G., Komarova, E.A., Kondratov, R.V., Christov-Tselkov, K., Coon, J.S., Chernov, M.V. & Gudkov, A.V. (1999) *Science*, 285, 1733-7.

67.   Foster, B.A., Coffey, H.A., Morin, M.J. & Rastinejad, F. (1999) *Science*, 286, 2507-10.

68.   Takimoto, R., Wang, W., Dicker, D.T., Rastinejad, F., Lyssikatos, J. & el-Deiry, W.S. (2002 *Cancer Biol Ther*, 1, 47-55.

69.   Wang, W., Takimoto, R., Rastinejad, F. & El-Deiry, W.S. (2003) *Mol Cell Biol*, 23, 2171-81.

70.   Bykov, V.J., Issaeva, N., Selivanova, G. & Wiman, K.G. (2002) *Carcinogenesis*, 23, 2011-8.

71.   Peng, Y., Li, C., Chen, L., Sebti, S. & Chen, J. (2003) *Oncogene*, 22, 4478-87.

72.   Friedler, A., Hansson, L.O., Veprintsev, D.B., Freund, S.M., Rippin, T.M., Nikolova, P.V., Proctor, M.R., Rudiger, S. & Fersht, A.R. (2002) *Proc Natl Acad Sci U S A*, 99, 937-42.

73.   Sohn, T.A., Bansal, R., Su, G.H., Murphy, K.M. & Kern, S.E. (2002). *Carcinogenesis*, 23, 949-57.

74.   Sax, J.K., Fei, P., Murphy, M.E., Bernhard, E., Korsmeyer, S.J. & El-Deiry, W.S. (2002). *Nat Cell Biol*, 4, 842-9.

75.   Chene, P. (2003). *Expert Opin Ther Targets*, 7, 453-61.

76.   Bottger, A., Bottger, V., Sparks, A., Liu, W.L., Howard, S.F. & Lane, D.P. (1997). *Curr Biol*, 7, 860-9.

77.   Bottger, V., Bottger, A., Howard, S.F., Picksley, S.M., Chene, P., Garcia-Echeverria, C., Hochkeppel, H.K. & Lane, D.P. (1996). *Oncogene*, 13, 2141-7.

78.   Midgley, C.A., Desterro, J.M., Saville, M.K., Howard, S., Sparks, A., Hay, R.T. & Lane, D.P. (2000). *Oncogene*, 19, 2312-23.

79.   Donehower, L.A. (1996). *Semin Cancer Biol*, 7, 269-78

80.   Donehower, L.A., Harvey, M., Slagle, B.L., McArthur, M.J., Montgomery, C.A., Jr., Butel, J.S. & Bradley, A. (1992). *Nature*, 356, 215-21.

81.   Jacks, T., Remington, L., Williams, B.O., Schmitt, E.M., Halachmi, S., Bronson, R.T. & Weinberg, R.A. (1994). *Curr Biol*, 4, 1-7.

82.   Roth, J., Dobbelstein, M., Freedman, D.A., Shenk, T. & Levine, A.J. (1998). *Embo J*, 17, 554-64.

RUNX PROTEIN SIGNALING IN HUMAN CANCERS

IAN ANGLIN AND ANTONIO PASSANITI

1. INTRODUCTION

The family of transcription factor genes (RUNX) that contain a conserved Runt DNA-binding domain regulate mammalian developmental events related to hematopoiesis[1] bone formation[2], and epithelial development [3]. These genes derive their names from the founding members of the family, the Drosophila *runt* genes (Runt and Lozenge), that are important in fly development, especially morphogenesis, sex determination, and eye formation [4]. Disruption of the Drosophila *runt* gene results in smaller flies because of impaired segmentation. Although translocation events and mutations in the *RUNX1* gene were positively implicated in human leukemias, the roles of the *RUNX2* and *RUNX3* genes in tumorigenesis have only recently been discovered. The RUNX proteins interact with a non-DNA binding factor, CBFß, through the Runt domain. Association of this factor with RUNX enhances RUNX DNA-binding [5] and reduces RUNX protein degradation through the proteasome pathway [6]. The importance of CBFß in hematopoiesis has long been appreciated because translocation events leading to a fusion of the *CBFß* and smooth muscle myosin heavy chain (SMMHC) gene also contribute to leukemogenesis [5].

The RUNX proteins function as strong transcriptional activators or repressors to regulate target gene expression [7]. A variety of binding partners determine whether activation or repression takes place and the expression and regulation of these cofactors have been the topics of much recent investigation [8]. The three mammalian RUNX proteins share some common features (Figure 1) [9]. In addition to the Runt domain in the N-terminus, which is highly homologous among all members of the group, the C-terminal region is the primary location for coactivator and corepressor binding.

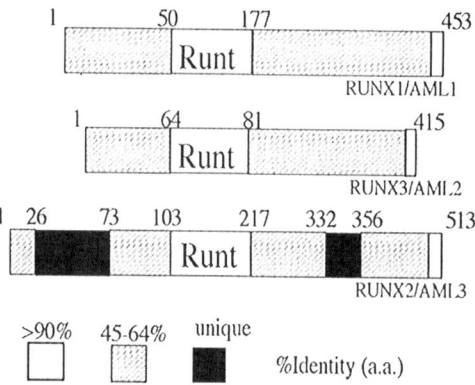

Figure 1: Comparison of Runt-related proteins
(from Westendorf and Hiebert[9])

The *RUNX1* gene is frequently the site of translocation events leading to expression of fusion proteins that convert RUNX1 from a transcriptional activator to a repressor [10]. This results in impaired hematopoiesis and contributes to leukemogenesis, eosinophilia, and blast crisis. RUNX2 is the largest member of the family. It contains two unique domains in the N- and C-termini. Targeted inactivation of the *Runx2* gene in mice leads to impaired bone formation and failure of vascularization. Several studies also implicate this factor in angiogenesis in adult tissues [11,12] and it can also cooperate with c-myc to enhance lymphomagenesis in transgenic mice [13]. In recent elegant studies, it was shown that *RUNX3* is important in development of the gastrointestinal tract and that this gene is frequently inactivated in human gastric carcinoma [3].

In this chapter, we will focus on the function of the human *RUNX* genes in tumorigenesis and how these genes regulate cancer and host cell function, including formation of the vascular endothelium. Reference to animal studies that implicate these genes in cancer will also be included. The upstream events that lead to expression or activation of *RUNX* genes and proteins and the regulation of the *RUNX* target genes will be described. Finally, some of the possible treatment strategies to address the tumorigenic events regulated by *RUNX* genes will be summarized.

2. RUNXGENESIN MAMMALIANDEVELOPMENT AND NON-CANCER PATHOLOGIES

Although this review will primarily emphasize the cancer-related functions of the *RUNX* genes, it is informative to consider some of the developmental and non-cancer pathologies that can result under conditions where these genes are inactivated, overexpressed, or mutated. For a more complete description of the role of the *RUNX* genes in normal development, the reader is referred to several recent reviews [7,10,14-17].

2.1. RUNX1

The *RUNX1* gene was cloned in 1991 [1] and soon after it was discovered that targeted inactivation of *Runx1* in mice resulted in a complete blockade in definitive hematopoiesis and embryonic lethality between E12.5 and E13.5 [18]. In this phase of hematopoiesis, hematopoietic stem cells (HSC) are produced, which give rise to both the myeloid and lymphoid progenitor cells in the adult. Runx1 was expressed in endothelial (hemogenic endothelium) and hematopoietic cells residing in the yolk sac and in endothelial cells within the vitelline and umbilical arteries [19] in E10.5 embryos. Endothelial cells within the aorta/gonad/mesonephros (AGM), a site of early hematopoiesis, and cells within the para-aortic splanchnopleure (PAS) also expressed Runx1. Takakura et al. [20] expanded on these observations to show that *Runx1* and HSC are essential in promoting angiogenesis in the embryo. *Runx1* null embryos lacked definitive hematopoiesis and exhibited defective angiogenesis in the head and pericardium. PAS explant cultures from *Runx1*-deficient mice, when grown on stromal cells, failed to generate HSCs and neovessels. However, angiogenesis could be rescued by the addition of HSCs or the angiogenic factor angiopoietin-1 (Ang1), which is produced by HSCs and which promoted endothelial cell migration *in vivo* and *in vitro*. However, HSCs from mice deficient in Ang1 could not rescue angiogenesis in the PAS explants from *Runx1*-deficient mice, suggesting a critical role for Ang1 derived from HSCs in the process. In other studies, Runx1 expression increased in murine blood vessels following stimulation of angiogenesis with FGF-2, while overexpression of dominant negative Runx resulted in reduction of Ang1 expression in mouse endothelial cells [11].

Changes in RUNX1 gene dosage in adults is apparently responsible for defects in hematopoiesis. Individuals with familial platelet disorder have been characterized with loss of function mutations in RUNX1, leading to thrombocytopenia and myelodysplastic syndrome, a pre-malignant condition [21]. Increased RUNX1 gene dosage occurs in individuals with Down's syndrome because of trisomy 21 [10] and may contribute to the observed myeloproliferative phenotype, which is characterized by increased myeloid, erythroid, and megakaryocyte cell populations with a propensity to develop AML.

Because RUNX1 regulates hematopoiesis and embryonic angiogenesis, these observations are relevant to understanding the role of human RUNX1 in neoplasia, a pathology that is promoted by the expression of angiogenic factors such as Ang1 [22] and vascular endothelial cell precursors from the bone marrow [23]. Therefore, the re-expression of

RUNX1 in tumor angiogenesis and the presence of mutations and translocations in leukemias (see section 3.1) certainly suggests an important role for RUNX1 in human cancer.

2.2. RUNX2

Expression of the *RUNX2* (Cbfa/PEBP2) gene was originally reported in T cells during thymic development [24], but it was also found to regulate expression of the osteoblast-specific gene, osteocalcin [2,25]. Expression of Runx2 in mouse embryos is first observed at embryonic day E12 in chondro-osteoprogenitor cells prior to ossification. Many matrix gene promoters in osteoblasts are activated by Runx2, including osteocalcin, type I collagen-alpha1 and alpha2 chains, bone sialoprotein, osteopontin [26] and, more recently, the marker of hypertrophic chondrocytes, collagen X [27]. In addition, Runx2 has been shown to mediate the expression of vascular endothelial growth factor (VEGF) in hypertrophic chondrocytes to activate angiogenesis necessary for bone formation [28]. Inactivation of *Runx2* in mice inhibits osteoblast development and ossification resulting in perinatal lethality from respiratory failure [29,30]. Heterozygote mice developed a pathology similar to that found in humans with cleidocranial dysplasia, a skeletal dysplasia resulting from mutations or heterozygous deletions in the *RUNX2* locus [31]. Runx2 was also found to be elevated in mouse [11] and human [12] models of angiogenesis suggesting a possible role for RUNX2 in neovascularization of adult tissues.

2.3. RUNX3

Although the human *RUNX3* gene was mapped to the short arm of chromosome 1 (1p36), a region that is deleted in certain well-differentiated gastric cancers [32], the role of the *RUNX3* gene in development has only recently been elucidated from studies on the C. elegans homologue, *run* [33], and some definitive studies in mice in which the *Runx3* gene was inactivated [3]. RUNX3 also contains a Runt domain that mediates DNA binding and thus, like other members of the family, plays a role in cell fate determination during development [34]. In C. elegans, the *run* gene is expressed in seam cells, which provide cells for the developing hypodermis and neuronal system, and in the intestine. Inhibition of run expression using RNAi resulted in defects in the mature hypodermis and gut, suggesting that run regulates growth of intestinal epithelial cells. Indeed, these results are consistent with mouse studies which show that *Runx3* null mice develop gut hyperplasia as a result of reduced epithelial cell apoptosis and increased proliferation. The epithelial cells from these mice were resistant to transforming growth factor-ß (TGFß)-induced growth inhibition and failed to undergo apoptosis after TGFß treatment. The *Runx3-/-* mice exhibited hyperplastic gastric epithelium and died soon after birth as a result of starvation. The role of the human *RUNX3* gene in tumorigenesis is consistent with these observations and will be summarized below (section 3.2).

2.4. CBFß

The role of CBFß in normal development is closely associated with the function of the *Runx1* gene [5]. In fact, targeted deletion of *CBFß* in mice results in loss of definitive hematopoietic precursor cells, as is observed in *Runx1* deficient mice [35]. CBFß forms a heterodimer with Runx1 to increase DNA binding and recent structural data also supports the necessary interaction between CBFß and the Runt domain of Runx1 in mediating DNA-binding and target gene regulation [36]. The *RUNX1* gene is a substrate of the ubiquitin/proteasome pathway. However, CBFß has been shown to protect RUNX1 from proteolytic degradation [6]. This suggests that one of the mechanisms by which Runx1 is able to maintain normal differentiation of hematopoietic cells is because CBFß heterodimerization results in transcriptionally active Runx1. Data from *CBFß*-deficient mice also support these conclusions since the levels of Runx1 in these mice was extremely low. Although CBFß protected Runx2 and Runx3 from proteolytic degradation, it is not clear whether this has a physiologically relevant effect on Runx2 or Runx3 biology. However, the interaction of Cbfß with Runx2 appears to be required for skeletal development [37]. *CBFß-/-* mice die at midgestation. Therefore, the role of Cbfß in bone formation cannot be ascertained in these mice. However, introduction of *CBFß* into *CBFß-/-* mice using a Gata1 promoter that directs expression to primitive and definitive erythroid cells allowed hematopoiesis to proceed, but mice displayed delayed bone formation and died soon after birth from respiratory failure. Immature osteoblasts were detected, but no endochondral bones were formed. The mechanism for these defects appeared to be reduced Runx2 DNA binding and transcriptional activation in the absence of CBFß.

3. CANCERS AND HOST RESPONSES

Developmentally active genes are often re-expressed in cancers and many of these genes are proto-oncogenes that may be overexpressed or are the dysregulated because of mutations in their coding sequence [38,39]. The Runx genes are examples of developmental genes that continue to be expressed in the adult and that regulate hematopoiesis (RUNX1), bone formation (RUNX2), and gut epithelial homeostasis (RUNX3), but which may also contribute to neoplasia.

3.1. Leukemia/Lymphoma

The RUNX1 (acute myeloid leukemia-1, AML-1) and CBFß genes, are the targets of chromosomal translocations, inversions, and mutations that are associated with the development of leukemias [40]. The most frequent alterations include the TEL-AML1 t(12;21) fusion, which accounts for 20% of acute lymphoblastic leukemias and AML1-ETO t(8;21) that is found in 12% of AML cases. The CBFß/SMMHC inversion event on chromosome 16 is associated with the M4 AML phenotype that is characterized by eosinophilia [10]. It is interesting to note that the mechanisms by which these common fusion events contribute to leukemias involve (1) a dominant negative activity of the fusion constructs that inhibits wild-type RUNX1 function, and (2) the recruitment of transcriptional repressors by the fusion partner to the RUNX1 promoter elements that are normally activated. The AML-ETO [41], CBFß-SMMHC [42], and the TEL-RUNX1 [43] fusion proteins have all been shown to exhibit dominant inhibitory activity for RUNX1. This mode of regulation is supported by experiments in which knock-in mice expressing RUNX1-ETO or CBFß-SMMHC exhibit the same phenotype as Runx1 null mice [44]. The AML-ETO gene codes for a strong repressor protein at the C-terminus that recruits the SIN3 corepressor and histone deacetylases (HDACs) responsible for target gene repression [45,46]. The TEL-RUNX1 fusion protein contains a pointer (PTN) motif and a repression domain at the N-terminus [47]. TEL is an ETS family member gene which recruits nuclear corepressor complexes (N-Cor) to RUNX1 DNA-binding promoter elements. This fusion gene, however, still contains the normal activator regions of RUNX1 that may function to balance the effects of the fused repressor domain. Therefore, the net effect of most of the RUNX1 fusion genes is to convert strong transcriptional activators, such as c/EBP-alpha, that is necessary for hematopoiesis, to proteins that mediate transcriptional repression of genes [48].

Runx2 expression has been shown to enhance T-cell lymphoma development in mice by cooperating with c-myc, p53 and Pim1 [49,50], while Runx3 is a common insertional site in retroviral-induced leukemias in mice [34]. Although these studies implicate Runx2 and Runx3 as oncogenic factors in mice, there is no definitive evidence for their roles in human hematopoietic malignancies.

3.2. Solid Tumors

Since the identification of the human RUNX transcription factors, there has been an increasing number of publications documenting the altered expression of these factors in a variety of solid tumors and cancer cell lines. RUNX1 expression was linked to regulation of acetylcholinesterase gene expression and correlated with increasing tumor grade in human brain tumors [51].

Accumulation of RUNX1 mRNA in astrocytomas and melanomas correlated with tumor aggressiveness. Co-expression of RUNX1 and the C-terminal domain of acetylcholinesterase-S in COS1 cells increased RUNX1 nuclear localization, suggesting that post-translational regulation of RUNX1 might be involved in increased tumorigenesis. Interestingly, overexpression of RUNX1 in fibroblasts enhanced neoplastic transformation, a process that is dependent on a conserved cysteine residue in the runt homology domain [52]. Endogenous RUNX2 in malignant breast cancer cells was found to regulate bone sialoprotein (BSP) expression via a RUNX2 binding site located within the proximal −110bp site of this gene [53]. However, normal human mammary epithelial cells showed extremely low RUNX2 expression and BSP promoter activity. Since breast cancers preferentially metastasize to bone and BSP has been significantly associated clinically with skeletal metastases [54], it is apparent that RUNX2 may play a key role in a critical stage of breast tumorigenesis. Another study using immunohistochemical analysis showed that RUNX2 mRNA was expressed in two malignant melanoma cell lines *in vitro* and RUNX2 protein was expressed in higher grade (Clarke's level IV) human melanomas [55]. These studies suggest that BSP expression may be linked to tumor cell invasion and are consistent with the known regulation of matrix metalloproteinases (matrix metalloproteinase-13, MMP13; membrane type-1 matrix metalloproteinase, MT_1MMP) by RUNX2 [56,57]. In addition, expression of RUNX2 has been linked to the expression of the osteopontin gene and activation of its promoter in breast cancer cells [58], events that may be involved in mediating tumor cell migration or survival.

RUNX2 is expressed in clinical prostate cancer specimens and in the metastatic, androgen independent PC-3 cell line [59,60]. Interestingly, RUNX2 expression was not detected in the less aggressive LNCaP cell line. This evidence raises the possibility that RUNX2 may play a role in hormone-independent prostate tumor growth and may likely offer another target for therapeutic intervention in these progressive tumors. Therefore, these findings support evidence from other experimental models that RUNX2 has oncogenic properties.

While increased expression of RUNX2 is associated with tumorigenesis, it is the loss of RUNX3 that contributes to tumor progression. Approximately 30% of 46 primary human gastric tumors analyzed by Li et al. [3] were hemizygous for the *RUNX3* gene. Furthermore, RUNX3 expression was reduced with increasing stage of the cancers and up to 90% of late stage metastatic tumors exhibited reduced RUNX3 levels [3]. Although only one of the examined tumors contained a mutation within the RUNX3 coding sequence, the presence of hypermethylation in the promoter region of RUNX3 correlated with gene silencing. In addition, wild-type RUNX3 transfected into gastric carcinoma cells implanted into nude mice reduced tumorigenicity while transfection of RUNX3 with a mutation in the DNA-binding Runt domain increased tumor formation, thus overcoming the tumor-suppressive effect of RUNX3. Therefore, the loss of RUNX3 appears to be causally related to the progression of human gastric carcinoma. A clue to the mechanistic basis of the tumor suppressive activity of RUNX3 is provided from studies with gastric epithelial cells from *Runx3* null mice [3]. As noted in section 2.3, these cells appear to be resistant to the growth inhibitory activity of TGFß and display apoptotic defects in response to TGFß. Since all RUNX proteins bind receptor-Smads (R-Smads) [61], it appears that the loss of RUNX3 may promote tumorigenesis, in part, via increased cell survival through loss of a Smad-mediated apoptotic pathway.

3.3. Tumor Angiogenesis

While it is now well established that tumors require the development of new blood vessels to survive and spread, the precise mechanisms underlying this complex process are not well defined. Several reviews discuss the possible models that regulate this "angiogenic switch" and a consistent feature of these models is that the alteration in the balance of specific angiogenic factors and inhibitors expressed by either tumor or endothelial cells (EC) regulates the angiogenic phenotype [62,63]. Clearly, any factor(s) that can regulate this balance will play a vital role in tumorigenesis. The molecular mechanisms through which transcription factors, including the Runx genes, might regulate angiogenesis have been reviewed recently [16,64].

Runx1 mRNA can be detected by *in situ* hybridisation in vascular EC and tumor cells of human gliomas with significant correlation observed with increasing tumor grade [51]. Similar results were obtained from sections of brain metastases of malignant melanoma. RUNX1 was observed in the vascular EC associated with these tumor specimens [51]. Several reports have suggested a role for the Runx genes in EC migration, invasion, and angiogenesis [11,12]. Fibroblast growth factor (FGF) and vascular endothelial growth factor (VEGF) were found to increase RUNX expression in mouse EC. EC expressed angiopoietin-1, a regulator of vasculogenesis and angiogenesis that mediates VEGF responses [65]. Expression of the dominant negative CBFß-SMMHC fusion construct, however, inhibited expression of angiopoietin-1 and cell migration and invasion suggesting that RUNX genes may respond to angiogenic factors to regulate target genes that are important in angigogenesis [11]. In other studies [12], RUNX2 mRNA and protein expression in human bone marrow EC were found to increase in cells that were activated for *in vitro* angiogenesis (tube formation) when cultured on basement membrane proteins. Angiogenesis and RUNX2 expression were induced by treatment with the angiogenic factor insulin-like growth factor-1 (IGF-1). Although endothelium is a tissue that expresses RUNX genes, it is not clear what the transcriptional targets are in EC. Expression of a dominant negative Runt DNA-binding domain inhibited EC migration and invasion and expression of the genes encoding the proteolytic enzymes urokinase plasminogen activator (uPA) and MT_1MMP, which mediate cell invasion

[12]. In addition, comparison of angiogenesis and extracellular matrix gene expression by microarray analysis in human umbilical vein EC shows that expression of the well-characterized RUNX2 target gene, MMP13 [66], is increased 20-fold after over-expression of RUNX2 (*I. Anglin and A. Passaniti, unpublished observations*). These results suggest that RUNX2 is involved in IGF-1 and extracellular matrix-induced angiogenesis and that one function of RUNX2 in angiogenesis may be to control EC migration and/or invasion through the regulation of protease expression.

4. PATHWAYS INVOLVED IN INDUCTION OR DOWNREGULATION OF RUNX

While several studies have now described the expression and function of RUNX genes in a variety of biological situations and pathological events, the pathways that lead to RUNX activation or inhibition have only recently been examined. Overall, several factors have been shown to increase RUNX mRNA or protein or to increase RUNX activity via regulation of post translational modifications. In this section, a summary of the current knowledge of upstream events leading to RUNX induction or downregulation will be presented. Although many of these studies have not linked RUNX regulation to tumorigenesis, the signaling pathways that regulate RUNX expression are also involved in cancer. It is noteworthy that several pathways leading to RUNX activation converge on the mitogen-activated protein kinase (MAPK) cascade, which is the subject of much therapeutic interest [67].

4.1. RUNX Activation

Few upstream events have been described that are important in regulating Runx1 or Runx3 activity [17]. However, the RUNX1 gene was upregulated in U937 cells during retinoic acid-induced cell differentiation [68] and RUNX3 was regulated in hematopoietic cells by the retinoic acid receptor-alpha [69]. RUNX3 expression in hematopoietic cells was also activated by vitamin D3. Other hormone receptor interactions may regulate RUNX2 expression, since tamoxifen, estrogen receptor modulators, and glucocorticoids can increase RUNX2 expression [70,71].

Several intracellular signaling pathways have been shown to activate RUNX2 [17,72]. Bone morphogenic proteins (BMPs) and TGFß exhibit some common effects including induction of osteoblast and chondrocyte differentiation, cell cycle arrest, angiogenesis and tumor metastasis. BMPs and TGFß stimulate cells by ligand activation of type I and II receptors and subsequent binding to R-Smads, which interact with Co-Smads to increase DNA-binding to the promoters of specific response genes [73]. BMPs were shown to increase Runx2 expression in osteoblasts and myoblasts and recently BMP7 was reported to upregulate Runx2 mRNA expression in osteoblasts and DNA binding activity in C2C12 myoblast cells [74-76]. Since BMPs are expressed by tumor cells at metastatic sites [77], they might induce cellular Runx2 expression and/or activity in a paracrine or autocrine fashion to further enhance cellular processes required for tumor maintenance and progression. Conversely, TGFß suppresses RUNX2 expression in osteosarcoma cells [78] but increases it in C2C12 myoblasts [79].

The extracellular matrix (ECM) plays an important role in cell differentiation, development and tumorigenesis. Changes in cellular interactions with the ECM lead to alterations in signaling and gene expression. The integrins are cell membrane-bound proteins that interact with the ECM and initiate intracellular signal transduction to the nucleus to activate nuclear factors such as RUNX2. In fact, ECM secretion by MC3T3 pre-osteoblasts increased expression of the osteocalcin gene, a RUNX2 target [80], which required the presence of the osteoblast-specific element (OSE2) in the osteocalcin promoter. RUNX2 DNA-binding activity was enhanced by interaction with the alpha-2 integrin without much change in RUNX2 protein or mRNA expression [81]. Interestingly, over-expression of constitutively active MEK1 (an ERK1/2 pathway kinase) increased RUNX2 phosphorylation demonstrating that the MAPK signaling pathway may activate and phosphorylate RUNX2 and may explain the increased ECM-induced activation without changes in synthesis of RUNX2 protein [82]. These studies suggested that phosphorylation may occur in a Pro/Ser/Thr-rich region in the C-terminal region of RUNX2. Metabolic labeling of RUNX2 with provided further evidence for post-translational phosphorylation of RUNX2 [83], but the kinases responsible for these modifications were not identified.

Inflammatory cytokines are expressed in pathological situations such as osteoarthritis, angiogenesis, and cancer and regulate the expression of MMPs, often through activation of the MAPK signal transduction pathway [84]. In osteoarthritis, chondrocytes secrete MMP13 in response to interleukin-1 (IL-1) [85], leading to collagen II degradation. Overexpression of RUNX2 in human chondrosarcoma cells resulted in increased MMP13 expression through the p38 MAPK pathway. A stably integrated MMP13 promoter-reporter vector was activated 5 to 6-fold after IL-1 treatment, a process requiring a proximal AP1 site, a RUNX site, and activation of the p38 MAPK pathway. Mutation of the RUNX binding site in the MMP13 promoter reduced reporter gene activation, suggesting that RUNX2 is required for MMP13 activation by IL-1 in these cells.

FGF-2 is a potent angiogenic growth factor and activator of endothelial and tumor cell proliferation [86, 87] that is synergistic with the alphaVbeta3 integrin pathway [88]. FGF-2 can also regulate skeletal development [89,90] and increase osteocalcin gene expression in pre-osteoblastic MC3T3-E1 cells [91]. FGF-2 increased phosphorylation of

RUNX2 in pre-osteoblast cells, while treatment with the MEK/ERK inhibitor, U0126, prevented this increased phosphorylation [92]. FGF-2 increased RUNX2-mediated osteocalcin promoter activation, which was also inhibited by UO126, suggesting that a MAPK pathway was responsible for transcriptional activity. This FGF-2 activated response was synergistically enhanced by the PKA pathway after forskolin treatment [57].

Parathyroid hormone (PTH) also regulated RUNX2 expression via activation of the PKA or PKC pathways [93,94]. Although PTH did not alter RUNX2 protein levels, it was suggested that activation occurred through a PKA-dependent phosphorylation pathway. Overexpression of RUNX2 in osteoblasts enhanced PTH-mediated stimulation of the MMP13 promoter [95], a mechanism that required a Runt domain interaction with the leucine zipper domain of AP-1 [96]. Therefore, PKC activation of RUNX2 appears to be the result of at least two pathways that involve the PTH/PTHRP and FGF2/RTK signal transduction cascades [72].

IGF-1 is a potent growth factor for many cells and an angiogenic factor that can stimulate EC proliferation and differentiation through selective crosstalk between the alphaVbeta5 integrin and PKC signaling pathways [97]. RUNX2 mRNA and protein expression in a human bone marrow EC line (HBME-1) were shown to be inducible after IGF-1 treatment, with an IC_{50} of 14pM [12]. RUNX2 expression was associated with a dose-dependent IGF-1 activation of EC tube formation on extracellular matrix, which could be inhibited by incubation with neutralizing IGF-1 receptor antibodies. It now appears that the PI3K signaling pathway may mediate both IGF-1-dependent RUNX2 synthesis and DNA-binding activity (*M.Qiao and A. Passaniti, unpublished results*). In summary, many extraceullular events may activate specific signal transduction pathways such as the MAPK and IGF-1 pathways to regulate RUNX protein phosphorylation and activity in a variety of cells.

4.2. Autoregulation of RUNX2 Promoter

In most physiological situations, such as bone formation and skeletal development, RUNX2 expression must be precisely regulated [15]. Although it is well-known that several cytokine signaling pathways can activate RUNX2 expression or activity, as summarized above, much less is known about the events that inhibit RUNX2 expression. Recent studies of the rat Cbfa1 (Runx2) promoter [98] suggest that Runx2 binding sites regulate repression of the Runx2 gene. At least three Runx2 recognition motifs are present in the rat promoter and three tandemly repeated sites are also present in the 5'UTR. Over-expression of Runx2 protein was found to downregulate promoter activity and this repression was dependent on a single Runx2 binding site in the proximal promoter. These results show that the Runx2 gene can be autoregulated by a negative feedback mechanism. The promoters of the human RUNX1 [99] and RUNX3 [100] genes also contain Runx binding sites, which may be important in autoregulation of expression. Whether RUNX-interacting factors (see sections 5.2, 5.3) regulate RUNX gene expression is not known, although some transcription factors, including NF1-A, FosB, Msx2, Bapx1, Hoxa2, and PPARg2 have been implicated in regulation of Runx2 expression [101-105].

4.3. Negative Regulation of RUNX Activity

Although vitamin D3 activates RUNX3 expression (section 4.1), vitamin D3 inhibited expression of RUNX2 in MC3T3 and ROS osteosarcoma cells [56], an effect not observed in ROS cells lacking functional vitamin D receptors. Consistent with these observations, the RUNX2 proximal promoter contains a vitamin D-responsive element, which when mutated, did not respond to vitamin D [106].

The cAMP pathway has been shown to promote RUNX2 proteolytic degradation in proteasomes and, therefore, to inhibit RUNX2 activity in osteoblast-like cells [107]. Increasing cAMP levels resulted in downregulation of RUNX2-regulated or osteoblast-specific genes such as alkaline phosphatase, bone sialoprotein, and osteocalcin. RUNX2 protein levels were notably lower when cAMP was elevated and polyubiquitinated forms of RUNX2 accumulated in the cells.

The signal transducer and activators of transcription (Stats) are targets of janus-activated kinase (Jak) phosphorylation and dimerization events that respond to inflammatory and viral signals in a variety of pathological situations, including cancer, where they function as onco-suppressor genes [108]. Stat1 plays a central role in regulating osteoclast differentiation in response to interferon-ß or interferon-γ [109]. Stat1-deficient mice exhibit excessive osteoclastogenesis while also showing increased bone mass, a process that appears to involve inhibition of osteoblast differentiation [110]. This inhibition was the result of interference with Runx2 transcriptional activity by interaction of Stat1 with Runx2 in the cytoplasm. This interaction did not require Stat1 phosphorylation. Therefore, the latent Stat1 protein interferes with Runx2 activation by sequestering Runx2 in the cytoplasm to prevent nuclear localization.

TNF-alpha mediates apoptotic or angiogenic signals [111] to regulate tumor cell apoptosis through caspase [112]. Therefore, it is interesting that TNF-alpha represses expression of the RUNX2 gene, which appears to have a role in solid tumor function (section 3.2). TNF-alpha was found to suppress RUNX2 steady-state mRNA levels of two major isoforms [113]. RUNX2 protein in the nucleus and DNA-binding activity were also reduced. The effect of TNF-alpha was observed at the promoter level since a RUNX2 promoter-luciferase reporter vector was inhibited by up to 50% in the presence of TNF-alpha. These results suggest that TNF-alpha regulates RUNX2 expression by altering the stability of the RNA and by suppression of transcription. In conclusion, negative regulation of RUNX activity can occur through hormone or cytokine-activated signaling pathways or through interaction with other transcription factors that alter RUNX subcellular localization.

5. DOWNSTREAM EVENTS: NUCLEAR LOCALIZATION AND TRANSCRIPTION

The Runx proteins contain a conserved nuclear localization signal (NLS) C-terminal to the Runt DNA-binding domain and a nuclear matrix targeting sequence (NMTS) [114] that localize expressed and activated Runx proteins to the cell nucleus and specific chromatin domains, respectively. Several Runx target genes have been identified to date that are expressed in hematopoietic, skeletal, or tumor tissues. For a summary of recent data and of interacting factors that regulate Runx target promoter activity, the reader is referred to the review by Otto [17].

5.1. RUNX Target Genes

Since the first report of RUNX transcription factors, an increasing number of genes that are positively or negatively regulated by these factors have been identified [7]. Some of these genes also have important functions in tumorigenesis, cell cycle progression, or angiogenesis. These include osteopontin (OPN) and bone sialoprotein (BSP), which function in adhesion and bone metastases [53, 55, 58], osteocalcin, which is important in bone mineralization and metastasis [55], BMP2 and VEGF, important in angiogenesis [28, 115], MMP13, a protease involved in invasion and tumorigenesis [66, 116], and the cell cycle regulators cyclin D1 and the cyclin-dependent kinase inhibitor p21^{CIP1} [40, 117]. The AML-ETO fusion RUNX protein also represses transcription of the p14^{ARF} tumor suppressor in acute myeloid leukemia [118]. In addition, the IgC-alpha promoter has been defined as a RUNX3 target gene [61]. Therefore, the genes regulated by a particular Runx factor are diverse. Initial studies with Runx2 suggested that it only regulated bone specific-genes such as osteocalcin. However, subsequent studies have shown that Runx2 can modulate the expression of other genes including VEGF and MMP-13 that are important in tumorigenesis and angiogenesis.

5.2. RUNX Proteins as Transcriptional Activators

The mechanisms by which Runx transcription factors can activate gene expression involve the recruitment of coactivator proteins such as p300/CBP and CBFß that do not interact with DNA or DNA-binding proteins such as Ets-1 that also interact with Runx to mediate transactivation [7]. RUNX proteins contain an autoinhibitory domain that masks the DNA-binding Runt domain. One of the functions of the CBFß cofactor is to enhance RUNX DNA-binding by interacting with the Runt domain and displacing the autoinhibitory domain. Similarly, RUNX1 and Ets-1 contain autoinhibitory domains that limit DNA binding of each factor to their respective target sequences [119]. However, interaction of RUNX1 and Ets-1 autoinhibitory domains leads to exposure of their DNA-binding domains. Subsequent transactivation of gene expression has been shown for T-cell receptor (TCR-alpha, TCR-beta) and murine leukemia virus core promoters [119-121]. Interestingly, the Ets-1 factor also plays a central role in the regulation of neovascularization during tumorigenesis [16, 122]. In fact, expression of an Ets-1 dominant-negative construct *in vivo* inhibited FGF-2-induced angiogenesis while expression of dominant negative Ets-1 in tumor and stromal cells inhibited tumor-induced angiogenesis [123].

Runx2 binding sites were recently identified between two androgen receptor sites in the promoter of the mouse sex-limited protein (Slp) gene [124]. Gel shift assays demonstrated that Runx2 bound to this promoter and GST pull down analyses showed that Runx2 interacts with androgen receptor (AR) and glucocorticoid receptor. These data suggest that Runx2 can be converted into an activator of gene expression via an AR-dependent mechanism. Since AR continues to be expressed in hormone-independent prostate tumors and *in vitro* studies show that AR and RUNX2 can be activated in prostate tumor cell lines with growth factors in a hormone-independent manner [60, 125], the cooperation of Runx2 and AR to activate gene expression may be an event that underlies androgen independent gene expression and prostate cancer cell growth. Recent studies also implicate the estrogen receptor (ER) in the stimulation of Runx2-dependent gene transcription [126]. Runx2 was found to interact with ER to increase Runx2 transcriptional activity, which was independent of changes in Runx2 protein or DNA binding.

Another class of important cofactors that interact with each of the RUNX proteins is the TGFß-activated Smad family of transcriptional activators [61, 73]. Mutations in the TGFß pathway are also highly implicated in tumorigenesis (TGFR, Smad4). TGFß is a multi-functional cytokine that either stimulates or inhibits angiogenesis [127, 128] and inhibits EC growth through activation of the Smad signaling pathway [73, 29-132]. Chromosomal translocations in leukemia give rise to a fusion protein consisting of AML1 (RUNX1) and the eight-twenty-one co-repressor (AML1-ETO) (See section 3.1). Interestingly, AML1-ETO interacts with Smads, but fails to activate transcriptional responses upon TGFß treatment [133], suggesting that possible coactivator interactions with the C-terminal domain of RUNX1 are important in transcriptional activation. The IgC-alpha promoter contains both Smad and Runx binding sites and has been used as a model to study the mechanisms of promoter activation by these factors [61]. RUNX3 and Smad respond to TGFß treatment by cooperative binding to the IgC-alpha promoter elements. Recently, the normal function of RUNX3 in gut epithelial cells was found to be the suppression of uncontrolled growth during development [3, 134]. Cells from mice in which the *RUNX3* gene was inactivated failed to exhibit growth suppression or apoptosis in response to TGFß. It will be interesting to see which target gene promoters are regulated by Smads and RUNX3 in these cells.

Expression of the extracellular matrix protein BSP correlates with the presence of skeletal metastases in breast cancer [54]. Regulation of BSP expression by RUNX genes appeaed to take place at the level of the BSP promoter and occured in a context-dependent fashion [135]. Interestingly, most of the distal BSP promoter sequences were found to repress the *BSP* gene, while a proximal Runx binding site was important in activation of promoter activity [53]. This activation was found to occur in response to the MRIPV, but not the MASN isoform of RUNX2. RUNX2 DNA-binding activity was detected in metastatic breast tumor cells, but not in normal human mammary epithelial cells. In similar studies, RUNX2 expression was observed in epithelial cells derived from normal mammary glands, in non-metastatic tumor cells, and in primary mammary epithelial cells, where it was found to regulate expression of the osteoblast gene, osteopontin [58]. Osteopontin is important in mammary gland development [136-138] and EC migration [139]. RUNX2 was found to bind to a consensus binding site in the osteopontin promoter, which regulated expression of a reporter gene in mammary epithelial cells. Dominant-negative RUNX proteins inhibited activation of the promoter gene and expression of endogenous osteopontin. Since RUNX2 appears to regulate osteopontin expression in normal mammary epithelial cells, the results suggest a possible role for RUNX2 in mammary gland development.

The peroxisome proliferator-activated receptor (PPAR) family of genes are important regulators of transformed cell growth in a variety of tumors, including colon, prostate, and breast cancers [140]. Targeted PPAR-gamma ligands express anti-proliferative activity against many transformed cells and have been used in adjuvant and chemoprevention protocols [140, 141]. PPAR-gamma activation is known to suppress osteoblast differentiation and osteocalcin expression [103, 142]. Recent studies found that PPAR-gamma activation can inhibit RUNX2-mediated transcription of the osteocalcin gene in osteoblasts [143]. A PPAR-gamma activator ligand inhibited RUNX2 and osteocalcin mRNA production. Activation of PPAR-gamma also directly inhibited osteocalcin promoter-reporter gene expression, a process that was mediated by a physical interaction between RUNX2 and PPAR-gamma. In light of reports showing that RUNX2 is expressed in breast and prostate tumors that metastasize to the bone, these studies may suggest some important connections between RUNX2 and PPAR-gamma in tumor regulation.

5.3. RUNX Proteins as Transcriptional Repressors

In contrast to activation of gene expression, Runx-mediated gene repression involves the recruitment and direct interaction with corepresssor proteins including mSin3a, histone deacetylases (HDAC) and the TLE1/Groucho factor through conserved motifs within specific domains of the Runx protein [134]. Although the majority of RUNX target genes are activated, the transcriptional repression of some target genes suggests that RUNX factors can act as both activators and repressors of transcription [8]. Target genes in hematopoietic, skeletal, and fibroblast cells that are repressed by RUNX proteins include the leukocyte antigen CD53 [144], the multi-drug resistance, MDR1 gene [135], several primitive myeloid cell differentiation genes [144], BSP [2, 135], and the cyclin-dependent kinase inhibitor $p21^{CIP1}$ [145, 146].

RUNX1 (AML1) was found to repress the $p21^{CIP1}$ native promoter even in the absence of the Groucho corepressor binding domain (WRPY). A region 5' to a second consensus Runx binding site on the promoter was necessary to mediate promoter repression. It was found that RUNX1 could interact with mSin3A, a member of the mSin3 family of transcriptional corepressors that often act in cooperation with nuclear hormone corepressors N-CoR and SMRT and histone deacetylases [147]. Deletion mutants of RUNX1 defined the mSin3A interaction site to a region near the C-terminus, but distinct from the Groucho/TLE interaction domain. Endogenous RUNX1 associated with endogenous mSin3A. In addition, transcriptional repression was reduced after treatment with histone deacetylase inhibitors, although a direct association of RUNX1 and histone deacetylase-1 and –2 was not demonstrated. The results suggest that recruitment of corepressors such as mSin3A may represent a common pathway for RUNX1 and the RUNX1 chromosomal translocation fusion genes that contribute to leukemogenesis.

RUNX2 repression mechanisms appear to be the result of multiple corepressor interactions [148]. RUNX2 interacts with mSin3A and TLE/Groucho, although the WRPY (TLE-binding) domain is not necessary for repression of engrailed, $p21^{CIP1}$, or bone sialoprotein gene promoters by RUNX2. The class II family of histone deacetylases (HDACs), which include HDAC6, has been shown to be important in transcriptional repression and HDAC6 was found to mediate $p21^{CIP1}$ promoter repression through a C-terminal domain distinct from the TLE-binding C-terminal domain [146]. The C-terminal domain alone could repress a heterologous Gal-promoter reporter and repression was abrogated by treatment with the HDAC inhibitor trichostatin A (inhibits HDAC6) but not trapoxin B (does not inhibit HDAC6). RUNX2 interaction with HDAC6 was localized to a C-terminal region near the nuclear matrix targeting signal. In addition, RUNX2 appears to exhibit strong repression activity in response to the monocyte leukemia zinc finger protein homologue, MORF, in a Gal-reporter luciferase assay in HEK293 cells [149].

In conclusion, it appears that RUNX proteins function as both transcriptional activators and repressors of gene expression. Target gene promoters and other DNA-binding or non-DNA-binding transcriptional regulators that are expressed by specific cells determine the context-dependent expression of target genes. RUNX transcriptional complexes will, therefore, be assembled on enhanceosomal or repressosomal molecular structures to mediate specific transcriptional activity.

6. TARGETING STRATEGIES AND THERAPEUTIC OPTIONS

Cancer cells acquire a variety of properties that regulate tumor progression and metastasis [150]. Runx genes may be involved in mediating the ability of tumor cells to acquire growth factor independence, become insensitive to anti-growth signals (such as TGFß), acquire invasive properties and become metastatic, avoid apoptosis, or sustain angiogenesis. However, therapeutic options may differ for each of the Runx genes. Whereas it is logical to inhibit the function of oncogenic proteins, the reduction in activity of suppressor proteins needs to be prevented or the genes need to be re-expressed. Runx1 translocation fusion genes appear to be oncogenic in the context of their transcriptional repression of certain differentiation-specific genes. Therefore, strategies to inhibit transcriptional repression (such as the use of HDAC inhibitors) may be a good strategy to inactivate these RUNX oncogenes [10]. In addition, since several specific corepressors, such as mSin3A interact with RUNX1, inhibitors of these corepressors may also be appropriate targets to treat acute leukemias [145]. Another possibility lies in the development of small molecule inhibitors that may interact specifically with the new epitopes created by the fusion of RUNX1 with transcriptional repression domains of ETO or TEL. Since CBFß is a necessary cofactor in RUNX1 DNA-binding, identification of molecules that interfere with CBFß/RUNX1 interactions may also be possible.

RUNX2 re-expression in solid tumors and angiogenesis appears to be related to growth-promoting functions [16, 53]. Therefore, dominant negative, RNA interference, or anti-sense approaches to inhibit RUNX2 expression have been suggested to inhibit tumorigenesis. Some of the RUNX2 target genes in tumors and EC regulate extracellular matrix (osteopontin, BSP, osteocalcin) content or protease expression (MMP13, MT_1MMP, uPA). Targeting of these gene products to reduce tumor or EC migration or invasion has been an established approach to treat a variety of tumors [150]. However, because of the many possible redundant target genes, this strategy may not be very effective. Another approach consists in activating or inhibiting the function of RUNX2-interacting proteins such as PPAR-gamma to alter gene expression in specific tumors [143]. PPAR-gamma is an attractive target because of the existence of many activator ligands that are already in clinical trials [140, 141]. As described for RUNX1, targeting of specific RUNX2 coactivators or corepressors may also be another strategy to inhibit RUNX2 transcriptional activation or repression, respectively.

The RUNX3 gene is inactivated in gastric carcinoma and thus acts as a putative suppressor gene [3]. Strategies to increase expression of its suppressor function by treatment with differentiation-inducing agents such as retinoids or vitamin D3 would, therefore, be reasonable [151]. In addition, demethylation strategies to re-express the normal gene or the use of HDAC inhibitors to reduce repression of the RUNX3 gene promoter are both plausible approaches [3].

Ultimately, the ability to develop therapeutic strategies for RUNX gene inhibition will depend on the existence of appropriate models to evaluate possible compounds. Since mice in which the Runx genes have been inactivated die in utero or soon after birth, many tumor or angiogenesis-related mechanisms cannot be tested. However, strategies to rescue Runx gene functions [37] or to introduce specific Runx mutant genes into hematopoietic stem cells [152] or conditionally into embryonic stem cells [153] have been developed. Introduction of the human t(8:21) AML1-ETO gene by retroviral expression in mouse hematopoietic stem cells was used to reconstitute animals expressing AML1-ETO and to recapitulate the developmental abnormalities seen in the human bone marrow [152]. Since these animals developed multiple hematopoietic abnormalities, but not leukemia, additional mutations would need to be engineered to mimic the human disease. Another approach used the cre-lox system to activate expression of AML1-ETO after birth. Cre-mediated deletion of a transcriptional stop cassette avoided embryonic lethality, which results from constitutive expression of AML1-ETO [153]. In this model, enhanced replating efficiency of myeloid progenitor cells was observed, but not leukemia. Induction of cooperating mutations, however, did result in the development of acute myeloid disease that expressed features similar to human leukemia. For RUNX2, transgenic mice have been generated that express RUNX2 in T-cells, leading to T-cell lymphomas in cooperation with c-myc, pim1, and p53 [49]. Similar strategies for RUNX3 heterozygote knockout mice or conditional inactivation of RUNX3 may prove to be useful models to test anti-tumor agents to treat gastric cancers [3].

7. CONCLUSIONS

Research into the role that the Runx genes play in development has provided valuable insight into hematopoietic and skeletal differentiation. Some of these studies identified a clear relationship between disruption of normal Runx gene expression and hematopoietic malignancies and identified chromosomal translocations and inversions specific to the RUNX1 gene that contribute to human leukemias [10]. Understanding the function of RUNX2 and RUNX3 in cancer development has only recently been possible and has resulted from several cumulative studies implicating RUNX2 in hematopoietic and solid tumors and RUNX3 in gastric carcinoma [3]. Related to these developments has been the renewed interest in understanding how vascular development contributes to tumorigenesis and how RUNX genes may play an important role in this process [16, 154]. However, many of these studies are in the early stages of research and much more needs to be done to unravel the mechanisms through which the RUNX genes regulate angiogenesis and solid tumor development. Future studies will most likely focus on specific pathways to identify molecular targets for therapeutic purposes. In general, development of murine models should generate powerful tools to identify the mechanisms that regulate tumorigenesis by RUNX family members and their interacting proteins. In addition, mutations that cooperate with RUNX genes can also be modeled in these systems. These models could have great utility in assessing novel therapeutic strategies that target the RUNX genes, proteins, and components of their downstream signaling pathways.

Note on nomenclature and citations: In this review, RUNX refers to the human proteins whereas Runx refers to both human and/or rodent proteins. Where possible, the most current publications related to *Runx* genes in cancer are cited, but because of space limitations, some references to Runx target genes have not been included.

Ian Anglin and Antonino Passaniti
University of Maryland at Baltimore, Greenebaum Cancer Center, Baltimore, Maryland

8. REFERENCES

1. Miyoshi, H. et al. t(8;21) breakpoints on chromosome 21 in acute myeloid leukemia are clustered within a limited region of a single gene, AML1. *Proc Natl Acad Sci U S A* 88, 10431-4 (1991).
2. 2.Ducy, P., Zhang, R., Geoffroy, V., Ridall, A.L. & Karsenty, G. Osf2/Cbfa1: a transcriptional activator of osteoblast differentiation [see comments]. *Cell* 89, 747-54 (1997).
3. Li, Q.L. et al. Causal relationship between the loss of RUNX3 expression and gastric cancer. *Cell* 109, 113-124 (2002).
4. Canon, J. & Banerjee, U. Runt and Lozenge function in Drosophila development. *Semin Cell Dev Biol* 11, 327-36 (2000).

208

5. Wang, Q. et al. The CBFbeta subunit is essential for CBFalpha2 (AML1) function in vivo. *Cell* 87, 697-708 (1996).
6. Huang, G. et al. Dimerization with PEBP2beta protects RUNX1/AML1 from ubiquitin-proteasome-mediated degradation. *Embo J* 20, 723-33 (2001).
7. Wheeler, J.C., Shigesada, K., Gergen, J.P. & Ito, Y. Mechanisms of transcriptional regulation by Runt domain proteins. *Semin Cell Dev Biol* 11, 369-75 (2000).
8. Canon, J. & Banerjee, U. In vivo analysis of a developmental circuit for direct transcriptional activation and repression in the same cell by a Runx protein. *Genes Dev* 17, 838-43 (2003).
9. Westendorf, J.J. & Hiebert, S.W. Mammalian runt-domain proteins and their roles in hematopoiesis, osteogenesis, and leukemia. *J Cell Biochem* 32/33, 51-8 (1999).
10. Speck, N.A. & Gilliland, D.G. Core-binding factors in haematopoiesis and leukaemia. *Nat Rev Cancer* 2, 502-13 (2002).
11. Namba, K. et al. Indispensable role of the transcription factor PEBP2/CBF in angiogenic activity of a murine endothelial cell MSS31. *Oncogene* 19, 106-14 (2000).
12. Sun, L., Vitolo, M. & Passaniti, A. Runt-related Gene 2 in Endothelial Cells: Inducible Expression and Specific Regulation of Cell Migration and Invasion. *Cancer Res* 61, 4994-5001. (2001).
13. Vaillant, F. et al. A full-length Cbfa1 gene product perturbs T-cell development and promotes lymphomagenesis in synergy with myc. *Oncogene* 18, 7124-34 (1999).
14. Cohen, M.M., Jr. RUNX genes, neoplasia, and cleidocranial dysplasia. *Am J Med Genet* 104, 185-8 (2001).
15. Karsenty, G. Minireview: transcriptional control of osteoblast differentiation. *Endocrinology* 142, 2731-3 (2001).
16. Sato, Y. Molecular mechanism of angiogenesis transcription factors and their therapeutic relevance. *Pharmacol Ther* 87, 51-60 (2000).
17. Otto, F., Lubbert, M. & Stock, M. Upstream and downstream targets of RUNX proteins. *J Cell Biochem* 89, 9-18 (2003).
18. Okuda, T., van Deursen, J., Hiebert, S.W., Grosveld, G. & Downing, J.R. AML1, the target of multiple chromosomal translocations in human leukemia, is essential for normal fetal liver hematopoiesis. *Cell* 84, 321-30 (1996).
19. Tracey, W.D. & Speck, N.A. Potential roles for RUNX1 and its orthologs in determining hematopoietic cell fate. *Semin Cell Dev Biol* 11, 337-42 (2000).
20. Takakura, N. et al. A role for hematopoietic stem cells in promoting angiogenesis. *Cell* 102, 199-209. (2000).
21. Ho, C.Y. et al. Linkage of a familial platelet disorder with a propensity to develop myeloid malignancies to human chromosome 21q22.1-22.2. *Blood* 87, 5218-24 (1996).
22. Yancopoulos, G.D. et al. Vascular-specific growth factors and blood vessel formation. *Nature* 407, 242-8 (2000).
23. Rafii, S. Circulating endothelial precursors: mystery, reality, and promise. *J Clin Invest* 105, 17-9. (2000).
24. Satake, M. et al. Expression of the Runt domain-encoding PEBP2 alpha genes in T cells during thymic development. *Mol Cell Biol* 15, 1662-70 (1995).
25. Ducy, P. & Karsenty, G. Two distinct osteoblast-specific cis-acting elements control expression of a mouse osteocalcin gene. *Mol Cell Biol* 15, 1858-69 (1995).
26. Kern, B., Shen, J., Starbuck, M. & Karsenty, G. Cbfa1 contributes to the osteoblast-specific expression of type I collagen genes. *J Biol Chem* 276, 7101-7 (2001).
27. Zheng, Q. et al. Type X collagen gene regulation by Runx2 contributes directly to its hypertrophic chondrocyte-specific expression in vivo. *J Cell Biol* 162, 833-42 (2003).
28. Zelzer, E. et al. Tissue specific regulation of VEGF expression during bone development requires Cbfa1/Runx2. *Mech Dev* 106, 97-106 (2001).

29. Komori, T. et al. Targeted disruption of Cbfa1 results in a complete lack of bone formation owing to maturational arrest of osteoblasts [see comments]. *Cell* 89, 755-64 (1997).
30. Otto, F. et al. Cbfa1, a candidate gene for cleidocranial dysplasia syndrome, is essential for osteoblast differentiation and bone development [see comments]. *Cell* 89, 765-71 (1997).
31. Lee, B. et al. Missense mutations abolishing DNA binding of the osteoblast-specific transcription factor OSF2/CBFA1 in cleidocranial dysplasia. *Nat Genet* 16, 307-10 (1997).
32. Ezaki, T. et al. Deletion mapping on chromosome 1p in well-differentiated gastric cancer. *Br J Cancer* 73, 424-8 (1996).
33. Nam, S. et al. Expression pattern, regulation, and biological role of runt domain transcription factor, run, in Caenorhabditis elegans. *Mol Cell Biol* 22, 547-54 (2002).
34. Lund, A.H. & Van Lohuizen, M. RUNX: A trilogy of cancer genes. *Cancer Cell* 1, 213-215 (2002).
35. Sasaki, K. et al. Absence of fetal liver hematopoiesis in mice deficient in transcriptional coactivator core binding factor beta. *Proc Natl Acad Sci U S A* 93, 12359-63 (1996).
36. Bravo, J., Li, Z., Speck, N.A. & Warren, A.J. The leukemia-associated AML1 (Runx1)-CBFbeta complex functions as a DNA-induced molecular clamp. *Nat Struct Biol* 8, 371-378. (2001).
37. Yoshida, C.A. et al. Core-binding factor beta interacts with Runx2 and is required for skeletal development. *Nat Genet* 32, 633-8 (2002).
38. Hanahan, D. Signaling vascular morphogenesis and maintenance [comment]. *Science* 277, 48-50 (1997).
39. Baltzinger, M., Mager-Heckel, A.M. & Remy, P. Xl erg: expression pattern and overexpression during development plead for a role in endothelial cell differentiation. *Dev Dyn* 216, 420-33 (1999).
40. Lutterbach, B. & Hiebert, S.W. Role of the transcription factor AML-1 in acute leukemia and hematopoietic differentiation. *Gene* 245, 223-35 (2000).
41. Meyers, S., Lenny, N. & Hiebert, S.W. The t(8;21) fusion protein interferes with AML-1B-dependent transcriptional activation. *Mol Cell Biol* 15, 1974-82 (1995).
42. Hiebert, S.W. et al. The t(12;21) translocation converts AML-1B from an activator to a repressor of transcription. *Mol Cell Biol* 16, 1349-55 (1996).
43. Lutterbach, B., Hou, Y., Durst, K.L. & Hiebert, S.W. The inv(16) encodes an acute myeloid leukemia 1 transcriptional corepressor. *Proc Natl Acad Sci U S A* 96, 12822-7 (1999).
44. Castilla, L.H. et al. Failure of embryonic hematopoiesis and lethal hemorrhages in mouse embryos heterozygous for a knocked-in leukemia gene CBFB-MYH11. *Cell* 87, 687-96 (1996).
45. Lutterbach, B. et al. ETO, a target of t(8;21) in acute leukemia, interacts with the N-CoR and mSin3 corepressors. *Mol Cell Biol* 18, 7176-84 (1998).
46. Gelmetti, V. et al. Aberrant recruitment of the nuclear receptor corepressor-histone deacetylase complex by the acute myeloid leukemia fusion partner ETO. *Mol Cell Biol* 18, 7185-91 (1998).
47. Fenrick, R. et al. Both TEL and AML-1 contribute repression domains to the t(12;21) fusion protein. *Mol Cell Biol* 19, 6566-74 (1999).
48. Westendorf, J.J. et al. The t(8;21) fusion product, AML-1-ETO, associates with C/EBP-alpha, inhibits C/EBP-alpha-dependent transcription, and blocks granulocytic differentiation. *Mol Cell Biol* 18, 322-33 (1998).
49. Blyth, K. et al. Runx2: a novel oncogenic effector revealed by in vivo complementation and retroviral tagging. *Oncogene* 20, 295-302 (2001).

50. Stewart, M. et al. Proviral insertions induce the expression of bone-specific isoforms of PEBP2alphaA (CBFA1): evidence for a new myc collaborating oncogene. *Proc Natl Acad Sci U S A* 94, 8646-51 (1997).
51. Perry, C. et al. Complex regulation of acetylcholinesterase gene expression in human brain tumors. *Oncogene* 21, 8428-41 (2002).
52. Kurokawa, M. et al. A conserved cysteine residue in the runt homology domain of AML1 is required for the DNA binding ability and the transforming activity on fibroblasts. *J Biol Chem* 271, 16870-6 (1996).
53. Barnes, G.L. et al. Osteoblast-related Transcription Factors Runx2 (Cbfa1/AML3) and MSX2 Mediate the Expression of Bone Sialoprotein in Human Metastatic Breast Cancer Cells. *Cancer Res* 63, 2631-7 (2003).
54. Waltregny, D. et al. Increased expression of bone sialoprotein in bone metastases compared with visceral metastases in human breast and prostate cancers. *J Bone Miner Res* 15, 834-43 (2000).
55. Riminucci, M. et al. Coexpression of Bone Sialoprotein (BSP) and the Pivotal Transcriptional Regulator of Osteogenesis, Cbfa1/Runx2, in Malignant Melanoma. *Calcif Tissue Int* (2003).
56. Jimenez, M.J. et al. A regulatory cascade involving retinoic acid, Cbfa1, and matrix metalloproteinases is coupled to the development of a process of perichondrial invasion and osteogenic differentiation during bone formation. *J Cell Biol* 155, 1333-44 (2001).
57. Selvamurugan, N., Pulumati, M.R., Tyson, D.R. & Partridge, N.C. Parathyroid hormone regulation of the rat collagenase-3 promoter by protein kinase A-dependent transactivation of core binding factor alpha1. *J Biol Chem* 275, 5037-42 (2000).
58. Inman, C.K. & Shore, P. The osteoblast transcription factor Runx2 is expressed in mammary epithelial cells and mediates osteopontin expression. *J Biol Chem* (2003).
59. Brubaker, K.D., Vessella, R.L., Brown, L.G. & Corey, E. Prostate cancer expression of runt-domain transcription factor Runx2, a key regulator of osteoblast differentiation and function. *Prostate* 56, 13-22 (2003).
60. Yeung, F. et al. Regulation of human osteocalcin promoter in hormone-independent human prostate cancer cells. *J Biol Chem* 277, 2468-76 (2002).
61. Hanai, J. et al. Interaction and functional cooperation of PEBP2/CBF with Smads. Synergistic induction of the immunoglobulin germline Calpha promoter. *J Biol Chem* 274, 31577-82 (1999).
62. Hanahan, D. & Folkman, J. Patterns and emerging mechanisms of the angiogenic switch during tumorigenesis. *Cell* 86, 353-64. (1996).
63. Bergers, G. & Hanahan, D. Combining antiangiogenic agents with metronomic chemotherapy enhances efficacy against late-stage pancreatic islet carcinomas in mice. *Cold Spring Harb Symp Quant Biol* 67, 293-300 (2002).
64. Gorski, D.H. & Walsh, K. The role of homeobox genes in vascular remodeling and angiogenesis. *Circ Res* 87, 865-72 (2000).
65. Maisonpierre, P.C. et al. Angiopoietin-2, a natural antagonist for Tie2 that disrupts in vivo angiogenesis [see comments]. *Science* 277, 55-60 (1997).
66. Jimenez, M.J. et al. Collagenase 3 is a target of Cbfa1, a transcription factor of the runt gene family involved in bone formation. *Mol Cell Biol* 19, 4431-42 (1999).
67. Chang, F. et al. Signal transduction mediated by the Ras/Raf/MEK/ERK pathway from cytokine receptors to transcription factors: potential targeting for therapeutic intervention. *Leukemia* 17, 1263-93 (2003).
68. Tanaka, K. et al. Increased expression of AML1 during retinoic-acid-induced differentiation of U937 cells. *Biochem Biophys Res Commun* 211, 1023-30 (1995).
69. Le, X.F. et al. Regulation of AML2/CBFA3 in hematopoietic cells through the retinoic acid receptor alpha-dependent signaling pathway. *J Biol Chem* 274, 21651-8 (1999).

70. Prince, M. et al. Expression and regulation of Runx2/Cbfa1 and osteoblast phenotypic markers during the growth and differentiation of human osteoblasts. *J Cell Biochem* 80, 424-40 (2001).

71. Tou, L., Quibria, N. & Alexander, J.M. Regulation of human cbfa1 gene transcription in osteoblasts by selective estrogen receptor modulators (SERMs). *Mol Cell Endocrinol* 183, 71-9 (2001).

72. Franceschi, R.T. & Xiao, G. Regulation of the osteoblast-specific transcription factor, Runx2: Responsiveness to multiple signal transduction pathways. *J Cell Biochem* 88, 446-54 (2003).

73. Massague, J. & Wotton, D. Transcriptional control by the TGFß/Smad signaling system. *EMBO J* 19, 1745-1754 (2000).

74. Tsuji, K., Ito, Y. & Noda, M. Expression of the PEBP2alphaA/AML3/CBFA1 gene is regulated by BMP4/7 heterodimer and its overexpression suppresses type I collagen and osteocalcin gene expression in osteoblastic and nonosteoblastic mesenchymal cells. *Bone* 22, 87-92 (1998).

75. Tou, L., Quibria, N. & Alexander, J.M. Transcriptional regulation of the human Runx2/Cbfa1 gene promoter by bone morphogenetic protein-7. *Mol Cell Endocrinol* 205, 121-9 (2003).

76. Banerjee, C. et al. Differential regulation of the two principal Runx2/Cbfa1 n-terminal isoforms in response to bone morphogenetic protein-2 during development of the osteoblast phenotype. *Endocrinology* 142, 4026-39 (2001).

77. Reddi, A.H., Roodman, D., Freeman, C. & Mohla, S. Mechanisms of tumor metastasis to the bone: challenges and opportunities. *J Bone Miner Res* 18, 190-4 (2003).

78. Lee, K.S. et al. Runx2 is a common target of transforming growth factor beta1 and bone morphogenetic protein 2, and cooperation between Runx2 and Smad5 induces osteoblast-specific gene expression in the pluripotent mesenchymal precursor cell line C2C12. *Mol Cell Biol* 20, 8783-92 (2000).

79. Alliston, T., Choy, L., Ducy, P., Karsenty, G. & Derynck, R. TGF-beta-induced repression of CBFA1 by Smad3 decreases cbfa1 and osteocalcin expression and inhibits osteoblast differentiation. *Embo J* 20, 2254-72 (2001).

80. Xiao, G., Cui, Y., Ducy, P., Karsenty, G. & Franceschi, R.T. Ascorbic acid-dependent activation of the osteocalcin promoter in MC3T3-E1 preosteoblasts: requirement for collagen matrix synthesis and the presence of an intact OSE2 sequence. *Mol Endocrinol* 11, 1103-13 (1997).

81. Xiao, G., Wang, D., Benson, M.D., Karsenty, G. & Franceschi, R.T. Role of the alpha2-integrin in osteoblast-specific gene expression and activation of the Osf2 transcription factor. *J Biol Chem* 273, 32988-94 (1998).

82. Xiao, G. et al. MAPK pathways activate and phosphorylate the osteoblast-specific transcription factor, Cbfa1. *J Biol Chem* 275, 4453-9 (2000).

83. Wee, H.J., Huang, G., Shigesada, K. & Ito, Y. Serine phosphorylation of RUNX2 with novel potential functions as negative regulatory mechanisms. *EMBO Rep* 3, 967-74 (2002).

84. Liacini, A. et al. Induction of matrix metalloproteinase-13 gene expression by TNF-alpha is mediated by MAP kinases, AP-1, and NF-kappaB transcription factors in articular chondrocytes. *Exp Cell Res* 288, 208-17 (2003).

85. Mengshol, J.A., Vincenti, M.P., Coon, C.I., Barchowsky, A. & Brinckerhoff, C.E. Interleukin-1 induction of collagenase 3 (matrix metalloproteinase 13) gene expression in chondrocytes requires p38, c-Jun N-terminal kinase, and nuclear factor kappaB: differential regulation of collagenase 1 and collagenase 3. *Arthritis Rheum* 43, 801-11 (2000).

212

86. LaVallee, T.M., Prudovsky, I.A., McMahon, G.A., Hu, X. & Maciag, T. Activation of the MAP kinase pathway by FGF-1 correlates with cell proliferation induction while activation of the Src pathway correlates with migration. *J Cell Biol* 141, 1647-58 (1998).
87. Folkman, J. Angiogenesis in cancer, vascular, rheumatoid and other disease. *Nat Med* 1, 27-31 (1995).
88. Friedlander, M. et al. Definition of two angiogenic pathways by distinct alpha v integrins. *Science* 270, 1500-2 (1995).
89. Coffin, J.D. et al. Abnormal bone growth and selective translational regulation in basic fibroblast growth factor (FGF-2) transgenic mice. *Mol Biol Cell* 6, 1861-73 (1995).
90. Liang, H., Pun, S. & Wronski, T.J. Bone anabolic effects of basic fibroblast growth factor in ovariectomized rats. *Endocrinology* 140, 5780-8 (1999).
91. Boudreaux, J.M. & Towler, D.A. Synergistic induction of osteocalcin gene expression: identification of a bipartite element conferring fibroblast growth factor 2 and cyclic AMP responsiveness in the rat osteocalcin promoter. *J Biol Chem* 271, 7508-15 (1996).
92. Xiao, G., Jiang, D., Gopalakrishnan, R. & Franceschi, R.T. Fibroblast growth factor 2 induction of the osteocalcin gene requires MAPK activity and phosphorylation of the osteoblast transcription factor, Cbfa1/Runx2. *J Biol Chem* 277, 36181-7 (2002).
93. Selvamurugan, N., Chou, W.Y., Pearman, A.T., Pulumati, M.R. & Partridge, N.C. Parathyroid hormone regulates the rat collagenase-3 promoter in osteoblastic cells through the cooperative interaction of the activator protein-1 site and the runt domain binding sequence. *J Biol Chem* 273, 10647-57 (1998).
94. Karaplis, A.C. & Goltzman, D. PTH and PTHrP effects on the skeleton. *Rev Endocr Metab Disord* 1, 331-41 (2000).
95. Hess, J., Porte, D., Munz, C. & Angel, P. AP-1 and Cbfa/runt physically interact and regulate parathyroid hormone-dependent MMP13 expression in osteoblasts through a new osteoblast-specific element 2/AP-1 composite element. *J Biol Chem* 276, 20029-38 (2001).
96. D'Alonzo, R.C., Selvamurugan, N., Karsenty, G. & Partridge, N.C. Physical interaction of the activator protein-1 factors c-Fos and c-Jun with Cbfa1 for collagenase-3 promoter activation. *J Biol Chem* 277, 816-22 (2002).
97. Brooks, P.C. et al. Insulin-like growth factor receptor cooperates with integrin alpha v beta 5 to promote tumor cell dissemination in vivo. *J Clin Invest* 99, 1390-8 (1997).
98. Drissi, H. et al. Transcriptional autoregulation of the bone related CBFA1/RUNX2 gene. *J Cell Physiol* 184, 341-50 (2000).
99. Ghozi, M.C., Bernstein, Y., Negreanu, V., Levanon, D. & Groner, Y. Expression of the human acute myeloid leukemia gene AML1 is regulated by two promoter regions. *Proc Natl Acad Sci U S A* 93, 1935-40 (1996).
100. Bangsow, C. et al. The RUNX3 gene--sequence, structure and regulated expression. *Gene* 279, 221-32 (2001).
101. Zambotti, A., Makhluf, H., Shen, J. & Ducy, P. Characterization of an osteoblast-specific enhancer element in the CBFA1 gene. *J Biol Chem* 277, 41497-506 (2002).
102. Kanzler, B., Kuschert, S.J., Liu, Y.H. & Mallo, M. Hoxa-2 restricts the chondrogenic domain and inhibits bone formation during development of the branchial area. *Development* 125, 2587-97 (1998).
103. Lecka-Czernik, B. et al. Inhibition of Osf2/Cbfa1 expression and terminal osteoblast differentiation by PPARgamma2. *J Cell Biochem* 74, 357-71 (1999).
104. Tribioli, C. & Lufkin, T. The murine Bapx1 homeobox gene plays a critical role in embryonic development of the axial skeleton and spleen. *Development* 126, 5699-711 (1999).
105. Satokata, I. et al. Msx2 deficiency in mice causes pleiotropic defects in bone growth and ectodermal organ formation. *Nat Genet* 24, 391-5 (2000).

106. Drissi, H. et al. 1,25-(OH)2-vitamin D3 suppresses the bone-related Runx2/Cbfa1 gene promoter. *Exp Cell Res* 274, 323-33 (2002).

107. Tintut, Y., Parhami, F., Le, V., Karsenty, G. & Demer, L.L. Inhibition of osteoblast-specific transcription factor Cbfa1 by the cAMP pathway in osteoblastic cells. Ubiquitin/proteasome-dependent regulation. *J Biol Chem* 274, 28875-9 (1999).

108. Calo, V. et al. STAT proteins: from normal control of cellular events to tumorigenesis. *J Cell Physiol* 197, 157-68 (2003).

109. Takayanagi, H. et al. T-cell-mediated regulation of osteoclastogenesis by signalling cross-talk between RANKL and IFN-gamma. *Nature* 408, 600-5 (2000).

110. Kim, S. et al. Stat1 functions as a cytoplasmic attenuator of Runx2 in the transcriptional program of osteoblast differentiation. *Genes Dev* 17, 1979-91 (2003).

111. Ruegg, C., Dormond, O. & Foletti, A. Suppression of tumor angiogenesis through the inhibition of integrin function and signaling in endothelial cells: which side to target? *Endothelium* 9, 151-60 (2002).

112. Aggarwal, B.B. Signalling pathways of the TNF superfamily: a double-edged sword. *Nat Rev Immunol* 3, 745-56 (2003).

113. Gilbert, L. et al. Expression of the osteoblast differentiation factor RUNX2 (Cbfa1/AML3/Pebp2alpha A) is inhibited by tumor necrosis factor-alpha. *J Biol Chem* 277, 2695-701 (2002).

114. Stein, G.S. et al. Temporal and spatial parameters of skeletal gene expression: targeting RUNX factors and their coregulatory proteins to subnuclear domains. *Connect Tissue Res* 44 Suppl 1, 149-53 (2003).

115. Crosier, P.S. et al. Pathways in blood and vessel development revealed through zebrafish genetics. *Int J Dev Biol* 46, 493-502 (2002).

116. Selvamurugan, N. & Partridge, N.C. Constitutive expression and regulation of collagenase-3 in human breast cancer cells. *Mol Cell Biol Res Commun* 3, 218-23 (2000).

117. Strom, D.K. et al. Expression of the AML-1 oncogene shortens the G(1) phase of the cell cycle. *J Biol Chem* 275, 3438-45 (2000).

118. Linggi, B. et al. The t(8;21) fusion protein, AML1 ETO, specifically represses the transcription of the p14(ARF) tumor suppressor in acute myeloid leukemia. *Nat Med* 8, 743-50 (2002).

119. Kim, W.Y. et al. Mutual activation of Ets-1 and AML1 DNA binding by direct interaction of their autoinhibitory domains. *Embo J* 18, 1609-20 (1999).

120. Gu, T.L., Goetz, T.L., Graves, B.J. & Speck, N.A. Auto-inhibition and partner proteins, core-binding factor beta (CBFbeta) and Ets-1, modulate DNA binding by CBFalpha2 (AML1). *Mol Cell Biol* 20, 91-103 (2000).

121. Wotton, D., Ghysdael, J., Wang, S., Speck, N.A. & Owen, M.J. Cooperative binding of Ets-1 and core binding factor to DNA. *Mol Cell Biol* 14, 840-50 (1994).

122. Sato, Y. Transcription factor ETS-1 as a molecular target for angiogenesis inhibition. *Hum Cell* 11, 207-14 (1998).

123. Pourtier-Manzanedo, A. et al. Expression of an Ets-1 dominant-negative mutant perturbs normal and tumor angiogenesis in a mouse ear model. *Oncogene* 22, 1795-806 (2003).

124. Ning, Y.M. & Robins, D.M. AML3/CBFalpha1 is required for androgen-specific activation of the enhancer of the mouse sex-limited protein (Slp) gene. *J Biol Chem* 274, 30624-30 (1999).

125. Culig, Z. et al. Androgen receptor activation in prostatic tumor cell lines by insulin-like growth factor-I, keratinocyte growth factor, and epidermal growth factor. *Cancer Res* 54, 5474-8 (1994).

126. McCarthy, T.L., Chang, W.Z., Liu, Y. & Centrella, M. Runx2 integrates estrogen activity in osteoblasts. *J Biol Chem* (2003).

127. Risau, W. & Flamme, I. Vasculogenesis. *Annu Rev Cell Dev Biol* 11, 73-91 (1995).

128. Pepper, M.S., Belin, D., Montesano, R., Orci, L. & Vassalli, J.D. Transforming growth factor-beta 1 modulates basic fibroblast growth factor-induced proteolytic and angiogenic properties of endothelial cells in vitro. *J Cell Biol* 111, 743-55 (1990).

129. Koff, A., Ohtsuki, M., Polyak, K., Roberts, J.M. & Massague, J. Negative regulation of G1 in mammalian cells: inhibition of cyclin E-dependent kinase by TGF-beta. *Science* 260, 536-9 (1993).

130. Lyons, R.M. & Moses, H.L. Transforming growth factors and the regulation of cell proliferation. *Eur J Biochem* 187, 467-73 (1990).

131. Laiho, M., DeCaprio, J.A., Ludlow, J.W., Livingston, D.M. & Massague, J. Growth inhibition by TGF-beta linked to suppression of retinoblastoma protein phosphorylation. *Cell* 62, 175-85 (1990).

132. Newton, L.K., Yung, W.K., Pettigrew, L.C. & Steck, P.A. Growth regulatory activities of endothelial extracellular matrix: mediation by transforming growth factor-beta. *Exp Cell Res* 190, 127-32 (1990).

133. Jakubowiak, A. et al. Inhibition of the transforming growth factor beta 1 signaling pathway by the AML1/ETO leukemia-associated fusion protein. *J Biol Chem* 275, 40282-7 (2000).

134. Ito, Y. & Miyazono, K. RUNX transcription factors as key targets of TGF-beta superfamily signaling. *Curr Opin Genet Dev* 13, 43-7 (2003).

135. Javed, A. et al. runt homology domain transcription factors (Runx, Cbfa, and AML) mediate repression of the bone sialoprotein promoter: evidence for promoter context-dependent activity of Cbfa proteins. *Mol Cell Biol* 21, 2891-905 (2001).

136. Senger, D.R., Perruzzi, C.A. & Papadopoulos, A. Elevated expression of secreted phosphoprotein I (osteopontin, 2ar) as a consequence of neoplastic transformation. *Anticancer Res* 9, 1291-9 (1989).

137. Brown, L.F. et al. Expression and distribution of osteopontin in human tissues: widespread association with luminal epithelial surfaces. *Mol Biol Cell* 3, 1169-80 (1992).

138. Nemir, M. et al. Targeted inhibition of osteopontin expression in the mammary gland causes abnormal morphogenesis and lactation deficiency. *J Biol Chem* 275, 969-76 (2000).

139. Shijubo, N., Uede, T., Kon, S., Nagata, M. & Abe, S. Vascular endothelial growth factor and osteopontin in tumor biology. *Crit Rev Oncog* 11, 135-46 (2000).

140. Koeffler, H.P. Peroxisome proliferator-activated receptor gamma and cancers. *Clin Cancer Res* 9, 1-9 (2003).

141. Badawi, A.F. & Badr, M.Z. Chemoprevention of breast cancer by targeting cyclooxygenase-2 and peroxisome proliferator-activated receptor-gamma (Review). *Int J Oncol* 20, 1109-22 (2002).

142. Jackson, S.M. & Demer, L.L. Peroxisome proliferator-activated receptor activators modulate the osteoblastic maturation of MC3T3-E1 preosteoblasts. *FEBS Lett* 471, 119-24 (2000).

143. Jeon, M.J. et al. Activation of peroxisome proliferator-activated receptor-gamma inhibits the Runx2-mediated transcription of osteocalcin in osteoblasts. *J Biol Chem* 278, 23270-7 (2003).

144. Shimada, H. et al. Analysis of genes under the downstream control of the t(8;21) fusion protein AML1-MTG8: overexpression of the TIS11b (ERF-1, cMG1) gene induces myeloid cell proliferation in response to G-CSF. *Blood* 96, 655-63 (2000).

145. Lutterbach, B. et al. A mechanism of repression by acute myeloid leukemia-1, the target of multiple chromosomal translocations in acute leukemia. *J Biol Chem* 275, 651-6 (2000).

146. Westendorf, J.J. et al. Runx2 (Cbfa1, AML-3) interacts with histone deacetylase 6 and represses the p21(CIP1/WAF1) promoter. *Mol Cell Biol* 22, 7982-92 (2002).

147. Torchia, J., Glass, C. & Rosenfeld, M.G. Co-activators and co-repressors in the integration of transcriptional responses. *Curr Opin Cell Biol* 10, 373-83 (1998).

148. Aronson, B.D., Fisher, A.L., Blechman, K., Caudy, M. & Gergen, J.P. Groucho-dependent and -independent repression activities of Runt domain proteins. *Mol Cell Biol* 17, 5581-7 (1997).

149. Pelletier, N., Champagne, N., Stifani, S. & Yang, X.J. MOZ and MORF histone acetyltransferases interact with the Runt-domain transcription factor Runx2. *Oncogene* 21, 2729-40 (2002).

150. Hanahan, D. & Weinberg, R.A. The hallmarks of cancer. *Cell* 100, 57-70. (2000).

151. Balmain, A. Cancer: new-age tumour suppressors. *Nature* 417, 235-7 (2002).

152. de Guzman, C.G. et al. Hematopoietic stem cell expansion and distinct myeloid developmental abnormalities in a murine model of the AML1-ETO translocation. *Mol Cell Biol* 22, 5506-17 (2002).

153. Higuchi, M. et al. Expression of a conditional AML1-ETO oncogene bypasses embryonic lethality and establishes a murine model of human t(8;21) acute myeloid leukemia. *Cancer Cell* 1, 63-74 (2002).

154. Oettgen, P. Transcriptional regulation of vascular development. *Circ Res* 89, 380-8 (2001).

SIGNAL TRANSDUCTION MEDIATED BY CYCLIN D1: FROM MITOGENS TO CELL PROLIFERATION: A MOLECULAR TARGET WITH THERAPEUTIC POTENTIAL

CHENGUANG WANG, ZHIPING LI, MAOFU FU, TOULA BOURAS
AND RICHARD G. PESTELL

1. INTRODUCTION

Cyclin D1 integrates extracellular signals by coupling signals from cell surface receptors to transcription factors, thereby regulating diverse gene expression networks. Cyclin D1 has well defined roles in cell cycle progression that are aberrantly activated in many cancers, thus, it is an appropriate pathway to target for therapeutic intervention. This review will outline the basic regulatory machinery responsible for cell cycle control and describe the latest advances made in the field. Strategies for targeting cyclin D1 and cyclin D1-mediated signaling pathways as means of developing novel and perhaps more effective anticancer treatments will be discussed. Examples of novel cell cycle-targeting molecules that are currently being tested in clinical trials will be discussed as well.

2. OVERVIEW OF CYCLIN D1'S FUNCTION

2.1. Role in Cell Cycle Process

The cyclin-dependent kinase holoenzymes are a family of serine/threonine kinases that play a pivotal role in controlling progression through the cell cycle [1-3]. The cyclins encode regulatory subunits of the kinases, which phosphorylate specific proteins, including the retinoblastoma (pRb) protein, to promote transition through specific cell cycle check points [3-6] (Figure 1). Cyclin D1 plays a pivotal role in G_1/S phase cell-cycle progression in fibroblasts and is rate limiting in growth factor or estrogen-induced mammary epithelial cell proliferation [7,8].

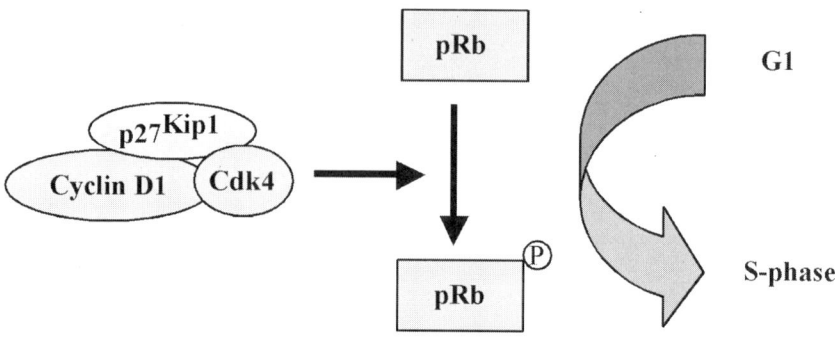

Figure 1. Cyclin D1 drives cell cycle from G1 to S phase through phosphorylating pRb.

2.2. New Properties

In addition to binding cyclin-dependent kinases 4 and 6 (CDK4/6) and pRb, cyclin D1 forms physical associations with various transcriptional regulators including P/CAF (p300/CBP associated factor), Myb, MyoD and the cyclin D1 myb-like binding protein (DMP1) [9-13]. The ability of cyclin D1 to differentially regulate nuclear receptor activity can lead to distinct transcription outputs, depending on the transcription factor being studied. For example, binding of cyclin D1 to the estrogen receptor α (ERα) enhances ligand-independent reporter gene activity, whereas liganded androgen receptor (AR) reporter gene activity is inhibited by cyclin D1 [9,14-16]. The *in vivo* or genetic evidence indicating a requirement for cyclin D1 in nuclear receptor function until recently had remained to be determined [17].

Table 1. Cyclin D1 Associated Proteins

Proteins	Functional Relationship with Cyclin D1	References
Cell cycle machinery		
CDK4/6	Cyclin D1 forms complex with CDK4/6 and facilitates CDK4 phosphorylation of pRb by CAK	[18-20]
p21(CIP1)	p21 represses cyclin D1/CDK4 kinase activity, and promotes cyclin D1 nuclear accumulation	[19,21-23]
p27	p27 assembles cyclin D1/CDK4 kinase complex and represses cyclin D1/CDK4 kinase activity	[22,24]
p57(Kip2)	p57 inhibits cyclin D1-CDK4 kinase activity	[22,25,26]

pRb	Cyclin D1/CDKs phosphorylates pRb and releases E2F from an inhibitory complex	[1,2]
PCNA	Forms multiple kinase complexes	[19,21]
Hsc70	Hsc70 promotes cyclin D1 and cyclin D1-dependent kinase maturation	[27]
Hsp90	Hsp90 promotes cyclin D1 nuclear accumulation	[28]
MCM3/7	Cyclin D1 promotes dissociation of inhibitory pRb/MCM7 complex	[23]
GSK-3β	Phosphorylates cyclin D1	[29]
CRM1	Promotes cyclin D1 nuclear export	[30]
Acetylase/Deacetylase		
p300/CBP	Cyclin D1 represses HAT activity	Unpublished observation
P/CAF	Cyclin D1 represses HAT activity	[9]
SRC-1	Cyclin D1 recruits SRC-1 to ERα	[15]
HDAC1	Cyclin D1 recruits HDAC1 to AR	Unpublished observation
HDAC3	Cyclin D1 recruits HDAC3 to TR to form ternary complexes	[31]
Transcriptional Factor		
ERα	Cyclin D1 recruits SRC-1 to ERα and activates unliganded ERα.	[14,16]
AR	Cyclin D1 represses ligand-bound AR activity	[9,32]
PPARγ	Cyclin D1 represses PPARγ-mediated transcription and differentiation	[17]
TR	Cyclin D1 represses both the unliganded TR and liganded TR activity	[31]
Myb	Cyclin D1 antagonizes B-Myb activity	[33,34]
DMP1	Cyclin D1 antagonizes DMP1 transactivation and overrides DMP1-mediated growth arrest	[13]
Myo D	Cyclin D1 represses muscle differentiation and MyoD-mediated transcription	[35,36]
Stat3	Cyclin D1 represses STAT3 activation	[37]
Sp1	Cyclin D1 represses Sp1-mediated transactivation	[38]
BETA2/NeuroD	Cyclin D1 represses the bHLH transcription factor, BETA2/NeuroD.	[39]
bHLH	Cyclin D1 inhibits the activity of myogenic bHLH regulator	[40]
Others		
TAF(II)250	Cyclin D1 represses Sp1-mediated transcription	[41]
DIP1	Repression	[42]
BRCA1	Cyclin D1 rescues BRCA1-mediated ERα repression	[43] and Unpublished observation
GCIP	Cyclin D1 inhibits cyclin D1/CDK4 activity	[44]

2.3. Cyclin D1 in Cancer

Disregulation of the cell-cycle control apparatus is an almost uniform aberration in tumorigenesis [3]. Cyclin D1 overexpression is found in carcinomas of varying tissue origins, being particularly common in human breast carcinomas, with more than 30% of cases displaying cyclin D1 overexpression, correlating with poor prognosis. Several different oncogenic signals induce cyclin D1 expression including the ErbB2, Ras and Wnt/APC/β-catenin signaling pathways [45-48]. Mammary-targeted expression of cyclin D1 is sufficient for the induction of mammary adenocarcinoma [49] and cyclin $D1^{-/-}$ mice are resistant to Ras- and ErbB2-induced tumorigenesis [50], indicating the strict requirement for cyclin D1 by these specific oncogenic pathways.

3. REGULATION OF CYCLIN D1

Cyclin D1 abundance is under strict control during the G1-phase of cell cycle progression. Multiple DNA sequences recognized by different transcriptional factors in the cyclin D1 promoter region have been identified (Figure 1) [48,51-59]. These distinct response elements binding many transcriptional complexes are thought to be responsible for mediating the transcriptional responsiveness of the cyclin D1 gene to a broad range of mitogenic stimuli. In some, but not all circumstances, transcription factors binding to these elements have been confirmed. Disregulation in the signaling pathways, which impinge on the activity of transcription complexes recruited to the cyclin D1 promoter, may lead to aberrant expression of cyclin D1 and thus, a disordered cell cycle. Because of the critical role for cyclin D1 in coupling extracellular or cytoplasmic signals to nuclear responses it was well predicted that cyclin D1 would be frequently targeted, amplified or overexpressed in various types of cancer.

Figure 2. The Transcription Responsive Elements of the Cyclin D1 Promoter

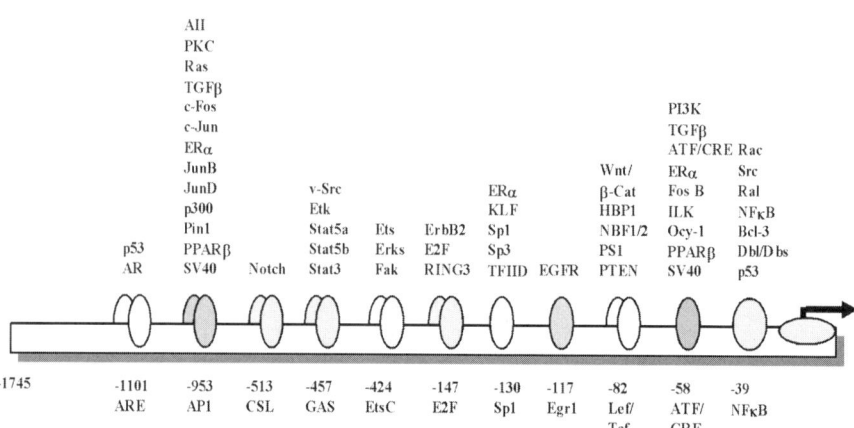

Table 2. Proteins regulating cyclin D1 abundance through transcription.

Regulator	Promoter Sequence Involved	Effect	References
AIB1	unknown	Induction	[60]
ATF/CRE	ATF/CRE	Induction	[55,57,61,62]
AP-1 family:	AP1		[51,52,63-65]
c-jun, c-Fos		Induction	
JunB, JunD		Repression	
Bcl-2	unknown	Induction	[66]
BRCA1	AP1,CRE	Repression	[67]
β-Catenin	Lef/Tcf	Induction	[48,68]
Caveolin-1	Lef/Tcf	Repression	[69]
CSL	CSL	Induction	[70]
Dbl/Dbs	NFκB	Induction	[53,71]
E2Fs	E2F/Sp-1	Induction/Repression	[72]
EGFR	Egr	Induction	[73]
ERα	AP1, Sp1, CRE	Induction	[74-76]
Etk/BMX	STAT5	Induction	[77]
Ets	Ets	Induction	[51]
Fak	Ets	Induction	[78]
FKHR	unknown	Repression	[79]
Galectin	Sp1,CRE	Induction	[80]
GKLF	Sp1	Induction	[81]
HBP1	Lef/Tcf	Repression	[82]
HGF/SF	unknown	Induction	[83]
Id2	unknown	Repression	[84]
IKKα	Lef/Tcf	Induction	[68]
ILK	ATF/CRE	Induction	[62]
INI1/hSNF5	unknown	Induction	[85]
Neu	E2F	Induction	[45]
Oct-1	ATF/CRE	Induction	[86]
NFκB	NFκB	Induction	[58,59]
Notch	CSL	Induction	[70]
p53/Bcl-3	NFκB	Induction	[87,88]
p300	AP1	Induction	[89]
PKC	AP1	Induction	[90]
PI3K/Akt	ATF/CRE	Induction	[68,91]
Pin1	AP1	Induction	[92]
PPARγ	AP1,CRE	Repression	[93]
Presenilin 1 (PS1)	Lef/Tcf	Repression	[94]
PTEN	Lef/Tcf	Repression	[95]
Ral	NFκB	Induction	[96]
Ras	AP1	Induction	[51]
c-Rel	NFκB	Induction	[97]
RING3	E2F	Induction	[98]
Rac	NFκB	Induction	[59]
Sp1/3	Sp1	Induction	[57]
v-Src	STAT3	Induction	[57,99]
STAT5	GAS	Induction	[56]
STAT3	GAS	Induction	[100]
SV40 small t	AP1, CRE	Induction	[52]
TGFβ/PTHrP	AP1,CRE	Induction	[101]

TFIID	Sp1	Repression	[102]
Tob	unknown	Repression	[103]
UBF1/2	LEF/TCF	Induction	[104]

3.1. Oncogenic Signaling Pathways Targeting Cyclin D1

3.1.1. ErbB2

ErbB2 (also known as neu, Her-2) is a member of the epidermal growth factor family of receptors, which includes the epidermal growth factor receptor (EGFR). Oncogenic activation by ErbB2 can occur through its overexpression, point mutation within the transmembrane domain or deletion of the extracellular domain, which culminate in constitutive ErbB2 signaling [105]. ErbB-2 is sufficient for the induction of mammary tumorigenesis in transgenic mice [45] and cyclin D1 was identified as a downstream target of ErbB2 and required for ErbB2-induced mammary tumor growth *in vivo* [45]. In subsequent studies, genetically engineered mice generated by crossing ErbB-2 transgenic mice with cyclin D1 knockout mice demonstrated that cyclin D1-deficient mice are relatively resistant to breast cancers induced by the neu and ras oncogenes [50,106]. In mammary epithelial cells, the cyclin D1 dependency of the neu and ras pathways for malignant transformation suggests that therapeutic strategies targeting cyclin D1 might be highly specific in treating human breast cancers with activated Neu-Ras pathways.

3.1.2. Ras

The product of the *ras* gene plays a critical role in the regulation of cellular proliferation and differentiation [107-110]. Activated Ras induces cyclin D1 expression and anti-sense cyclin D1 reduces the cell proliferation rate in Ras-transformed cells [46]. Receptor-mediated Ras signaling promotes cyclin D1 expression and activation through the Raf/Mek/ErK kinase cascade, which targets the cyclin D1 promoter [51]. Mammary-targeted overexpression of activated Ras is sufficient for the induction of mammary tumorigenesis, which is associated with the induction of cyclin D1 [69,111]. *In vivo* evidence of cyclin D1 dependency in ras-mediated tumorigenesis has been provided by studies showing lack of ras induced mammary tumorigenesis in cyclin D1-deficient mice [50,112]. Cyclin D1 expression and cyclin D1/CDK kinase are also up-regulated in keratinocytes in response to oncogenic Ras. Furthermore, cyclin D1 deficiency decreases the development of squamous tumors generated through either grafting of retroviral Ras-transduced keratinocytes, phorbol ester treatment of ras transgenic mice, or two-stage carcinogenesis by 80% [112].

3.1.3. Src

The pp60[c-src] is a cytoplasmic tyrosine kinase that is sufficient to initiate cellular transformation. In NIH3T3 cells, v-Src overexpression resulted in cellular transformation and an acceleration of G1 progression is associated with an induction of cyclin D1 protein [113]. Activating c-src induces mammary gland tumor formation when introduced into mice under the control of the murine mammary tumor virus (MMTV) promoter [114,115]. Activation of v-Src leads to induction of cyclin D1 expression in the mammary gland [57] and induction of cyclin D1 occurs through CRE sequences in the cyclin D1 promoter [59].

3.1.4. Wnt/APC/β-Catenin

The Wingless/Wnt pathway plays a crucial role in development and cell cycle control [116-118]. Disregulation of the Wnt/β-catenin/Tcf pathway has been implicated in tumorigenesis of diverse tissue types [119]. A role for cyclin D1 in β-catenin/Tcf-mediated signaling and cell transformation *in vivo* is corroborated by correlative studies in human breast carcinomas showing coexpression of cyclin D1 and β-catenin that is correlated with poor prognosis [48,120,121]. β-catenin activates the transcription of cyclin D1 through the single functional TCF-binding site within the promoter [47,48]. G1-phase arrest of the cell cycle by expression of the dominant-negative TCF mutant was overcome by ectopic expression of cyclin D1 [47], although the requirement of cyclin D1 in tumorgenesis by β-catenin may be cell-type specific [122].

3.1.5. STATs

Signal transducers and activators of transcription (STATs) are transcription factors that mediate cytokine and growth factor induced signals, culminating in various biological responses, including proliferation and differentiation. Disregulation of the STAT signaling pathway is commonly found in tumor samples and alteration of STAT activation can be found in various human cancers and transformed cell lines [123]. Therefore, an improved understanding of the mechanisms underlying STAT regulation of cell survival may lead to successful strategies for targeting STATs in cancer therapy. In response to cytokine stimulation, Stat proteins become phosphorylated. Phosphorylated STATs accumulate in the nucleus and activate transcription through binding to a specific responsive element. Many genes are known to be downstream targets of STATs including antiapoptotic proteins Mcl-1, Bcl-xL, Bcl-2, caspases, Fas, TRAIL [124-129], and the proliferation-associated proteins such as cyclin D1 [56,100,126,130], p21 [131], and c-Myc [132]. Cellular transformation by activated Stat3 occurs through the transcriptional regulation of specific genes [100,133]. Cyclin D1 was shown as a crucial target of STAT5 in cytokine-dependent proliferation. Overexpression of cyclin D1 overcomes the growth inhibitory effects of dominant negative mutant STAT5 [56].

4. NUCLEAR RECEPTOR REGULATION OF CYCLIN D1 EXPRESSION

PPARγ agonists inhibit the growth of human colorectal cancer cells [134], promote fibroblast, prostate, and breast epithelial cell differentiation [135-137], and according to one study, inhibit the growth of implanted colonic tumors with APC mutations [138]. The molecular mechanisms by which 15d-PGJ2 regulates cellular proliferation has been assessed [93]. In MCF-7 cells, 15d-PGJ2 selectively inhibits S-phase entry and both the abundance and kinase activity of cyclin D1. Cyclin D1 Mrna and promoter activity are repressed by 15d-PGJ2, while cyclin D1 overexpression reversed 15d-PGJ2-mediated S-phase inhibition [93].

Induction of cyclin D1 protein is a critical step in estrogen-induced cell proliferation in ERα-positive cells. A number of non-classical estrogen responsive elements in the cyclin D1 promoter region have been identified as mediators of ERα responsiveness [74-76], including the Camp response-like element (CRE), stimulating factor 1 (Sp-1), and activating protein-1 (AP-1).

4.1. Cyclin D1 polymorphism

Betticher et al identified an alternatively spliced transcript of the cyclin D1 gene using Northern blotting and PCR analysis of cDNA derived from cell lines and tissues [139]. The variant transcript shows no splicing at the downstream of exon 4 and encodes a protein with an altered carboxy-terminal domain. Two forms of mRNA (transcript a and b) are generated in variety of cells, and in normal and tumor tissues. Although the function of this polymorphism remains controversial, polymorphisms of cyclin D1 are found to correlate with non-small cell cancer of the lung [139], squamous cell carcinoma of the head and neck [140], epithelial ovarian cancer [141], hereditary nonpolyposis colorectal cancer [142], increased risk of urinary bladder cancer [143] and prostate cancer [144], esophageal and gastric cardiac carcinoma [145], and early onset of lung cancer in men, especially squamous cell cancer[146]. The protein encoded by transcript b was constitutively localized in the nucleus, possibly due to the loss of threonine 286, the phosphorylation site of transcript a required for nuclear export [29,147].

5. CYCLIN D1 REGULATION OF NUCLEAR RECEPTOR FUNCTION

Nuclear receptors coordinate diverse physiological roles in metabolism and development through ligand-dependent and -independent mechanisms. Nuclear receptors form multiprotein complexes with coactivator and corepressor proteins to orchestrate dynamic transcriptional events in response to ligand. In the absence of ligand, nuclear receptors repress transcription through a dominant association with corepressor complexes with histone deacetylase activity [148]. Conformational changes induced upon nuclear receptor ligand binding release corepressors, with subsequent transient association of coactivator proteins [148-150].

5.1. Estrogen Receptor

It has been demonstrated that overexpression of cyclin D1 mRNA correlates with poor prognosis within human ERα-positive breast carcinomas [151]. The causal relationship between ERα and cyclin D1 is an area of intense investigation. The induction of cellular proliferation induced by estrogen in breast cancer cell lines correlates with increased cyclin D1 expression [152,153]. Upon ligand binding, ERα undergoes conformational changes recruiting coactivators (p160, p300/CBP) and disengaging corepressors [148]. ERα then binds in a transient and cyclical manner to target DNA sequences in the promoters of estrogen responsive genes to activate gene transcription [149]. Estrogen-independent mechanisms of ERα activation also exist. For example, ERα is also activated by both ERK [154] and PI3K [155], which are induced by ErbB2. Recent studies also showed that overexpression of cyclin D1 renders breast cancer cells less mitogen dependent [8,156]. We, and others, have found that cyclin D1 enhances transcription of ER-responsive genes without the assistance of estrogens by stimulating ER binding to ERE-sequences and ligand-independent recruitment of coactivators [14-16].

5.2. Androgen Receptor

Androgens induce differentiation of male reproductive organs [157]. Our studies demonstrate that cyclin D1 selectively and specifically inhibits liganded AR activity [9]. Cyclin D1 can inhibit differentiation and either promote or inhibit cellular proliferation in a cell type-specific manner [3]. Ectopic expression of cyclin D1 results in reduced cell cycle progression in androgen-dependent LNCaP cells, which is independent of CDK4 association [32,158].

5.3. Peroxisome Proliferator-activated Receptor

PPARγ was originally cloned as a transcription factor involved in fat cell differentiation and is required for the induction of adipocyte differentiation [159,160]. The PPARγ ligands include eicosanoids, such as 15-deoxy-D12,

226

14-prostaglandin J2 (15d-PGJ2) and synthetic ligands of the thiazolidinedione (TZD) class.

PPARγ protein is widely expressed in various tumors and cell lines, and it has also become a target for developing new anticancer drugs that take advantage of its antiproliferative effects [161]. Reduced PPARγ gene expression has also been reported, for example, in lung cancer and esophageal cancer correlating with poor prognosis in patients [162,163]. Our recent studies indicate that reduced PPARγ expression together with increased cyclin D1 may be a genetic feature of the transition from normal breast epithelium to a subset of adenocarcinomas. PPARγ immunopositivity was decreased in benign breast disease compared with normal mammary epithelium and was reduced further in adenocarcinomas. Cyclin D1 immunopositivity increased from normal epithelium to benign disease and adenocarcinomas. Furthermore, the reduction in PPARγ expression in cyclin D1 infected MEFs, together with the finding of increased levels of PPARγ mRNA and protein in *cyclin D1$^{-/-}$* livers by microarray and Western blotting, suggested cyclin D1 inhibits PPARγ expression and function. The overexpression of cyclin D1 with ERα reflects poor prognosis in some human breast cancers. Given the repression of PPARγ function and expression by cyclin D1 and the cyto inhibitory role of PPARγ in breast epithelium, these studies raise the question of whether reduced PPARγ may contribute to poor prognosis in a subset of patients.

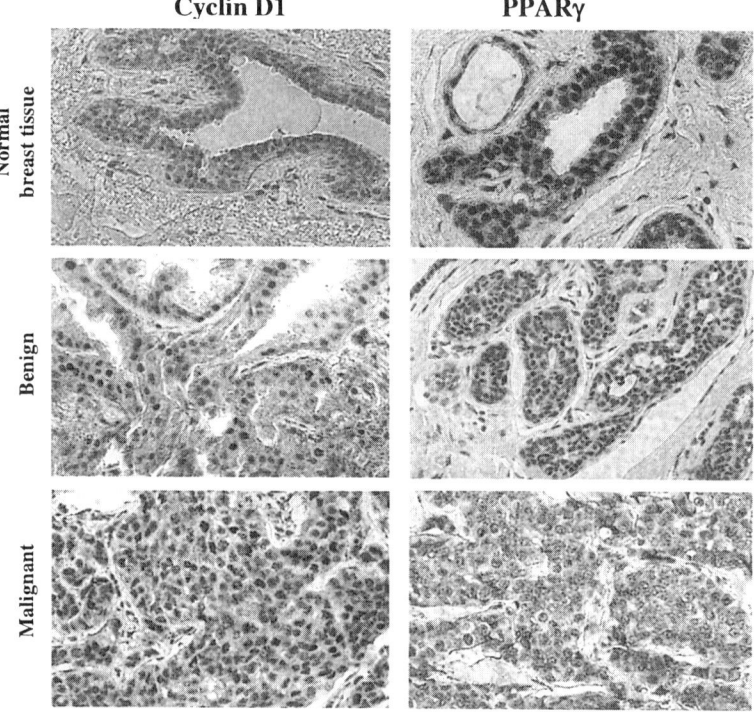

Figure 3. *PPARγ and cyclin D1 abundance in human benign and malignant breast disease (reproduced with permission of MCB)*.

6. KINASE INHIBITOR

Transition through the cell-cycle is induced by a family of protein kinase holoenzymes, the cyclin-dependent kinases (CDKs) and their heterodimeric cyclin partner. Sequential and ordered progression through the cell-cycle involves coordinated activation of the CDKs, which in the presence of an associated CDK-activating kinase, phosphorylate target substrates including members of the 'pocket protein' family. This family includes the product of the retinoblastoma susceptibility gene (the pRb protein) and the related p107 and p130 proteins. Activity of CDKs is tightly controlled by many regulatory mechanisms that either permit or restrain its progression. The regulatory proteins that play key roles in regulating CDK activity are cyclin-dependent kinase inhibitors (CKIs).

6.1. *Cyclin-dependent Kinase Inhibitors (CKIs)*

There are a number of endogenous proteins that have been identified to be CKIs, including members of the p27, the p21CIP1/WAF1 family, and the p16 family. Cellular neoplastic transformation is accompanied by dysregulation of cell cycle checkpoints in conjunction with aberrant expression of CDKs and/or cyclins and/or the loss or mutation of the negative regulators, particularly the CKIs. One strategy to inhibit malignant cellular proliferation involves inhibiting CDK activity or enhancing the function of the CKIs.

6.1.1. p27

$p27^{Kip1}$ was cloned and described as a cyclin-dependent kinase inhibitor that associates with cyclin D1-CDK4 and cyclin E/CDK2 complexes *in vivo* and *in vitro*, and inhibits their kinase activity [164-166]. p27 is required for TGF-β-mediated cell arrest of cells in the G1 phase and overexpression of $p27^{Kip1}$ blocks cell entry into the S phase, indicating p27 is a negative regulator of G1 progression. Overexpressing p27 in fibroblasts delays cell cycle progression, and the ability of specific mitogens to allow transit through the restriction point is correlated with their ability to down regulate p27. Anti-sense p27 can overcome cell-cycle arrest in response to mitogen depletion, indicating a critical role for p27 in the establishment or maintenance of cellular quiescence [3,167]. $p27^{Kip1}$ is bound to p21 and shares sequence similarity to p21.

6.1.2. p21

Harper et al [168], using a two-hybrid screen, isolated a human gene encoding CDK-interacting protein 1 (CIP1), a novel 21kd protein found in cyclin A, cyclin D1, cyclin E, and CDK2 immunoprecipitates. It is a potent, tight-binding inhibitor of CDKs and can inhibit the phosphorylation of the retinoblastoma (pRb) protein by cyclin A-CDK2, cyclin E/CDK2, cyclin D1/CDK4, and cyclin D2/CDK4 complexes. At the same time, by using a subtractive hybridization approach, EI-Deiry et al [169] identified a gene named WAF1 (for wild type p53-activated fragment 1), the induction of which was associated with wild type but not mutant p53 gene expression. Introduction of the WAF1 cDNA suppressed the growth of human brain, lung, and colon tumor cells in culture. A p53-binding site occurs 2.4 kb upstream of WAF1 coding sequences. The sequence of CIP1 was identical to that of WAF1. Combined, these independent results provide a dramatic example of the interplay between tumor suppressor genes and the cell cycle.

6.1.3. p16

Inactivation of the INK4a/ARF locus on human chromosome 9p21 is a common and critical genetic event in the development of human cancers. This locus engages the pRb and p53 tumor suppressor pathways through its capacity to encode two distinct gene products, p16INK4a and p19ARF (p14ARF in humans) [170]. Inactivation of INK4a/ARF occurs through mutation, promoter silencing through methylation and deletion [171]. p16INK4a was identified as a protein binding partner of CDK4 that acts upstream of pRb to cause G1 arrest [172]. p19ARF is also capable of inducing cell-cycle arrest [173].

Altered p16 gene expression occurs in 20% to 50% of certain types of cancer, including breast, bladder and esophageal cancers, and melanomas. High throughput advanced screening methods and structure based approaches have been used to develop small molecule inhibitors that "switch off" inappropriate cell division through mimicking the effects of the p16 gene or by blocking the active site of the cyclin D1 associated kinase [50].

7. DESIGNED SMALL MOLECULAR INHIBITOR OF CDK

The list of designed small molecular CDK inhibitors is growing rapidly [50]. The rationale of these approaches hinges on the role of the kinase activity of cyclin holoenzymes rather than the abundance of cyclin itself. Flavopiridol and UCN-01 are novel inhibitors of CDKs that have been tested in human clinical trials [174,175]. Flavopiridol is a synthetic flavone that is derived from the flavanoid rohitukine, originally isolated from a plant *Dysoxylum binectaisriferum*. Flavopiridol functions through multiple mechanisms. It inhibits CDKs activity, represses cyclin D1 [176,177] and vascular endothelial growth factor expression [178-180], induces apoptosis [181,182], and blocks most RNA polymerase II-dependent transcription by inhibiting positive elongation factor b [183]. It also inhibits the activity of various kinases including epidermal growth factor receptor tyrosine kinase [178], and serine/threonine kinases. Administration of flavopiridol blocks pRb phosphorylation and increased E2F1 levels by repressing cyclin D1 [176,184]. UCN-01, another CDK modulator that has been evaluated in clinical trials, UCN-01 [174,175] inhibits protein kinase C (PKC) activity, promotes cell-cycle arrest by accumulation of p21/p27, and induces apoptosis.

8. CONCLUSIONS

The critical role for cyclin D1 as an integrator of mitogenic signals has led many investigators to examine its regulation.

8.1. *Understanding CKI*

The CDK inhibitors p21CIP1, p27Kip1, and p57Kip2 [22] have been shown to promote the association of CDK4 with the D-type cyclins. Both p21CIP1 and p27Kip1 promote the nuclear accumulation of cyclin D1/CDK4 complexes [20,22,185]. Alt et al (2002) demonstrated that p21CIP1 promotes the nuclear accumulation of cyclin D1/CDK4 through its ability to inhibit GSK-3β-triggered cyclin D1 nuclear export. p27 also has the capacity to enhance cyclin D-dependent activity by acting as an "assembly factor" [24,185]. The cyclin/CDK complex to which p27 is bound determines its functional activity. p27 is found associated with cyclin E in a variety of cell types during quiescence. Overexpression of cyclin D1/CDK4/6 complexes sequesters p27(Kip1) from the cyclin E/CDK2 complex and reduces the inhibitory potential through a stoichiometric mechanism [24]. Understanding how CKI function to promote as assembly factors, or inhibit CDK activity, remains to be fully understood.

8.2. *Development of Small Molecules*

Cyclin D1 is expressed at high levels in a variety of tumors, including breast, prostate, and colon cancer. The functional requirement for cyclin D1 in cell cycle progression and cellular proliferation makes it an ideal target for molecular therapeutics. Several molecules including flavopiridol and rapamycin have been assessed as inhibitors of cyclin D1 expression. Rapamycin and herbimycin inhibit the translation of cyclin D1 mRNA. 17-allylamino-17-dimethoxy geldanamycin (17-AAG), a geldanamycin –related molecule, has been shown to inhibit cyclin D1 expression in breast cancer [186], is being tested in a phase I clinical trial of advanced epithelial carcinoma, malignant lymphoma or sarcoma.

8.3. Transgenic Approach

New experimental approaches have been developed to understand the distinct roles of the cell-cycle in differentiation and tumorigenesis. The role for cyclin D1 in tumorigenesis has been confounded by the mammary gland developmental defect in the *cyclin D1$^{-/-}$* mice. Such abnormal development may result from the loss of key cellular components. Tissue-specific inducible mammary gland-targeted transgenic mice have been developed to analyze events regulating the induction and maintenance of tumorigenesis *in vivo* [187,188]. This ecdysone system has recently been applied to regulate anti-sense cyclin D1 in the mammary gland. A caged ecdysteroid has been developed that can be activated upon exposure to brief illumination [188]. Caged beta-ecdysone is cell permeable and can be photo-uncaged. In combination with spot illumination, caged ponasterone can therefore be used to drive spatially discrete protein expression in a multicellular setting [188]. The combination of ponasterone-inducible transgenics and photo uncaging allows for single cell level gene targeting, and can thus shed light on tissue specific activities of genes regulating the cell-cycle in a temporally and spatially defined manner.

9. ACKNOWLEDGMENTS

This work was supported in part by awards from the Breast Cancer Alliance Inc., R01CA70896, R01CA75503, R01CA86072, R01CA86071 (R.G.P.), R.G.P was a recipient of the Irma T. Hirschl and Weil Caulier award and was the Diane Belfer Faculty Scholar in Cancer Research. Work conducted at the Lombardi Comprehensive Cancer Center was supported by the NIH Cancer Center Core grant.

Chenguang Wang, Zhiping Li, Maofu Fu, Toula Bouras and Richard G. Pestell,
Department of Oncology, Lombardi Comprehensive Cancer Center, Georgetown University, Washington, DC

10. REFERENCES

1. Kato, J., Matsushime, H., Hiebert, S. W., Ewen, M. E., and Sherr, C. J. (1993) *Genes Dev* 7, 331-342
2. Ewen, M. E., Sluss, H. K., Sherr, C. J., Matsushime, H., Kato, J., and Livingston, D. M. (1993) *Cell* 73, 487-497
3. Pestell, R. G., Albanese, C., Reutens, A. T., Segall, J. E., Lee, R. J., and Arnold, A. (1999) *Endocr Rev* 20, 501-534
4. Hunter, T., and Pines, J. (1994) *Cell* 79, 573-582

5. Sherr, C. J. (1996) *Science* 274, 1672-1677
6. Weinberg, R. A. (1995) *Cell* 81, 323-330
7. Baldin, V., Lukas, J., Marcote, M. J., Pagano, M., and Draetta, G. (1993) *Genes Dev* 7, 812-821
8. Zwijsen, R. M., Klompmaker, R., Wientjens, E. B., Kristel, P. M., van der Burg, B., and Michalides, R. J. (1996) *Mol Cell Biol* 16, 2554-2560
9. Reutens, A. T., Fu, M., Wang, C., Albanese, C., McPhaul, M. J., Sun, Z., Balk, S. P., Janne, O. A., Palvimo, J. J., and Pestell, R. G. (2001) *Mol Endocrinol* 15, 797-811
10. Cesi, V., Tanno, B., Vitali, R., Mancini, C., Giuffrida, M. L., Calabretta, B., and Raschella, G. (2002) *Cell Death Differ* 9, 1232-1239
11. Zhang, J. M., Wei, Q., Zhao, X., and Paterson, B. M. (1999) *Embo J* 18, 926-933
12. Bernards, R. (1999) *Biochim Biophys Acta* 1424, M17-22
13. Inoue, K., and Sherr, C. J. (1998) *Mol Cell Biol* 18, 1590-1600
14. Zwijsen, R. M., Wientjens, E., Klompmaker, R., van der Sman, J., Bernards, R., and Michalides, R. J. (1997) *Cell* 88, 405-415
15. Zwijsen, R. M., Buckle, R. S., Hijmans, E. M., Loomans, C. J., and Bernards, R. (1998) *Genes Dev* 12, 3488-3498
16. Neuman, E., Ladha, M. H., Lin, N., Upton, T. M., Miller, S. J., DiRenzo, J., Pestell, R. G., Hinds, P. W., Dowdy, S. F., Brown, M., and Ewen, M. E. (1997) *Mol Cell Biol* 17, 5338-5347
17. Wang, C., Pattabiraman, N., Zhou, J. N., Fu, M., Sakamaki, T., Albanese, C., Li, Z., Wu, K., Hulit, J., Neumeister, P., Novikoff, P. M., Brownlee, M., Scherer, P. E., Jones, J. G., Whitney, K. D., Donehower, L. A., Harris, E. L., Rohan, T., Johns, D. C., and Pestell, R. G. (2003) *Mol Cell Biol* 23, 6159-6173
18. Matsushime, H., Ewen, M. E., Strom, D. K., Kato, J. Y., Hanks, S. K., Roussel, M. F., and Sherr, C. J. (1992) *Cell* 71, 323-334
19. Xiong, Y., Zhang, H., and Beach, D. (1992) *Cell* 71, 505-514
20. Diehl, J. A., and Sherr, C. J. (1997) *Mol Cell Biol* 17, 7362-7374
21. Zhang, H., Xiong, Y., and Beach, D. (1993) *Mol Biol Cell* 4, 897-906
22. LaBaer, J., Garrett, M. D., Stevenson, L. F., Slingerland, J. M., Sandhu, C., Chou, H. S., Fattaey, A., and Harlow, E. (1997) *Genes Dev* 11, 847-862
23. Gladden, A. B., and Diehl, J. A. (2003) *J Biol Chem* 278, 9754-9760
24. Cheng, M., Sexl, V., Sherr, C. J., and Roussel, M. F. (1998) *Proc Natl Acad Sci U S A* 95, 1091-1096
25. Gomez Lahoz, E., Liegeois, N. J., Zhang, P., Engelman, J. A., Horner, J., Silverman, A., Burde, R., Roussel, M. F., Sherr, C. J., Elledge, S. J., and DePinho, R. A. (1999) *Mol Cell Biol* 19, 353-363
26. Reynaud, E. G., Guillier, M., Leibovitch, M. P., and Leibovitch, S. A. (2000) *Oncogene* 19, 1147-1152
27. Diehl, J. A., Yang, W., Rimerman, R. A., Xiao, H., and Emili, A. (2003) *Mol Cell Biol* 23, 1764-1774
28. Taules, M., Rius, E., Talaya, D., Lopez-Girona, A., Bachs, O., and Agell, N. (1998) *J Biol Chem* 273, 33279-33286
29. Diehl, J. A., Cheng, M., Roussel, M. F., and Sherr, C. J. (1998) *Genes Dev* 12, 3499-3511
30. Alt, J. R., Cleveland, J. L., Hannink, M., and Diehl, J. A. (2000) *Genes Dev* 14, 3102-3114
31. Lin, H. M., Zhao, L., and Cheng, S. Y. (2002) *J Biol Chem* 277, 28733-28741
32. Knudsen, K. E., Cavenee, W. K., and Arden, K. C. (1999) *Cancer Res* 59, 2297-2301
33. Ganter, B., Fu, S., and Lipsick, J. S. (1998) *Embo J* 17, 255-268
34. Horstmann, S., Ferrari, S., and Klempnauer, K. H. (2000) *Oncogene* 19, 298-306
35. Skapek, S., Rhee, J., Kim, P., Novitch, B., and Lassar, A. (1996) *Mol. Cell. Biol.* 16, 7043-7053
36. Skapek, S. X., Rhee, J., Spicer, D. B., and Lassar, A. B. (1995) *Science* 267, 1022-1024
37. Bienvenu, F., Gascan, H., and Coqueret, O. (2001) *J Biol Chem* 276, 16840-16847
38. Opitz, O. G., and Rustgi, A. K. (2000) *Cancer Res* 60, 2825-2830

39. Ratineau, C., Petry, M. W., Mutoh, H., and Leiter, A. B. (2002) *J Biol Chem* 277, 8847-8853
40. Rao, S. S., Chu, C., and Kohtz, D. S. (1994) *Mol Cell Biol* 14, 5259-5267
41. Adnane, J., Shao, Z., and Robbins, P. D. (1999) *Oncogene* 18, 239-247
42. Yao, Y., Doki, Y., Jiang, W., Imoto, M., Venkatraj, V. S., Warburton, D., Santella, R. M., Lu, B., Yan, L., Sun, X. H., Su, T., Luo, J., and Weinstein, I. B. (2000) *Exp Cell Res* 257, 22-32
43. Wang, H., Shao, N., Ding, Q. M., Cui, J., Reddy, E. S., and Rao, V. N. (1997) *Oncogene* 15, 143-157
44. Xia, C., Bao, Z., Tabassam, F., Ma, W., Qiu, M., Hua, S., and Liu, M. (2000) *J Biol Chem* 275, 20942-20948
45. Lee, R. J., Albanese, C., Fu, M., D'Amico, M., Lin, B., Watanabe, G., Haines, G. K., 3rd, Siegel, P. M., Hung, M. C., Yarden, Y., Horowitz, J. M., Muller, W. J., and Pestell, R. G. (2000) *Mol Cell Biol* 20, 672-683
46. Filmus, J., Robles, A. I., Shi, W., Wong, M. J., Colombo, L. L., and Conti, C. J. (1994) *Oncogene* 9, 3627-3633
47. Tetsu, O., and McCormick, F. (1999) *Nature* 398, 422-426
48. Shtutman, M., Zhurinsky, J., Simcha, I., Albanese, C., D'Amico, M., Pestell, R., and Ben-Ze'ev, A. (1999) *Proc Natl Acad Sci U S A* 96, 5522-5527
49. Wang, T. C., Cardiff, R. D., Zukerberg, L., Lees, E., Arnold, A., and Schmidt, E. V. (1994) *Nature* 369, 669-671
50. Yu, Q., Geng, Y., and Sicinski, P. (2001) *Nature* 411, 1017-1021
51. Albanese, C., Johnson, J., Watanabe, G., Eklund, N., Vu, D., Arnold, A., and Pestell, R. G. (1995) *J Biol Chem* 270, 23589-23597
52. Watanabe, G., Howe, A., Lee, R. J., Albanese, C., Shu, I. W., Karnezis, A. N., Zon, L., Kyriakis, J., Rundell, K., and Pestell, R. G. (1996) *Proc Natl Acad Sci U S A* 93, 12861-12866
53. Westwick, J. K., Lee, R. J., Lambert, Q. T., Symons, M., Pestell, R. G., Der, C. J., and Whitehead, I. P. (1998) *J Biol Chem* 273, 16739-16747
54. Brown, J. R., Nigh, E., Lee, R. J., Ye, H., Thompson, M. A., Saudou, F., Pestell, R. G., and Greenberg, M. E. (1998) *Mol Cell Biol* 18, 5609-5619
55. Beier, F., Lee, R. J., Taylor, A. C., Pestell, R. G., and LuValle, P. (1999) *Proc Natl Acad Sci U S A* 96, 1433-1438
56. Matsumura, I., Kitamura, T., Wakao, H., Tanaka, H., Hashimoto, K., Albanese, C., Downward, J., Pestell, R. G., and Kanakura, Y. (1999) *Embo J* 18, 1367-1377
57. Lee, R. J., Albanese, C., Stenger, R. J., Watanabe, G., Inghirami, G., Haines, G. K., 3rd, Webster, M., Muller, W. J., Brugge, J. S., Davis, R. J., and Pestell, R. G. (1999) *J Biol Chem* 274, 7341-7350
58. Guttridge, D. C., Albanese, C., Reuther, J. Y., Pestell, R. G., and Baldwin, A. S., Jr. (1999) *Mol Cell Biol* 19, 5785-5799
59. Joyce, D., Bouzahzah, B., Fu, M., Albanese, C., D'Amico, M., Steer, J., Klein, J. U., Lee, R. J., Segall, J. E., Westwick, J. K., Der, C. J., and Pestell, R. G. (1999) *J Biol Chem* 274, 25245-25249
60. Planas-Silva, M. D., Shang, Y., Donaher, J. L., Brown, M., and Weinberg, R. A. (2001) *Cancer Res* 61, 3858-3862
61. Watanabe, G., Lee, R. J., Albanese, C., Rainey, W. E., Batlle, D., and Pestell, R. G. (1996) *J Biol Chem* 271, 22570-22577
62. D'Amico, M., Hulit, J., Amanatullah, D. F., Zafonte, B. T., Albanese, C., Bouzahzah, B., Fu, M., Augenlicht, L. H., Donehower, L. A., Takemaru, K., Moon, R. T., Davis, R., Lisanti, M. P., Shtutman, M., Zhurinsky, J., Ben-Ze'ev, A., Troussard, A. A., Dedhar, S., and Pestell, R. G. (2000) *J Biol Chem* 275, 32649-32657
63. Westwick, J. K., Lambert, Q. T., Clark, G. J., Symons, M., Van Aelst, L., Pestell, R. G., and Der, C. J. (1997) *Mol Cell Biol* 17, 1324-1335
64. Bakiri, L., Lallemand, D., Bossy-Wetzel, E., and Yaniv, M. (2000) *Embo J* 19, 2056-2068
65. Weitzman, J. B., Fiette, L., Matsuo, K., and Yaniv, M. (2000) *Mol Cell* 6, 1109-1119

66. Lin, H. M., Lee, Y. J., Li, G., Pestell, R. G., and Kim, H. R. (2001) *Cell Death Differ* 8, 44-50
67. Welcsh, P. L., Lee, M. K., Gonzalez-Hernandez, R. M., Black, D. J., Mahadevappa, M., Swisher, E. M., Warrington, J. A., and King, M. C. (2002) *Proc Natl Acad Sci U S A* 99, 7560-7565
68. Albanese, C., Wu, K., D'Amico, M., Jarrett, C., Joyce, D., Hughes, J., Hulit, J., Sakamaki, T., Fu, M., Ben-Ze'ev, A., Bromberg, J. F., Lamberti, C., Verma, U., Gaynor, R. B., Byers, S. W., and Pestell, R. G. (2003) *Mol Biol Cell* 14, 585-599
69. Hulit, J., Bash, T., Fu, M., Galbiati, F., Albanese, C., Sage, D. R., Schlegel, A., Zhurinsky, J., Shtutman, M., Ben-Ze'ev, A., Lisanti, M. P., and Pestell, R. G. (2000) *J Biol Chem* 275, 21203-21209
70. Ronchini, C., and Capobianco, A. J. (2001) *Mol Cell Biol* 21, 5925-5934
71. Whitehead, I. P., Lambert, Q. T., Glaven, J. A., Abe, K., Rossman, K. L., Mahon, G. M., Trzaskos, J. M., Kay, R., Campbell, S. L., and Der, C. J. (1999) *Mol Cell Biol* 19, 7759-7770
72. Watanabe, G., Albanese, C., Lee, R. J., Reutens, A., Vairo, G., Henglein, B., and Pestell, R. G. (1998) *Mol Cell Biol* 18, 3212-3222
73. Lin, S. Y., Makino, K., Xia, W., Matin, A., Wen, Y., Kwong, K. Y., Bourguignon, L., and Hung, M. C. (2001) *Nat Cell Biol* 3, 802-808
74. Castro-Rivera, E., Samudio, I., and Safe, S. (2001) *J Biol Chem* 276, 30853-30861
75. Sabbah, M., Courilleau, D., Mester, J., and Redeuilh, G. (1999) *Proc Natl Acad Sci U S A* 96, 11217-11222
76. Liu, M. M., Albanese, C., Anderson, C. M., Hilty, K., Webb, P., Uht, R. M., Price, R. H., Jr., Pestell, R. G., and Kushner, P. J. (2002) *J Biol Chem* 277, 24353-24360
77. Wen, X., Lin, H. H., Shih, H. M., Kung, H. J., and Ann, D. K. (1999) *J Biol Chem* 274, 38204-38210
78. Zhao, J., Pestell, R., and Guan, J. L. (2001) *Mol Biol Cell* 12, 4066-4077
79. Schmidt, M., Fernandez de Mattos, S., van der Horst, A., Klompmaker, R., Kops, G. J., Lam, E. W., Burgering, B. M., and Medema, R. H. (2002) *Mol Cell Biol* 22, 7842-7852
80. Lin, H. M., Pestell, R. G., Raz, A., and Kim, H. R. (2002) *Oncogene* 21, 8001-8010
81. Shie, J. L., Chen, Z. Y., Fu, M., Pestell, R. G., and Tseng, C. C. (2000) *Nucleic Acids Res* 28, 2969-2976
82. Sampson, E. M., Haque, Z. K., Ku, M. C., Tevosian, S. G., Albanese, C., Pestell, R. G., Paulson, K. E., and Yee, A. S. (2001) *Embo J* 20, 4500-4511
83. Recio, J. A., and Merlino, G. (2002) *Oncogene* 21, 1000-1008
84. Lasorella, A., Iavarone, A., and Israel, M. A. (1996) *Mol Cell Biol* 16, 2570-2578
85. Zhang, Z. K., Davies, K. P., Allen, J., Zhu, L., Pestell, R. G., Zagzag, D., and Kalpana, G. V. (2002) *Mol Cell Biol* 22, 5975-5988
86. Boulon, S., Dantonel, J. C., Binet, V., Vie, A., Blanchard, J. M., Hipskind, R. A., and Philips, A. (2002) *Mol Cell Biol* 22, 7769-7779
87. Rocha, S., Martin, A. M., Meek, D. W., and Perkins, N. D. (2003) *Mol Cell Biol* 23, 4713-4727
88. Westerheide, S. D., Mayo, M. W., Anest, V., Hanson, J. L., and Baldwin, A. S., Jr. (2001) *Mol Cell Biol* 21, 8428-8436
89. Albanese, C., D'Amico, M., Reutens, A. T., Fu, M., Watanabe, G., Lee, R. J., Kitsis, R. N., Henglein, B., Avantaggiati, M., Somasundaram, K., Thimmapaya, B., and Pestell, R. G. (1999) *J Biol Chem* 274, 34186-34195
90. Soh, J. W., and Weinstein, I. B. (2003) *J Biol Chem*
91. Page, K., Li, J., Hodge, J. A., Liu, P. T., Vanden Hoek, T. L., Becker, L. B., Pestell, R. G., Rosner, M. R., and Hershenson, M. B. (1999) *J Biol Chem* 274, 22065-22071
92. Wulf, G. M., Ryo, A., Wulf, G. G., Lee, S. W., Niu, T., Petkova, V., and Lu, K. P. (2001) *Embo J* 20, 3459-3472
93. Wang, C., Fu, M., D'Amico, M., Albanese, C., Zhou, J. N., Brownlee, M., Lisanti, M. P., Chatterjee, V. K., Lazar, M. A., and Pestell, R. G. (2001) *Mol Cell Biol* 21, 3057-3070

234

94. Soriano, S., Kang, D. E., Fu, M., Pestell, R., Chevallier, N., Zheng, H., and Koo, E. H. (2001) *J Cell Biol* 152, 785-794

95. Persad, S., Troussard, A. A., McPhee, T. R., Mulholland, D. J., and Dedhar, S. (2001) *J Cell Biol* 153, 1161-1174

96. Henry, D. O., Moskalenko, S. A., Kaur, K. J., Fu, M., Pestell, R. G., Camonis, J. H., and White, M. A. (2000) *Mol Cell Biol* 20, 8084-8092

97. Romieu-Mourez, R., Kim, D. W., Shin, S. M., Demicco, E. G., Landesman-Bollag, E., Seldin, D. C., Cardiff, R. D., and Sonenshein, G. E. (2003) *Mol Cell Biol* 23, 5738-5754

98. Denis, G. V., Vaziri, C., Guo, N., and Faller, D. V. (2000) *Cell Growth Differ* 11, 417-424

99. Sinibaldi, D., Wharton, W., Turkson, J., Bowman, T., Pledger, W. J., and Jove, R. (2000) *Oncogene* 19, 5419-5427

100. Bromberg, J. F., Wrzeszczynska, M. H., Devgan, G., Zhao, Y., Pestell, R. G., Albanese, C., and Darnell, J. E., Jr. (1999) *Cell* 98, 295-303

101. Beier, F., Ali, Z., Mok, D., Taylor, A. C., Leask, T., Albanese, C., Pestell, R. G., and LuValle, P. (2001) *Mol Biol Cell* 12, 3852-3863

102. Hilton, T. L., and Wang, E. H. (2003) *J Biol Chem* 278, 12992-13002

103. Yoshida, Y., Nakamura, T., Komoda, M., Satoh, H., Suzuki, T., Tsuzuku, J. K., Miyasaka, T., Yoshida, E. H., Umemori, H., Kunisaki, R. K., Tani, K., Ishii, S., Mori, S., Suganuma, M., Noda, T., and Yamamoto, T. (2003) *Genes Dev* 17, 1201-1206

104. Grueneberg, D. A., Pablo, L., Hu, K. Q., August, P., Weng, Z., and Papkoff, J. (2003) *Mol Cell Biol* 23, 3936-3950

105. Bargmann, C. I., and Weinberg, R. A. (1988) *Embo J* 7, 2043-2052

106. Bowe, D. B., Kenney, N. J., Adereth, Y., and Maroulakou, I. G. (2002) *Oncogene* 21, 291-298

107. Feig, L. A., and Cooper, G. M. (1988) *Mol Cell Biol* 8, 3235-3243

108. Feig, L. A. (1993) *Science* 260, 767-768

109. Boguski, M. S., and McCormick, F. (1993) *Nature* 366, 643-654

110. McCormick, F. (1993) *Nature* 363, 15-16

111. Engelman, J. A., Lee, R. J., Karnezis, A., Bearss, D. J., Webster, M., Siegel, P., Muller, W. J., Windle, J. J., Pestell, R. G., and Lisanti, M. P. (1998) *J Biol Chem* 273, 20448-20455

112. Robles, A. I., Rodriguez-Puebla, M. L., Glick, A. B., Trempus, C., Hansen, L., Sicinski, P., Tennant, R. W., Weinberg, R. A., Yuspa, S. H., and Conti, C. J. (1998) *Genes Dev* 12, 2469-2474

113. Liu, J. J., Chao, J. R., Jiang, M. C., Ng, S. Y., Yen, J. J., and Yang-Yen, H. F. (1995) *Mol Cell Biol* 15, 3654-3663

114. Parker, R. C., Varmus, H. E., and Bishop, J. M. (1984) *Cell* 37, 131-139

115. Webster, M. A., Cardiff, R. D., and Muller, W. J. (1995) *Proc Natl Acad Sci U S A* 92, 7849-7853

116. Wong, G. T., Gavin, B. J., and McMahon, A. P. (1994) *Mol Cell Biol* 14, 6278-6286

117. Cadigan, K. M., and Nusse, R. (1997) *Genes Dev* 11, 3286-3305

118. Huelsken, J., and Behrens, J. (2002) *J Cell Sci* 115, 3977-3978

119. Polakis, P. (2000) *Genes Dev* 14, 1837-1851

120. Barker, N., and Clevers, H. (2000) *Bioessays* 22, 961-965

121. Barker, N., Morin, P. J., and Clevers, H. (2000) *Adv Cancer Res* 77, 1-24

122. Rowlands, T., Hatsell, S., Pechenkina, I., Pestell, R., and Cowin, P. (2003) *PNAS* In press

123. Bromberg, J. (2002) *J Clin Invest* 109, 1139-1142

124. Battle, T. E., and Frank, D. A. (2002) *Curr Mol Med* 2, 381-392

125. Epling-Burnette, P. K., Zhong, B., Bai, F., Jiang, K., Bailey, R. D., Garcia, R., Jove, R., Djeu, J. Y., Loughran, T. P., Jr., and Wei, S. (2001) *J Immunol* 166, 7486-7495

126. de Groot, R. P., Raaijmakers, J. A., Lammers, J. W., and Koenderman, L. (2000) *Mol Cell Biol Res Commun* 3, 299-305

127. Silva, M., Benito, A., Sanz, C., Prosper, F., Ekhterae, D., Nunez, G., and Fernandez-Luna, J. L. (1999) *J Biol Chem* 274, 22165-22169

128. Qin, J. Z., Zhang, C. L., Kamarashev, J., Dummer, R., Burg, G., and Dobbeling, U. (2001) *Blood* 98, 2778-2783

129. Catlett-Falcone, R., Landowski, T. H., Oshiro, M. M., Turkson, J., Levitzki, A., Savino, R., Ciliberto, G., Moscinski, L., Fernandez-Luna, J. L., Nunez, G., Dalton, W. S., and Jove, R. (1999) *Immunity* 10, 105-115

130. Brockman, J. L., Schroeder, M. D., and Schuler, L. A. (2002) *Mol Endocrinol* 16, 774-784

131. Bellido, T., O'Brien, C. A., Roberson, P. K., and Manolagas, S. C. (1998) *J Biol Chem* 273, 21137-21144

132. Bowman, T., Broome, M. A., Sinibaldi, D., Wharton, W., Pledger, W. J., Sedivy, J. M., Irby, R., Yeatman, T., Courtneidge, S. A., and Jove, R. (2001) *Proc Natl Acad Sci U S A* 98, 7319-7324

133. Darnell, J. E., Jr. (1997) *Science* 277, 1630-1635

134. Sarraf, P., Mueller, E., Smith, W. M., Wright, H. M., Kum, J. B., Aaltonen, L. A., de la Chapelle, A., Spiegelman, B. M., and Eng, C. (1999) *Mol Cell* 3, 799-804

135. Elstner, E., Muller, C., Koshizuka, K., Williamson, E. A., Park, D., Asou, H., Shintaku, P., Said, J. W., Heber, D., and Koeffler, H. P. (1998) *Proc Natl Acad Sci U S A* 95, 8806-8811

136. Mueller, E., Sarraf, P., Tontonoz, P., Evans, R. M., Martin, K. J., Zhang, M., Fletcher, C., Singer, S., and Spiegelman, B. M. (1998) *Mol Cell* 1, 465-470

137. Mueller, E., Smith, M., Sarraf, P., Kroll, T., Aiyer, A., Kaufman, D. S., Oh, W., Demetri, G., Figg, W. D., Zhou, X. P., Eng, C., Spiegelman, B. M., and Kantoff, P. W. (2000) *Proc Natl Acad Sci U S A* 97, 10990-10995

138. Sarraf, P., Mueller, E., Jones, D., King, F. J., DeAngelo, D. J., Partridge, J. B., Holden, S. A., Chen, L. B., Singer, S., Fletcher, C., and Spiegelman, B. M. (1998) *Nat Med* 4, 1046-1052

139. Betticher, D. C., Thatcher, N., Altermatt, H. J., Hoban, P., Ryder, W. D., and Heighway, J. (1995) *Oncogene* 11, 1005-1011

140. Matthias, C., Branigan, K., Jahnke, V., Leder, K., Haas, J., Heighway, J., Jones, P. W., Strange, R. C., Fryer, A. A., and Hoban, P. R. (1998) *Clin Cancer Res* 4, 2411-2418

141. Dhar, K. K., Branigan, K., Howells, R. E., Musgrove, C., Jones, P. W., Strange, R. C., Fryer, A. A., Redman, C. W., and Hoban, P. R. (1999) *Int J Gynecol Cancer* 9, 342-347

142. Kong, S., Wei, Q., Amos, C. I., Lynch, P. M., Levin, B., Zong, J., and Frazier, M. L. (2001) *J Natl Cancer Inst* 93, 1106-1108

143. Wang, L., Habuchi, T., Takahashi, T., Mitsumori, K., Kamoto, T., Kakehi, Y., Kakinuma, H., Sato, K., Nakamura, A., Ogawa, O., and Kato, T. (2002) *Carcinogenesis* 23, 257-264

144. Wang, L., Habuchi, T., Mitsumori, K., Li, Z., Kamoto, T., Kinoshita, H., Tsuchiya, N., Sato, K., Ohyama, C., Nakamura, A., Ogawa, O., and Kato, T. (2003) *Int J Cancer* 103, 116-120

145. Zhang, J., Li, Y., Wang, R., Wen, D., Sarbia, M., Kuang, G., Wu, M., Wei, L., He, M., Zhang, L., and Wang, S. (2003) *Int J Cancer* 105, 281-284

146. Qiuling, S., Yuxin, Z., Suhua, Z., Cheng, X., Shuguang, L., and Fengsheng, H. (2003) *Carcinogenesis* 24, 1499-1503

147. Solomon, D. A., Wang, Y., Fox, S. R., Lambeck, T. C., Giesting, S., Lan, Z., Senderowicz, A. M., and Knudsen, E. S. (2003) *J. Biol. Chem.*, M303969200

148. McKenna, N. J., Lanz, R. B., and O'Malley, B. W. (1999) *Endocr Rev* 20, 321-344

149. Shang, Y., Hu, X., DiRenzo, J., Lazar, M. A., and Brown, M. (2000) *Cell* 103, 843-852

150. Chen, H., Lin, R. J., Xie, W., Wilpitz, D., and Evans, R. M. (1999) *Cell* 98, 675-686

151. Kenny, F. S., Hui, R., Musgrove, E. A., Gee, J. M., Blamey, R. W., Nicholson, R. I., Sutherland, R. L., and Robertson, J. F. (1999) *Clin Cancer Res* 5, 2069-2076

152. Altucci, L., Addeo, R., Cicatiello, L., Dauvois, S., Parker, M. G., Truss, M., Beato, M., Sica, V., Bresciani, F., and Weisz, A. (1996) *Oncogene* 12, 2315-2324

153. Foster, J. S., and Wimalasena, J. (1996) *Mol Endocrinol* 10, 488-498

154. Kato, S., Endoh, H., Masuhiro, Y., Kitamoto, T., Uchiyama, S., Sasaki, H., Masushige, S., Gotoh, Y., Nishida, E., Kawashima, H., and et al. (1995) *Science* 270, 1491-1494

155. Campbell, R. A., Bhat-Nakshatri, P., Patel, N. M., Constantinidou, D., Ali, S., and Nakshatri, H. (2001) *J Biol Chem* 276, 9817-9824
156. Musgrove, E. A., Lee, C. S., Buckley, M. F., and Sutherland, R. L. (1994) *Proc Natl Acad Sci U S A* 91, 8022-8026
157. Heinlein, C. A., and Chang, C. (2002) *Endocr Rev* 23, 175-200
158. Petre, C. E., Wetherill, Y. B., Danielsen, M., and Knudsen, K. E. (2002) *J Biol Chem* 277, 2207-2215
159. Rosen, E. D., Sarraf, P., Troy, A. E., Bradwin, G., Moore, K., Milstone, D. S., Spiegelman, B. M., and Mortensen, R. M. (1999) *Mol Cell* 4, 611-617
160. Tontonoz, P., Hu, E., and Spiegelman, B. M. (1994) *Cell* 79, 1147-1156
161. Koeffler, H. P. (2003) *Clin Cancer Res* 9, 1-9
162. Sasaki, H., Tanahashi, M., Yukiue, H., Moiriyama, S., Kobayashi, Y., Nakashima, Y., Kaji, M., Kiriyama, M., Fukai, I., Yamakawa, Y., and Fujii, Y. (2002) *Lung Cancer* 36, 71-76
163. Terashita, Y., Sasaki, H., Haruki, N., Nishiwaki, T., Ishiguro, H., Shibata, Y., Kudo, J., Konishi, S., Kato, J., Koyama, H., Kimura, M., Sato, A., Shinoda, N., Kuwabara, Y., and Fujii, Y. (2002) *Jpn J Clin Oncol* 32, 238-243
164. Toyoshima, H., and Hunter, T. (1994) *Cell* 78, 67-74
165. Polyak, K., Kato, J. Y., Solomon, M. J., Sherr, C. J., Massague, J., Roberts, J. M., and Koff, A. (1994) *Genes Dev* 8, 9-22
166. Polyak, K., Lee, M. H., Erdjument-Bromage, H., Koff, A., Roberts, J. M., Tempst, P., and Massague, J. (1994) *Cell* 78, 59-66
167. Coats, S., Flanagan, W. M., Nourse, J., and Roberts, J. M. (1996) *Science* 272, 877-880
168. Harper, J. W., Adami, G. R., Wei, N., Keyomarsi, K., and Elledge, S. J. (1993) *Cell* 75, 805-816
169. el-Deiry, W. S., Tokino, T., Velculescu, V. E., Levy, D. B., Parsons, R., Trent, J. M., Lin, D., Mercer, W. E., Kinzler, K. W., and Vogelstein, B. (1993) *Cell* 75, 817-825
170. Sharpless, N. E., and DePinho, R. A. (1999) *Curr Opin Genet Dev* 9, 22-30
171. Serrano, M. (2000) *Carcinogenesis* 21, 865-869
172. Serrano, M., Hannon, G. J., and Beach, D. (1993) *Nature* 366, 704-707
173. Quelle, D. E., Zindy, F., Ashmun, R. A., and Sherr, C. J. (1995) *Cell* 83, 993-1000
174. Mani, S., Wang, C., Wu, K., Francis, R., and Pestell, R. (2000) *Expert Opin Investig Drugs* 9, 1849-1870
175. Senderowicz, A. M. (2003) *Cancer Chemother Pharmacol* 52, 61-73
176. Carlson, B., Lahusen, T., Singh, S., Loaiza-Perez, A., Worland, P. J., Pestell, R., Albanese, C., Sausville, E. A., and Senderowicz, A. M. (1999) *Cancer Res* 59, 4634-4641
177. Wu, K., Wang, C., D'Amico, M., Lee, R. J., Albanese, C., Pestell, R. G., and Mani, S. (2002) *Mol Cancer Ther* 1, 695-706
178. Melillo, G., Sausville, E. A., Cloud, K., Lahusen, T., Varesio, L., and Senderowicz, A. M. (1999) *Cancer Res* 59, 5433-5437
179. Rapella, A., Negrioli, A., Melillo, G., Pastorino, S., Varesio, L., and Bosco, M. C. (2002) *Int J Cancer* 99, 658-664
180. Nahta, R., Trent, S., Yang, C., and Schmidt, E. V. (2003) *Cancer Res* 63, 3626-3631
181. Parker, B. W., Kaur, G., Nieves-Neira, W., Taimi, M., Kohlhagen, G., Shimizu, T., Losiewicz, M. D., Pommier, Y., Sausville, E. A., and Senderowicz, A. M. (1998) *Blood* 91, 458-465
182. Arguello, F., Alexander, M., Sterry, J. A., Tudor, G., Smith, E. M., Kalavar, N. T., Greene, J. F., Jr., Koss, W., Morgan, C. D., Stinson, S. F., Siford, T. J., Alvord, W. G., Klabansky, R. L., and Sausville, E. A. (1998) *Blood* 91, 2482-2490
183. Chao, S. H., and Price, D. H. (2001) *J Biol Chem* 276, 31793-31799
184. Osuga, H., Osuga, S., Wang, F., Fetni, R., Hogan, M. J., Slack, R. S., Hakim, A. M., Ikeda, J. E., and Park, D. S. (2000) *Proc Natl Acad Sci U S A* 97, 10254-10259
185. Cheng, M., Olivier, P., Diehl, J. A., Fero, M., Roussel, M. F., Roberts, J. M., and Sherr, C. J. (1999) *Embo J* 18, 1571-1583
186. Basso, A. D., Solit, D. B., Munster, P. N., and Rosen, N. (2002) *Oncogene* 21, 1159-1166

187. Albanese, C., Reutens, A. T., Bouzahzah, B., Fu, M., D'Amico, M., Link, T., Nicholson, R., Depinho, R. A., and Pestell, R. G. (2000) *Faseb J* 14, 877-884
188. Lin, W., Albanese, C., Pestell, R. G., and Lawrence, D. S. (2002) *Chem Biol* 9, 1347-1353

SIGNAL TRANSDUCTION PATHWAYS IN BCR-ABL TRANSFORMED CELLS

RALPH ARLINGHAUS AND TONG SUN

1. INTRODUCTION

The formation of the Philadelphia chromosome (Ph) in pluripotent stem cells within the bone marrow leads to a lethal leukemia (Fig. 1). The Ph produces two main types of leukemia. One is a myeloid leukemia termed chronic myelogenous leukemia (CML) in which the levels of granulocytes are dramatically increased per unit volume of blood. This leukemia is frequently detected in the early stage of the disease, which is termed the chronic phase. If untreated, this stage is always followed by progression to an intermediate stage (accelerated phase) and then to a terminal stage termed blast crisis. The Ph fuses the 5' part of the BCR gene to most of the ABL gene (usually lacking the first alternate ABL first exon). In CML, the junction of BCR to ABL is almost always composed of either BCR exons 13 or 14 fused to ABL exon 2. This fusion leads to a hybrid Bcr-Abl protein of about 210,000 daltons (P210). The other form of leukemia caused by the Ph is an acute lymphocytic (B cell involvement) leukemia termed ALL. In this case the BCR-ABL fusion can either be the same as that in CML or one that involves fusion of exon 1 of BCR to ABL exon 2, yielding a 185,000 to 190,000 dalton protein (P185/190). A third and rare form of Bcr-Abl (P230) has been seen in chronic neutrophilic leukemia (Fig. 1). The crucial properties of the oncogenic Bcr-Abl proteins are their highly activated tyrosine kinase activity [1,2] and cytoplasmic location. In contrast to the Bcr-Abl oncoproteins, the normal Abl protein functions both in the nucleus and the cytoplasm. The fused Bcr sequences provide a coiled-coil domain at the amino terminus (Fig. 2) that allows the formation of tetramers [3] of the Bcr-Abl protein, which is known to be critical for both its tyrosine kinase activation [3] and leukemic activity in mouse models [4,5].

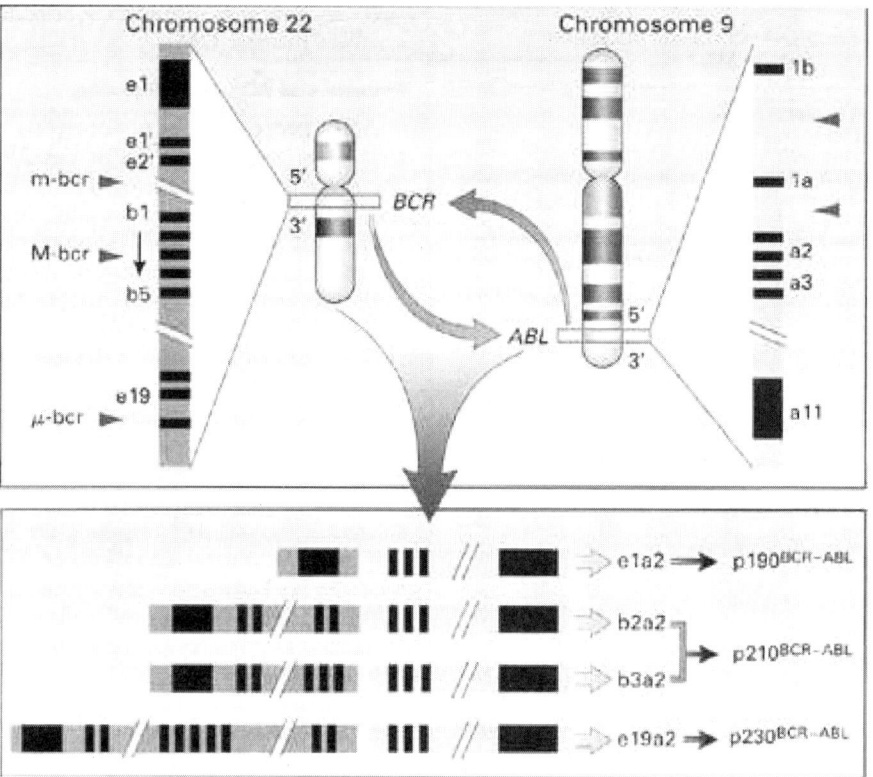

Figure 1. The Translocation of t(9;22)(q34;q11) in CML, the Philadelphia chromosome (Ph). (Taken from Faderl et al (72).

2. SIGNAL TRANSDUCTION PATHWAYS AFFECTED BY BCR-ABL.

A relatively large number of proteins have been observed to be tyrosine-phosphorylated in cells expressing Bcr-Abl. Several signal transduction pathways are altered in hematopoietic cells expressing Bcr-Abl (Fig. 3). These include activation of the Ras pathway, the Jak/Stat pathway, and pathways involved in blocking apoptosis. The latter effects are thought to be critical for development of disease, as CML cells are believed to have a longer half-life than normal blood cells [6]. Cytoskeletal protein are also involved in CML, as Bcr-Abl has an increased affinity for actin filaments compared to normal Abl [7]. Binding of c-Abl to actin filaments reduces c-Abl's tyrosine kinase activity [8]. Bcr-Abl despite its high affinity for actin filaments maintains its high level of tyrosine kinase activity. This difference between the two kinases is unknown but could relate to the ability of Bcr-Abl

241

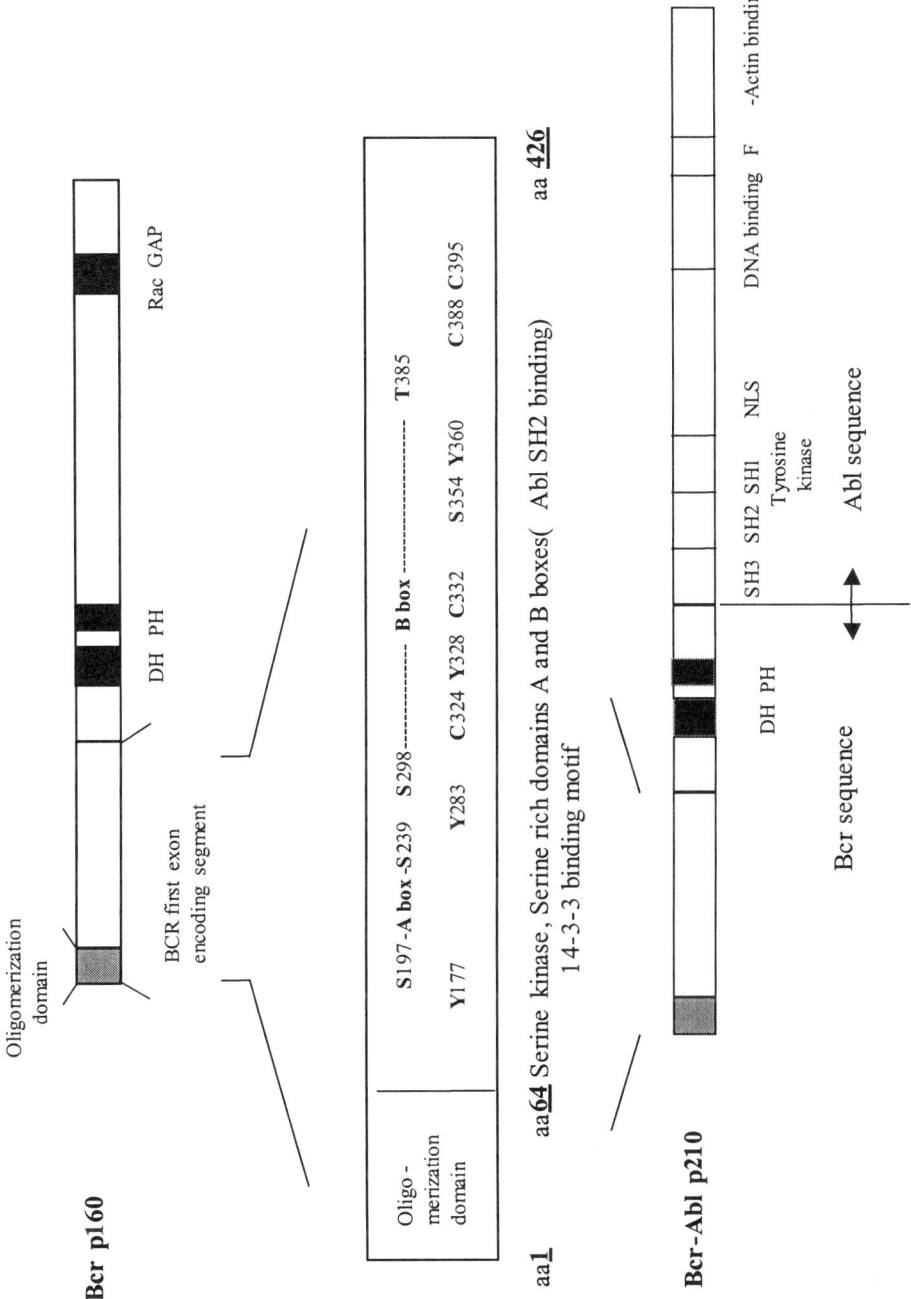

Figure 2. Structure of Bcr and Bcr-Abl proteins. The various domains are as follows: DH, Dbl

homology domain (a GDP exchange factor); PH, the pleckstrin homology domain (binds to phosphoinositol); Rac GAP, GTPase protein domain similar to Rac; SH3, Src homology domain 3; SH2, Src homology domain 2; SH1, Src homology domain 1 (the catalytic kinase domain); NLS, nuclear localization signal; not shown are a proline–rich domain for binding SH3 domains (overlapping with the DNA binding domain) and a nuclear export signal (NES) located between the NLS and F-Actin binding domain. Also several adaptor protein binding sites are located between the SH1 domain and the NLS.

to functionally down-regulate Bcr, one of the critical Abl kinase inhibitors (see section on Bcr). Paxillin and Cbl proteins are also targets of Bcr-Abl induced tyrosine phosphorylation [9]. In addition, adhesion proteins are also affected by Bcr-Abl expression, which is believed to induce changes in cell adhesion and cell mobility [10]. The latter effects correlate with the abnormal exit of leukemic cells from the marrow. Another property of Bcr-Abl is its cytoplasmic location. In contrast, the normal Abl protein is found in both the nucleus and cytoplasm [11], and shuttles back and forth between these two cellular compartments [12]. Interestingly, if Bcr-Abl is allowed to enter the nucleus and remain active as a tyrosine kinase, it causes cell death [13]. Thus, cytoplasmic Bcr-Abl is an oncogenic protein but nuclear Bcr-Abl can cause cell death of the leukemia cells [13]. These findings will likely prompt the search for new compounds that facilitate nuclear localization of Bcr-Abl as treatment for CML.

Adaptor proteins such as Shc and CrkL are also targets of the Bcr-Abl tyrosine kinase. Three forms of the Shc protein are expressed in cells; Bcr-Abl is known to phosphorylate one of the Shc proteins in the 52,000/53,000 dalton size class [14]. Shc is involved in activating the Ras pathway, as described below. CrkL is a protein related to c-Crk [15]. It contains one SH2 (Src Homology) domain followed by two SH3 domains. The role of CrkL in CML is not well understood, but it is widely used as a target of Bcr-Abl [16]. Early studies have identified CrkL as a major target of the Bcr-Abl oncoprotein in blood cells from CML patients [17].

The adaptor protein Grb2, which like CrkL also contains two SH3 domains and one SH2 domain, is not tyrosine-phosphorylated by Bcr-Abl but binds to a phosphotyrosine residue present in both Bcr and the Shc proteins (see below).

The Janus kinase (Jak) family of proteins are also targets of oncogenic Abl proteins [18,19,20]. Bcr-Abl appears to directly phosphorylate the Jak2 protein on a tyrosine that is involved in activating the Jak2 tyrosine kinase [18] (see below). Activation of Jak2 by Bcr-Abl appears to involve a pathway that is independent of the activation of Stat5 [18,20]. Thus, this Jak2 activation is quite different from the IL-3 pathway, which activates Jak2 which in turn activates Stat5, allowing Stat5 to translocate to the nucleus where it acts as a transcription factor. Bcr-Abl appears to directly activate Stat5 [21] independent of Jak2. However, a report from Skorski's lab

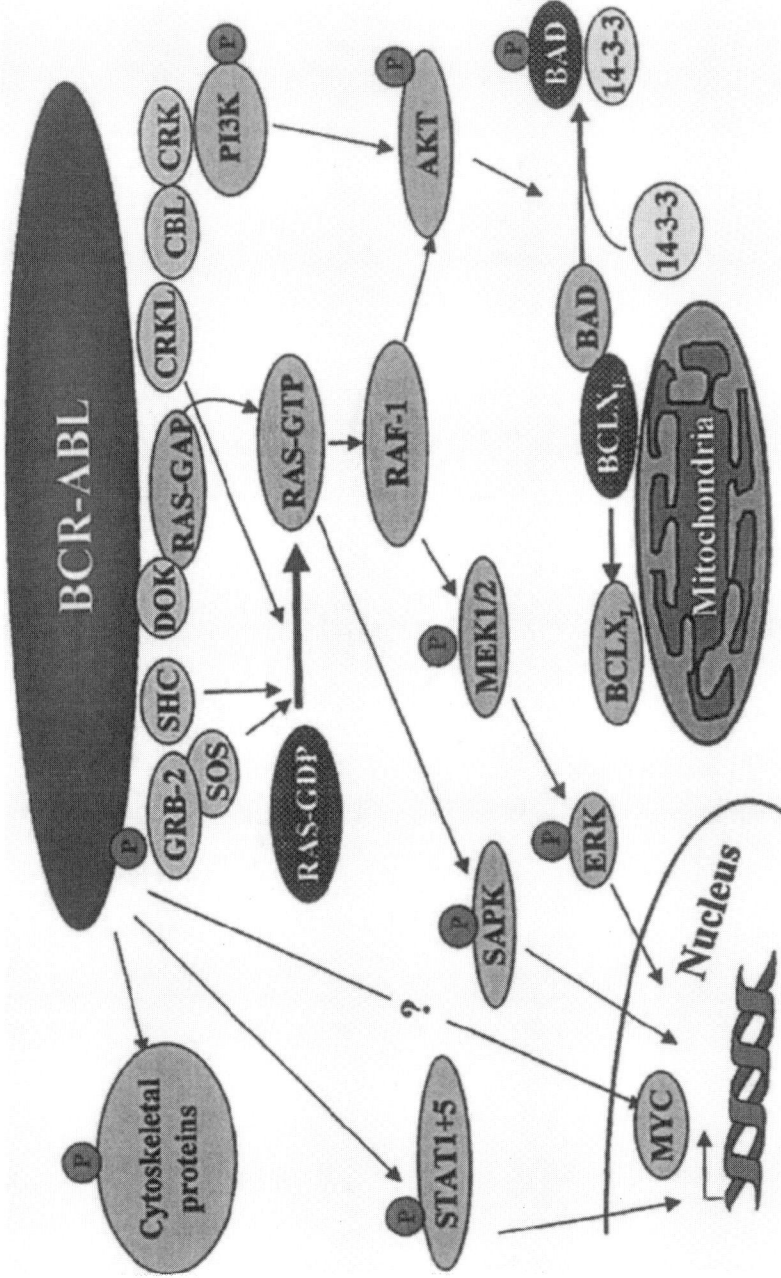

Figure 3. Signal transduction pathways regulated by the oncogenic Bcr-Abl protein. Modified from Figure 4 of Deninger et al (73). The asterisk (*) refers to the work of Klejman et al (22). Others have proposed a direct activation of Stat5 by Bcr-Abl (21).

[22] indicates that the Src kinase family member Hck mediates Stat5 activation in Bcr-Abl expression in myeloid leukemia cells.

3. RAS PATHWAY ACTIVATION

In the early 1990s my laboratory began exploring sites of autophosphorylation within the Bcr-Abl oncoprotein [23]. Surprisingly, we found that major sites of tyrosine phosphorylation reside in sequences encoded by the first exon of the BCR gene [23]. We explored the question of which tyrosine residues within the first exon sequences (there are 11 tyrosines) are targets for Bcr-Abl autophosphorylation. We identified four tyrosine residues as sites of phosphorylation: tyrosine 177, tyrosine 283, tyrosine 328 and tyrosine 360 [24, 25]. In collaboration with the Tony Pawson group, we found that tyrosine 177 of Bcr lies within the consensus binding site for Grb2 (pYVNV), an SH2/SH3 containing adaptor protein involved in receptor signaling. These studies showed that Grb2 bound to the P210 form of Bcr-Abl, specifically to the sequence pYVNV [14]. In the same study, the Shc protein became phosphorylated on a tyrosine within the YVNV sequence of Shc. Shc was shown to bind to Grb2. Based on the many studies performed in receptor-mediated signal transduction, Puil et al. [14] confirmed that Grb2 binding to either Bcr-Abl or Shc lead to binding of Sos (a GDP exchange factor), which in turn interacts with Raf (Fig. 3). This pathway eventually leads to activation of several genes by way of the Ras pathway. The importance of phosphorylation of tyrosine 177 was established by comparing the activity of the Y177F mutant of Bcr-Abl to wild type Bcr-Abl [26]. The Y177F mutant was defective in activating a gene driven by the Ras pathway.

Several studies established the biological importance of tyrosine 177 in mouse leukemia models [4,5]. The Y177F mutant has reduced leukemic activity in mouse marrow transplant model studies conducted by Ren and co-workers [5] and Pear and co-workers [4].

What remains to be determined is the mechanism by which Bcr-Abl has access to the Grb2/Sos Ras pathway. Adaptor molecules, such as Grb2, typically are involved in transmitting signals from cell surface receptors [27]. Thus, Grb2 is targeted to the inner side of the cell surface membrane by binding to phosphotyrosine sequences of a transmembrane receptor. In contrast to transmembrane receptors, there is no known association of Bcr-Abl with the inner surface of the cell membrane. Therefore, some targeting mechanism must exist to allow Bcr-Abl to have access of the Grb2/Sos-Ras signaling complex. We have proposed that Bcr plays that role [28]. In this type of scenario, Bcr binds 14-3-3 [29]; 14-3-3 is known to bind to Raf [30]; the Bcr amino terminus would form mixed oligomers with Bcr-Abl thereby allowing Bcr-Abl to accompany the Bcr/14-3-3/Raf to the inner surface of the cell membrane. Autophosphorylation of either Bcr-Abl or the complexed Bcr (which we have shown binds Grb2, ref. 31) or both would attract Grb2 to the complex and thereby would initiate activation of the Ras pathway (Fig. 3).

4. JAK2 INVOLVEMENT IN ONCOGENIC EFFECTS OF BCR-ABL

In human leukemia and transgene mouse models the oncogenic effects of Bcr-Abl are restricted to hematopoietic cells. Therefore, this review will focus on the effects of Bcr-Abl on hematopoietic cells in both mouse [32] and human systems. Earlier studies focused on the oncogenic effects of Bcr-Abl in fibroblasts like Rat-1 cells [33]. However, Bcr-Abl has difficulty transforming fibroblasts, and transgenic mice experiments re-inforce the specificity of Bcr-Abl towards hematopoietic cells, either the myeloid lineage or lymphoid lineage in human leukemia, as solid tumors were not observed in these mice.

The IL-3/GM-CSF receptor pathways are altered as a result of Bcr-Abl expression. Some years ago it was observed that Bcr-Abl expression in mouse hematopoietic cells abrogates their requirement for IL-3 in IL-3 dependent cell lines [34]. Since then, several examples of this effect of Bcr-Abl on the IL-3 receptor pathway were reported [35]. The lack of an IL-3 requirement in cell culture was either the result of autocrine stimulation of IL-3 production by Bcr-Abl expressing cells [36] or by activating the Jak2 and Stat5 members of the Jak/Stat pathway in an independent manner [18,21] (Fig. 3). In studies with a human megakaryocytic cell line (Mo7e, ref. 35) that requires either IL-3 or GM-CSF for survival/growth, Bcr-Abl expression did not induce autocrine effects involving IL-3 and GM-CSF in at least one cell clone (M3.16) [19]. It is likely that both types of effects take place in leukemia patients. That is, Bcr-Abl tyrosine kinase would persistently activate the Jak and Stat pathways but in some cell contexts, autocrine effects of Bcr-Abl on IL-3/GM-CSF might also be operable.

In the Mo7e system, Bcr-Abl expression induces tyrosine phosphorylation of Jak2 but not Jak1 [19]. These findings were confirmed in the 32D clone 3 cell line transformed by BCR-ABL [18]. Detailed studies in this system provided some very interesting new information. A Bcr-Abl/Jak2 pathway not involving Stat5 was described [18, 20]. This is of interest because in cells not transformed by Bcr-Abl, IL-3 induces Jak2 activation, which in turn activates translocation of Stat5 into the nucleus where it can affect the transcription of a battery of genes. In Bcr-Abl transformed hematopoietic cells, Stat5 is thought to be directly or indirectly activated by

Bcr-Abl. [21, 22]. It was assumed from these Stat5 results, that Jak2 was not involved in any way in the abrogation of the requirement of IL-3 for maintainance and survival of these IL-3 dependent cell lines. In contrast our findings, however, established that Jak2 plays a critical role in Bcr-Abl's oncogenic effects in cell culture systems [18, 20].

The steps involved in Jak2 activation involve binding of Bcr-Abl via the C-terminus of Abl to Jak2. This interaction does not require the kinase activity of either protein. Bcr-Abl phoshorylates Jak2 at the tyrosine residue known to activate the tyrosine kinase of Jak2. Jak2 complexes include Bcr-Abl and several other proteins, all of which become tyrosine phosphorylated [20]. Of interest, one of these proteins is SH2-Bβ. SH2-Bβ is involved in activating Jak2 and sustaining its activity [37]. Thus, Bcr-Abl directly activates the tyrosine kinase activity of Jak2 and also alters SH2-Bβ, a protein known to enhance Jak2 activity. How tyrosine phosphorylation of SH2-Bβ affects Jak2 kinase activation or other affects is not yet known.

One of the outcomes of Jak2 activation is enhancing the levels of c-Myc RNA and the c-Myc protein [20] (Fig. 3). The c-Myc transcription factor is known to be required for transformation by Bcr-Abl [38]. Two approaches were used to study the requirement of Jak2 activation. One involved use of an inhibitor of Jak2 tyrosine kinase, AG490. The other involved dominant-negative (DN) Jak2 effects. DN Jak2 treated K562 cells have reduced levels of c-Myc; those cells with reduced levels of c-Myc had reduced tumor-forming ability in nude mice [20]. The lower the level of c-Myc, the smaller the tumors in nude mice [20]. Lastly, a mutant of the adaptor protein, SH2-B□, having a defective SH2 domain also inhibited c-Myc elevation [20] in Bcr-Abl expressing mouse 32D cells, a myeloid hematopoietic cell. SH2-Bβ is involved in enhancing the activation of the Jak2 tyrosine kinase [39], but may have other functions in signal transduction.

4.1. Translational Control Effects of Bcr-Abl are Involved in Disease Progression in CML.

Calabretta and colleagues have published reports that describe two important translational control mechanisms that are involved in progression of CML from chronic stage to accelerated to blast crisis [40,41]. One involves the stimulation of La antigen expression [40]. La is a RNA binding protein that allows a by-pass of the normal translational repression of MDM2 mRNA. A net increase of Mdm2 protein would lower the level of p53, thereby contributing to an unstable genome. About 25% of blast crisis CML patients have a defect in the p53 gene [42]. Thus, Bcr-Abl also functionally inactivates p53 in earlier stages of the disease. Given the role of p53 as guardian of the genome, these effects could contribute to causing further mutations leading to disease progression. In another study [41] the level of HnRNP E2 was found to be increased in Bcr-Abl positive cells. This leads to suppression of transcription factor C/EBPα, which is involved in creating the cellular environment for granulocytic differentiation. HnRNP E2 binds to C-rich sequences located in either 5' or 3' UTR sequences in mRNAs, leading to a block of translation. Thus, C/EBPα expression is inhibited by the effects of increased levels of HnRNP E2. The thrust of this research suggests that Bcr-Abl inhibits differentiation of myeloid progenitors toward mature granulocytes as a direct result of suppression of C/EBPα translation.

5. BCR-ABL EXPRESSION ENHANCES THE DNA REPAIR ENVIRONMENT

Skorski and colleagues have explored the mechanisms of drug resistance in CML. Their findings indicate that Bcr-Abl activates pathways involved in DNA repair, lengthens the time of the G2/M phase, and increases the expression of Bcl-2 family members and thus greatly reducing mitototic driven apoptosis [43]. All three events allow cells expressing Bcr-Abl to be more resistant to the cell killing effects of cytotoxic drugs and radiation than normal cells.

6. ROLE OF BCR IN BCR-ABL POSITIVE CML

More than 10 years ago my laboratory began studies on the role of the Bcr protein in chronic myelogenous leukemia (CML) caused by the Bcr-Abl oncoprotein. The Bcr protein has a number of functional domains, which include a coiled-coil oligomerization domain, an Abl SH2 binding domain (serine-rich sequence), a serine/threonine kinase domain, and other domains as listed in Fig. 2. Initially, we characterized the Bcr protein by generating a number of site-directed anti-peptide Bcr antibodies. These antibodies proved to be very useful for isolating and characterizing the Bcr-Abl oncoprotein and the normal Bcr protein. Our first investigative use of these antibodies allowed us to detect the tyrosine kinase activity of the P210 BCR-ABL protein in CML patient cell line (W. Kloetzer,and R. Arlinghaus, Cold Spring Harbor RNA tumor Virus meeting, 1994; 1). This was about the same time that the Witte group demonstrated that Bcr-Abl has an activated tyrosine kinase activity. [2]. These antibodies were also useful in sorting out the structure of the P190 BCR-ABL tyrosine kinase [44]. The Bcr antibodies were also useful in detecting the serine/threonine kinase encoded by the Bcr protein [45]. We observed that the Bcr sequence contains a similar pattern of protein sequences found in all serine/threonine kinases, yet Bcr has significant deviations from the consensus kinase sequence, suggesting that Bcr is a novel kinase [45]. Subsequent studies by Maru and Witte provided convincing evidence that Bcr is indeed a novel serine/threonine kinase, and that the kinase domain has elements that resemble kinases like pyruvate kinase [46]. We investigated the role of the Bcr serine/threonine kinase in the oncogenic effects of the Bcr-Abl protein. We first showed that the Bcr protein itself is a target for tyrosine phosphorylation by Bcr-Abl [47]. However, these studies revealed that only a minor fraction of total Bcr was tyrosine-phosphorylated in a CML cell line. In addition, a minor fraction of Bcr was tightly associated with P210 BCR-ABL [48], most likely by way of co-oligomerization with Bcr-Abl [3]. It seems likely that these mixed tetramers of Bcr and Bcr-Abl play an important role in the leukemia syndrome driven by the Bcr-Abl oncoprotein (Fig. 2 and see above).

6.1. Bcr-Abl Autophosphorylates at Sites Encoded by the First Exon of BCR

The fusion of Bcr sequences activates the Abl tyrosine kinase causing autophosphorylation of the Bcr-Abl oncoprotein on tyrosine residues. Surprisingly, the predominant sites of tyrosine phosphorylation are located within Bcr protein sequences encoded by the first exon [23]. This was initially determined by comparing phosphotryptic peptide maps of c-Abl, P210 BCR-ABL and P185 BCR-ABL. Most of the phosphotryptic peptides resulting from in vitro autophosphorylation of Bcr-Abl are shared between the P210 and P185 forms of Bcr-Abl oncoprotein. The c-Abl specific peptides are only minor components of the Bcr-Abl autokinase reaction products. Since P185 BCR-ABL contains only Bcr sequences encoded by the first exon of BCR but not other BCR exons, we concluded that the major sites of Bcr-Abl autophosphorylation were present in sequences encoded by the first exon of BCR [23].

Peptide mapping studies of site-directed mutants of Bcr-Abl when compared to wild-type Bcr-Abl indicated that tyrosine residues (Y) 177, Y283, Y328 and Y360 are sites of tyrosine phosphorylation resulting from Bcr-Abl autophosphorylation [14,25, 26,49]. In agreement with our earlier findings [23], all of these sites are located within the first exon coding sequence of the BCR gene. Y177 [26] and Y360 [24] were also identified as sites of phosphorylation in P210 BCR-ABL expressed within cells.

Phosphotyrosine sequence-specific antibodies were also developed against phosphotyrosine (pTyr) 328 and pTyr 360 of Bcr [50] to determine whether these sites were phosphorylated within Bcr-Abl proteins and Bcr proteins. The Bcr protein co-expressed with Bcr-Abl is phosphorylated at these sites just as in the cis Bcr sequences in the Bcr-Abl oncoprotein, itself [25, 50]. Besides pTyr sites mentioned above, additional sites of Bcr tyrosine phosphorylation may exist but have not yet been identified. The known functional consequences of the sites of Bcr tyrosine phosphorylation are described below.

Activation of the Ras pathway. The role of Y177 phosphorylation was identified in a series of experiments published by Pendergast and her colleagues [26], and confirmed by Puil et al., [14]. Briefly, the wild-type Bcr-Abl oncoprotein binds tightly to the Grb2 adaptor protein through an SH2 domain of Grb2; Grb2 in turn binds to the GDP exchange protein Sos, and it in turn binds to Ras (Fig. 3), as described above.

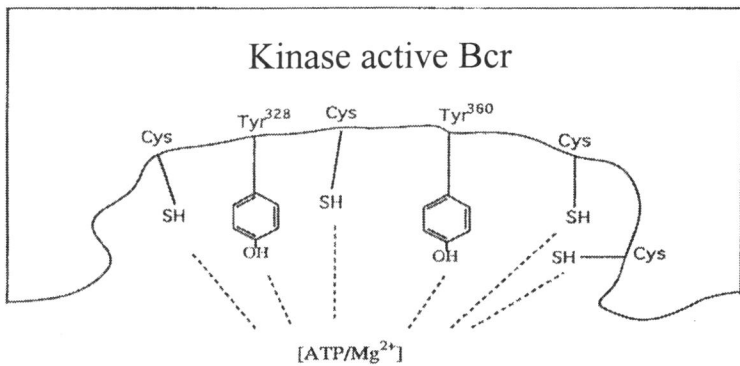

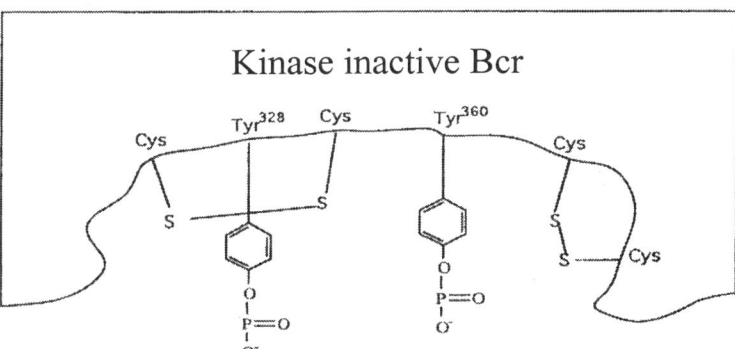

Figure 4. Model for inactivation of the Bcr serine/threonine protein kinase activity by tyrosine phosphorylation catalyzed by the Bcr-Abl oncoprotein.

6.2. *Tyrosine Phosphorylation Reduces the Serine/threonine Kinase Activity of the Bcr Protein*

We searched for other possible effects of tyrosine phosphorylation of Bcr first exon sequences. Surprisingly, tyrosine phosphorylation of the Bcr protein by the Bcr-Abl protein strongly reduced its serine/threonine kinase activity [24]. Since Y283, Y328 and Y360 were also phosphorylated by the Abl kinase domain within the Bcr-Abl oncoprotein, we performed experiments to determine whether any of these tyrosine residues were involved in inhibiting the Bcr serine/threonine kinase. Mutation of Y360 to phenylalanine (F) strongly inhibited the ability to the Bcr protein (P160 BCR) to phosphorylate an added substrate in kinase reactions. Similarly, the Y328F mutant also curtailed Bcr trans-kinase activity, but this mutation also reduced the ability of Bcr to autophosphorylate on serine residues. The double mutant Y328F/Y360F blocked both the trans- and auto-kinase activities of the Bcr protein [25]. We proposed a model indicating that the hydroxyl groups of tyrosine 328 and 360 are critically involved in regulating Bcr's serine/threonine kinase activity (Fig. 4) [51]. In this model, phosphorylation of these tyrosyl hydroxyl groups within Bcr would compromise its kinase function (e.g. proper conformation of the kinase domain). We further proposed that the phosphorylation of Y328 and Y360 alters the function of two cysteine pairs (e.g. possibly by oxidation to form S-S bonded pairs), because of the proposed role of the two cysteine pairs in Bcr kinase activity [46]. An important role for tyrosine residues has been proposed by those working on pyruvate kinase, another novel kinase that requires cysteine residues and tyrosine for its kinase activity [52,53]. We propose that removal of the hydroxyl groups of Y328 and Y360 by mutation of the tyrosine to phenylalanine has the same result as the phosphorylation of these hydroxyl groups (Fig.4).

6.3. Inhibitory Role of Phosphoserine Bcr in Bcr-Abl Leukemia

We reasoned that since reduction of the Bcr serine/threonine kinase activity resulted from tyrosine phosphorylation of Bcr sequences by the activated Abl tyrosine kinase within the Bcr-Abl oncoprotein, then Bcr's serine kinase must have an important role with regards to Bcr-Abl and that its oncogenic activity might require some level of Bcr serine kinase inhibition. By chance we discovered that removal of the oligomerization domain in a truncated form of Bcr [Bcr(64-413)] rendered it resistant to tyrosine phosphorylation by the activated Abl kinase [49]. We later found that although the oligomerization domain is required for the maximum level of Bcr serine kinase activity, Bcr(64-413) maintained a significant level of in vitro kinase activity [54]. Co-expression of Bcr(64-413) with over-expressed c-Abl revealed a modest reduction of tyrosine phosphorylation of c-Abl as compared to that observed with co-expression with Bcr(1-413) [49]; the latter being an excellent target for c-Abl and thus would have reduced serine/threonine kinase activity. Bcr(64-413) containing immune complexes also strongly inhibited the Bcr-Abl kinase auto-phosphorylation activity as judged by kinase assays whereas the heat-denatured Bcr(64-413) protein lost the inhibitory activity [49]. Also, co-expression of c-Abl with Bcr(64-413) formed complexes involving these proteins, and these complexes contained an inactive c-Abl tyrosine kinase. The biological activity of Bcr-Abl was also reduced by co-expression of Bcr(64-413), as expression of Bcr(64-413) in K562 cells reduced its ability to form colonies in soft agar when compared to vector only transfected cells [49]. In this paper, we also showed that rather high molar levels of a phosphoserine 354 Bcr peptide (but not the non-phosphorylated peptide) strongly inhibited the Bcr-Abl tyrosine kinase autophosphorylation reaction. These levels of the phosphoserine Bcr peptide did not inhibit the Src kinase, arguing that Bcr's inhibitory effects were specific for the Abl kinase. The importance of phosphoserine 354 was confirmed in later studies [54] reviewed below.

6.4. Bcr(64-413) Expression in CML Cell Lines and Primary Cultures Inhibits Their Growth and Survival

Because of the inhibitory effects of Bcr(64-413) on colony formation of the CML cell line K562 [49], we introduced this form of Bcr into other CML cell lines and primary cultures of blood cells of CML patients with active disease by adenovirus infection [55]. We used a replication-defective adenovirus encoding either BCR(64-413) or the β galactosidase (GAL) gene for these studies. Unfortunately, CML cell lines in log phase growth are resistant to adenovirus β GAL infection. We searched for conditions that would make hematopoietic cells susceptible to adenovirus infection. We found that serum-starvation combined with over-crowding rendered hematopoietic cells susceptible to adenovirus infection, as 70-80% of the cells exhibited expression of the β GAL gene following infection under these conditions with adenovirus β GAL [55]. Under these same conditions, K562 cells (myeloid blast cells expressing P210 BCR-ABL), BV173 cells (lymphoid blast cells expressing P210 BCR-ABL), primary blood cell cultures from CML patients with active disease, SMS-SB (a pre-B cell lacking BCR-ABL expression), and normal marrow cells had similar levels of infection with the β GAL adenovirus. Infection of these cell lines with adenovirus encoding BCR(64-413) inhibited cell proliferation of K562 cells but not that of SMS-SB cells. In addition, Bcr(64-413) expression induced high levels of apoptosis in primary CML cells but not normal marrow cells, as measured by annexin V/propidium iodide staining [55]. Thus, these findings indicate that Bcr(64-413) antagonizes the oncogenic effects of Bcr-Abl but had no detectable toxic effects on hematopoietic cells lacking Bcr-Abl expression at least in short-term tissue culture experiments. These findings raise the possibility that some form of Bcr can be used to treat CML.

6.5. Induction of BCR Gene Expression in Rat-1Cells Transformed by Bcr-Abl Reversed the Oncogenic Phenotype and Sharply Reduced Tyrosine Phosphorylation of Cellular Protein

Since Bcr(64-413) is inhibitory to the Bcr-Abl tyrosine kinase and blocks its oncogenic effects, we decided to test whether full length Bcr would have similar effects. Because Bcr is a target for tyrosine phosphorylation [47], we were concerned that it may not be able to antagonize the effects of Bcr-Abl. That is, tyrosine phosphorylation of Bcr would inhibit the Bcr serine kinase activity and thus neutralize Bcr's inhibitory effects directed towards Bcr-Abl. We repeatedly attempted to over-express human Bcr in Rat-1 cells transformed by Bcr-Abl expression (P185 BCR-ABL), but to our surprise during the selection process we produced Rat-1 cells lacking expression of both human Bcr and Bcr-Abl, as judged by anti-Bcr and anti-phosphotyrosine Western blotting experiments. We reasoned that the Bcr-Abl positive cells were counter-selected for as a result of BCR over-expression during the two to three week neomycin (neo) selection period. Thus, these findings suggested that Bcr expression in Bcr-Abl positive Rat-1 cells caused cell death. These findings suggested that excess BCR gene expression antagonizes the oncogenic effects of Bcr-Abl, and thereby induces cell death. Because of these findings, we searched for ways to introduce a transcriptionally silent form of the human BCR gene into Bcr-Abl cells, which would permit induction of BCR gene expression to determine the initial effects of Bcr on the Bcr-Abl oncogenic phenotype. We tried several induceable vector systems and chose the tetracycline (Tet) repression system [28]. In order to maximize the induction of Bcr and to insure a low level of synthesis of exogenous Bcr in the non-induced condition, we selected various clones of Bcr-Abl expressing clones in soft agar (as a means to select oncogenic Rat-1 cell clones) under neo selection conditions. We reasoned that the larger the size of the cell colony, the lower the level of Bcr expression in the presence of Tet. Subsequent induction of exogenous Bcr expression inhibited foci formation of Bcr-Abl transformed Rat-1 cells and reduced the phosphotyrosine content of cellular proteins [28]. However, the effects of BCR expression on a CML cell line (K562 cells) were considerably less drastic in culture than that observed in Bcr-Abl positive Rat-1 fibroblastic cells. Induction of BCR expression in K562 cells caused an increase in cell differentiation. Further studies are underway to probe further the characteristics of CML cell lines like K562 following induction of the Bcr protein. (Y. Wang and R. Arlinghaus, unpublished).

Despite these less drastic effects in a CML cell such as K562, we examined the phosphotyrosine/phosphoserine content of the Bcr protein following BCR induction. Before induction, the endogenous Bcr protein had a high level of phosphotyrosine compared to phosphoserine. After BCR induction, the level of Bcr protein increased several fold within 5-15 days. After increasing the Bcr concentration, the percentage of phosphotyrosine Bcr within the pool of total Bcr decreased dramatically whereas the content of phosphoserine Bcr increased to the point that Bcr was predominantly in the phosphoserine form [28, 56]. This was of interest to us, since Rat-1 fibroblastic cells transformed by Bcr-Abl were greatly inhibited by this phosphoserine form of Bcr but K562 cells, when maintained in tissue culture, were somehow able to survive these inhibitory effects in cell culture.

Of interest, Bcr-Abl positive hematopoietic 32D cells (a mouse myeloid blast cell line considered to have less defects than the K562 cell line) responded to Bcr over-expression in a manner similar to Rat-1 fibroblasts (X Ling and R. Arlinghaus, unpublished results). In these studies, induction of Bcr-Abl expression (Tet-off system) in the presence of human Bcr induced apoptosis, virtually eliminating cells that express both Bcr and Bcr-Abl from the culture. Bcr did not appear to be toxic to cells when Bcr-Abl expression was not induced (X. Ling and R. Arlinghaus, unpublished). These effects of Bcr-Abl positive mouse 32D cells over-expressing Bcr are similar to that seen in primary cultures of cells from CML patients expressing BCR(64-413) [55].

6.6. Bcr Expression Blocks Leukemia Formation in a NOD/scid Mouse Model

Despite the more subtle effects of enhanced BCR expression on CML cell line clones containing a Tet-repressed BCR gene maintained in culture compared to Rat-1 cells transformed by Bcr-Abl, we tested the effects of induced BCR expression on CML cell line clones following their injection into NOD/scid mice [56]. In a series of three experiments with a clone of K562 cells (K6, selected for colony formation in soft agar with low BCR expression), we tested the dose of CML cells necessary for adequate engraftment based on Southern blotting to detect human DNA. Under optimum conditions when BCR induction is prevented by feeding mice tetracycline (Tet) in the drinking water, 100% of mice died within 35 days (Fig. 5). Release of the block in BCR expression at the time of intravenous injection of the K6-clone, allowed 80% of the mice to survive for a 55-day period. The remaining 20% of mice died of the same leukemia syndrome (an extramedullary disease). These results were reproduced in two additional experiments with this K6 cell clone [56].

BCR induction after the leukemia process was underway, delayed the onset of death in most animals and allowed longer-term survival in 30% of the animals with what appeared to be a chronic leukemia syndrome (Fig. 5). Of interest, one of three mice had no detectable hematopoietic tumors in the spleen and marrow or any other types of tumors.

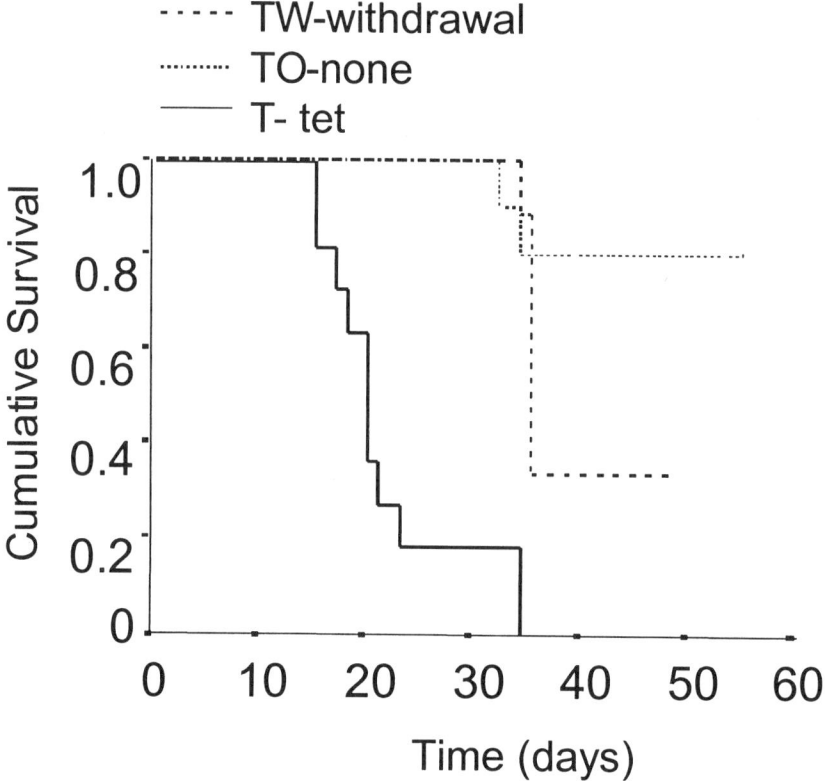

Figure 5. Induction of BCR gene expression inhibits death and disease caused by Bcr-Abl positive K562 cells in NOD/scid mice. Cumulative survival data as a Kaplan/Meier plot of mice injected intravenously in the tail vein with K6 K562 cells are shown. To suppress the expression of the BCR gene, the drinking water was supplemented with 1.0g of tetracycline per liter. Mice not injected with leukemia cells but maintained on Tet at these doses had no visible abnormal effects. At time zero Tet was withdrawn from the water of 10 mice (termed TO mice); at day 19 Tet was withdrawn from a second group of 10 mice (TW); a third group of 10 mice were maintained on Tet from day 0 (T). It must be emphasized that mice dying day 19 were included in the Tet group.

The leukemia syndrome was interesting in that the mice died of either a wasting syndrome with no detectable tumors in spleens, marrow and liver or died of neoplasia of these tissues either combined with or without a wasting syndrome [56]. Pathology studies revealed that mice with the wasting syndrome had either atrophy of cellular hematopoiesis in spleen/marrow or tumors or both. In another clone of K562 cells (K7), also selected for low expression of the Bcr protein under control of the Tet promoter, the atrophy of cellular hematopoiesis was also observed (but at a lower frequency, about 40%) in NOD/*scid* in mice in which BCR expression was blocked but death occurred at later times than the K6 clone. Induction of BCR expression reduced these leukemia effects in a dose-dependent manner (Wang,Y. , Sun, T., and R. Arlinghaus, unpublished).

These findings prompted us to propose a model that describes the relationship of Bcr protein levels in leukemic cells to the degree of aggressiveness of the disease in CML patients [28,56] (Fig. 6). CML patients are typically diagnosed with a benign phase of the leukemia, which if untreated progresses to a more aggressive state (e.g. blast crisis). In this model, the ratio of Bcr protein to the Bcr-Abl oncoprotein is proposed to be a major factor that determines the aggressiveness of the leukemia. Thus, those patients with lower Bcr/Bcr-Abl ratios per cell would progress to more aggressive disease (e.g. from benign phase to accelerated phase and/or blast crisis) in a shorter time frame than higher ratio patients, and be more resistant to therapy (e.g. interferon) than higher ratio patients.

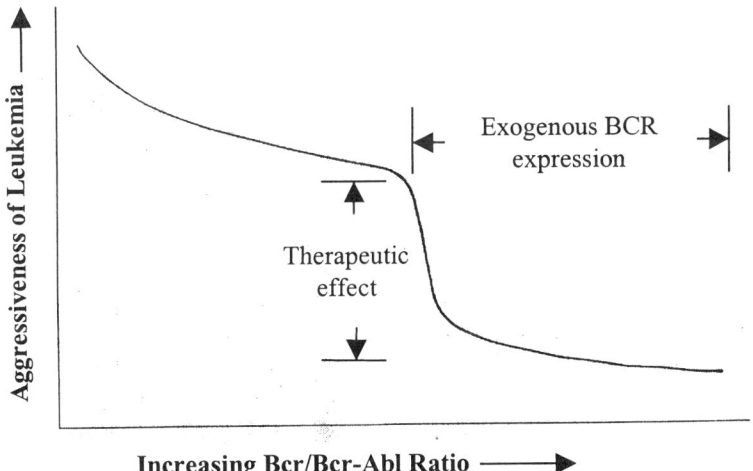

Figure 6. Model for the relationship of the ratio of Bcr to Bcr-Abl to the aggressiveness of the patient's chronic myelogenous leukemia. In this model, CML patients with low levels of Bcr relative to Bcr-Abl in their leukemia cells are proposed to have a more aggressive disease compared to those CML patients that have higher ratios of Bcr to Bcr-Abl protein. Further, it is proposed that BCR gene therapy will provide a significant therapeutic effect resulting in extended patient survival.

6.7. Serine 354 of Bcr Plays a Critical Role in the Inhibitory Effects of Bcr

Our earlier studies indicated that a Bcr peptide containing phosphoserine at residue 354 (350-SSRV**pSPS**PTTYRMFRDK-366) inhibits the tyrosine kinase activity of Bcr-Abl and c-Abl as judged by kinase assays (IC50=600 uM)[49]. The non-phosphoserine form of this peptide was less active as an inhibitor. Moreover, the Bcr phosphoserine peptide did not inhibit the c-Src tyrosine kinase. We have confirmed these results with a variety of phosphoserine Bcr peptides of varying lengths surrounding serine 354; mutant peptides conserving the phosphoserine-like structure (e.g. S354E) were also inhibitory (Jiaxin Liu, Jacki Lin and R. Arlinghaus, in preparation).

Using the adenovirus gene delivery system described above, we tested the effects of the S354A mutant of Bcr(64-413) on K562, BV173, and SMS-SB cells [54]. Bcr(64-413) expression had the expected result on Bcr-Abl expressing cells, causing growth inhibition and cell death. In contrast, the S354A mutant Bcr(64-413) had little effect on these Bcr-Abl positive cells; the effects of the S354A mutant were similar in magnitude to infection with the β GAL adenovirus (Fig. 7). We are exploring the mechanism by which Bcr(64-413) blocks the effects of the Bcr-Abl oncoprotein. Pendergast et al. [57] showed that the serine-rich sequences of Bcr bind firmly to the SH2 domain of c-Abl. Surprisingly, this binding was not dependent on the presence of phosphotyrosine sequences. This binding was confined to two serine-rich regions, termed A and B boxes encoded within the first exon of the BCR gene; the most efficient binding was observed from the B box domain that includes Bcr serine 354. Our findings indicate that phosphoserine forms of Bcr are important for Bcr's inhibitory effects [49, 54, 56]. Phosphoserine Bcr(64-413) as measured by Western blotting with a sequence specific phosphoserine 354 antibody bound tightly to the Abl SH2 domain in GST pull-down experiments conducted with GST-Abl SH2 [54]. Surprisingly, the S354A mutant of Bcr(64-413) had no detectable Abl SH2 binding ability. Thus, the ability of Bcr(64-413) to bind to the Abl SH2 domain correlated with its ability to inhibit Bcr-Abl's oncogenic activity.

We examined the nature of the Bcr(64-413) interaction with the Abl SH2 domain. Our findings at this point indicate that a unique structural form of Bcr interacts with the Abl SH2 domain [54]. We noticed that slower migrating forms of Bcr, either full length or deletion mutant forms containing the coiled-coil domain [3], are present in denaturing SDS polyacrylamide gels. We observed that Bcr(64-413) exists in two forms, a 45-47 kDa and a

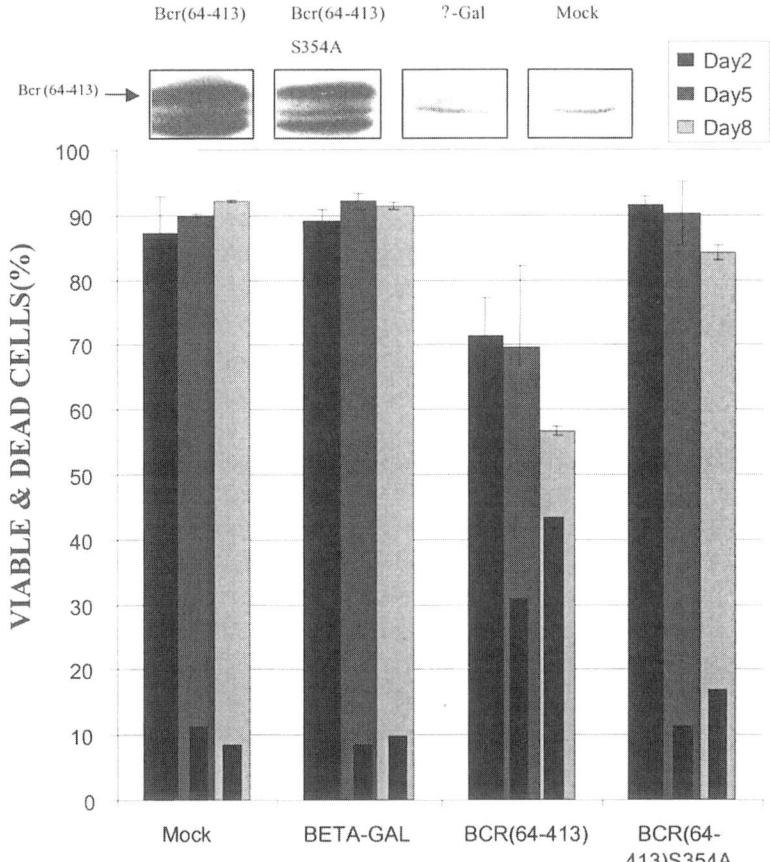

Figure 7. BCR(64-413) S354A is deficient in its ability to induce cell death in K562 cells. Viable K562 cells (stipled, hatched and gray bars) and dead K562 cells (narrow black bars) were measured by trypan blue staining. Cell counts were the average of triplicates. Each viability and dead cell value is the average of two separate adenovirus infection experiments. Inset: shows Western blot analysis of adenovirus infected K562 cells at eight days post infection with anti-Bcr(181-194). The arrow marks the intact form of the 45-47 kDa Bcr(64-413) protein.

55 kDa form. Both forms of BCR(64-413) are recognized by antibodies made against Bcr peptides 181-194 and 298-310. Importantly both forms are recognized by a sequence-specific phosphoserine 354 antibody [50,54]. Importantly, the S354A mutant generated only the 45-47 kDa form but not the 55 kDa form. Moreover, the phosphoserine-containing 55 kDa form but not the 45-47 kDa form was able to bind tightly to the Abl SH2 domain in GST-Abl SH2 pull-down experiments. Therefore, the lack of Abl SH2 binding by the S354A mutant of Bcr(64-413) and its inability to form the unique 55 kDa form of Bcr(64-413) strongly suggests that a unique form of phosphoserine Bcr is required for binding to/interaction with the Abl SH2 domain.

Importantly the inability of the S354A Bcr mutant to block Bcr-Abl's oncogenic effects [54], suggests that the formation of the unique form of Bcr appears to be a critical step in Bcr's inhibitory properties directed towards the Bcr-Abl oncoprotein.

The SH2 domain is well known to bind to phosphotyrosine-containing peptide sequences, with each SH2 domain having a different consensus phosphotyrosine peptide sequence [58]. The FLVRES sequence is highly conserved among SH2 domains and mutation of this sequence interferes with its phosphotyrosine binding ability [59]. We prepared a FLVRES mutant Abl SH2 (R to L within the FLVRES sequence) and compared Bcr(64-413) binding to both wild-type and mutant SH2 domains. No difference in Bcr(64-413) binding to either wild-type or mutant SH2 domain was observed [54]. These findings indicate that the phosphoserine form of Bcr and not phosphotyrosine Bcr that binds to or interacts with the Abl SH2 domain. Moreover, phosphoserine Bcr does not appear to bind to the site within the Abl SH2 domain that binds phosphotyrosine peptide sequences. Moreover, the role of serine 354 of Bcr in this binding is critical, since the S354A mutant of Bcr(64-413) does not bind to the Abl SH2 domain, nor does the S354A mutant produce the slower migrating form, which is the form of Bcr(64-413) that binds to the SH2 domain.

As discussed above, serine 354 of Bcr is a crucial amino acid required for inhibition of the Bcr-Abl kinase and for formation of the slower migrating form of Bcr(64-413) [54]. The slower migrating form of Bcr(64-413) with a mobility of about 55 kDa, we believe, is likely to be a hyperphosphorylated form of the 45 kDa form of Bcr(64-413), and that serine 354 plays a gate-keeper role, because the S354A mutant is unable to form the slower migrating form of Bcr(64-413). One possible interpretation of our findings is that phosphorylation of serine 354 is required for the phosphorylation of additional serine and or /threonine residues within Bcr. We believe that the observed serine-rich domains of Bcr (A & B boxes, ref. 57) play an important role in forming the hyperphosphorylated forms of Bcr. The kinase responsible for phosphorylation of serine 354 is likely to include Bcr itself. Kinase assays and peptide mapping experiments performed with Bcr(1-413) wild-type and the S354A mutant proteins obtained from BCR transfected cells detected the loss of a single phosphotryptic peptide spot from the mutant Bcr protein, as determined by 2-D mapping experiments (N. Hawk, Ph.D. thesis). Bcr is also phosphorylated at serine 354 in transfected COS1 cells, as determined by Western blotting with a sequence-specific phosphoserine 354 antibody [50]. These results indicate serine 354 is a site of phosphorylation within Bcr, and possibly a site of Bcr autophosphorylation. However, we cannot eliminate the possibility that a serine kinase other than Bcr can also phosphorylate serine 354.

We are currently exploring other sites of serine phosphorylation within Bcr. We have identified serine 356 of Bcr as a site of phosphorylation, using phosphoserine sequence-specific antibodies prepared from phosphoserine Bcr peptides [50]. Of interest, the sequence in this region of Bcr is 346-SSGQSSRV**SPSP**TTYR....; the bolded **serine-proline-serine-proline** sequence includes serines 354 and 356. Further studies are underway to provide information about the unique structural forms of Bcr and how they allow firm interaction with the Abl SH2 domain. Importantly, we believe that the interaction of the Abl SH2 sequence with phosphoserine Bcr perturbs the structure of the SH2 domain, thereby interfering with the kinase activity of the Abl tyrosine kinase.

6.8. Model for Bcr Inhibition of Bcr-Abl

Figure 8 depicts a model that fits our findings obtained thus far. Briefly, the phosphorylation of Bcr on serine residues including serine 354 creates a precursor structure, which upon further serine (or threonine) phosphorylation leads to a unique structural form of Bcr (**pSer Bcr****). The presence of serine 354 is required for the formation of the unique structural species that binds to the Abl SH2 domain. Of interest, this form of Bcr is associated with structures in the cell that make it difficult to solubilize with negatively charged strong detergents like sodium dodecyl sulfate (SDS). Also, this form of Bcr is resistant to tyrosine phosphorylation by Bcr-Abl, as shown in studies with Tet-off clones of K562 cells that over-express Bcr upon removal of the Tet block [28,56]. We believe that this resistance to tyrosine phosphorylation results from either its unique phosphoserine structure (e.g it is phosphorylated on either several or many serine residues) and therefore it adopts a conformation that is resistant to Abl kinase phosphorylation, or from its ability to bind to or react with and therefore perturb the SH2 domain of Bcr-Abl leading to inhibition of the tyrosine kinase, or both. The binding of, or interaction with, phosphoserine Bcr by the Abl SH2 domain apparently does not involve the same SH2 sequence necessary for phosphotyrosine peptide binding, as mutation of the conserved FLVRES sequence of the SH2 domain (R to L), which interferes with phosphotyrosine peptide binding, had no effect on phosphoserine Bcr binding [54]. This result suggests that the phosphoserine Bcr binding site is different from the pocket in the SH2 domain that binds phosphotyrosine. Thus by some new mechanism, the unique phosphoserine structure of Bcr binds to or interacts firmly with a different site within the Abl SH2 domain than the phosphotyrosine pocket.

Although the phosphorylation of serine 354 is required for the formation of the unique Bcr structure within Bcr(64-413), our unpublished

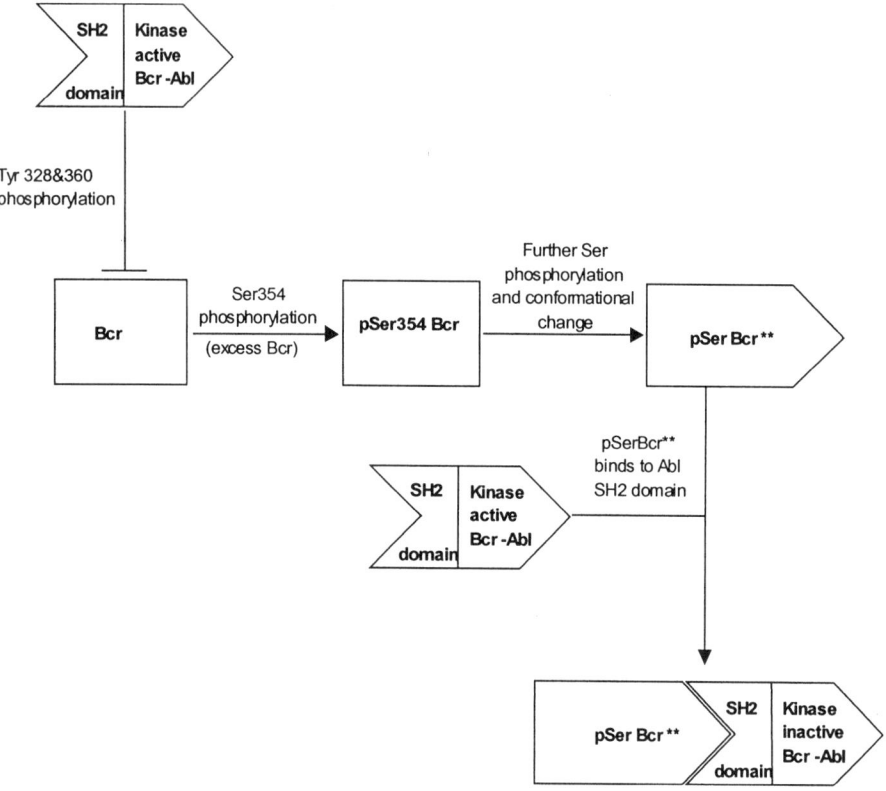

Figure 8. Model describing the interaction of phosphoserine Bcr with the SH2 domain of Bcr-Abl. The model illustrates a two-step serine phosphorylation process and subsequent conformational change of Bcr involving serine 354. Following phosphorylation of serine 354, a unique structural form of Bcr is produced by further serine phosphorylation. This form is identified as **pSer Bcr****. The **pSer Bcr**** structure is proposed to bind to a site within the SH2 domain of the Abl sequence causing perturbation of the catalytic domain, which leads to reduced tyrosine kinase activity. In essence the data predict a two-site model for the SH2 domain. One site binds phosphotyrosine, and a second site binds phosphoserine Bcr. As shown in the upper right portion of the figure, Bcr-Abl inactivates the Bcr serine/threonine kinase by tyrosine phosphorylation of Bcr tyrosines 328 and 360. Bcr-Abl inactivation of Bcr is proposed to occur when the ratio of Bcr-Abl to Bcr in moles is near 1. When the moles of Bcr is greater than Bcr-Abl, the excess Bcr molecules function to inhibit Bcr-Abl, as shown in the diagram.

studies indicate that Bcr proteins that contain the first 63 amino acids may have a back-up serine or serine residues that can serve as gate-keeper serine residues. The first 63 amino acids contain an oligomerization domain (coiled-coil structure, 3). Therefore, it is possible that serine 354, which is located in the serine-rich B box of Bcr, may be essential for Bcr(64-413) to form the unique structure but intact Bcr because of its higher kinase activity due to the tetramer structure might provide one or more back-up serine residues that can partially substitute for serine 354 and therefore allow the structural change to occur but at a lower frequency, when serine 354 is changed to alanine (Jiaxin Liu, Jacki Lin, and R. Arlinghaus, unpublished).

The phosphotyrosine-independent binding of proteins to SH2 domains has been reported in two other systems [60, 61] besides the Bcr/Abl SH2 system [57,62]. One describes the interaction of Raf-1 with the SH2 domains of Fyn and Src [60]. The other concerns the interaction of a 62,000 molecular weight protein with the SH2 domain of p56 Lck in a phosphotyrosine-independent manner [61]. Interestingly, inhibition of c-Abl kinase by actin requires the Abl SH2 domain but not its phosphotyrosine-binding activity [8].

These studies raise questions about how the binding of the phosphoserine Bcr structure affects the Abl catalytic domain and the adjacent SH2 domain. A recent study on the c-Abl protein provides evidence that the amino terminal 80 amino acids auto-inhibits the Abl tyrosine kinase activity [63]. A model describing the three-dimensional arrangement of the c-Abl protein and its regulatory motifs was presented. The N-terminal 80 residues is represented as a rod structure that binds to the catalytic domain (SH1) of the Abl protein, thereby stabilizing the down-regulated structural form (having reduced kinase activity) in which the SH3 domain is bound to the linker sequence between the SH2 domain and the SH1 domain. The resulting structure is viewed as a "closed structure" from the point of access to ATP and/or substrate.

In the Bcr-Abl oncoproteins, the amino terminal portion of the c-Abl sequence is of course deleted, which would remove the auto-inhibitory block. Moreover, the oligomerization of the Bcr-Abl protein would further facilitate the activation of the Abl tyrosine kinase by allowing cross-phosphorylation of key tyrosines in the Bcr-Abl protein. We view the binding of phosphoserine Bcr to the SH2 domain as an event that re-establishes the "closed structure" of the Abl catalytic complex of SH3/SH2/SH1 in a way that although different from that produced by the amino terminal Abl sequence, effectively down-regulates the tyrosine kinase activity of either Abl or Bcr-Abl.

6.9. Bcr is the Long-sought Inhibitor of the Cytoplasmic c-Abl Tyrosine Kinase

The cellular Abl protein or as it is termed the c-Abl protein (in contrast to the viral Abl protein encoded by the Abelson leukemia retrovirus) is a non-receptor tyrosine protein kinase that is localized both in the nucleus and cytoplasm [11,12]. Activation of c-Abl's oncogenic transforming activity is associated with cytoplasmic localization [11]. Recent studies have shown that the protein encoded by the ABL gene becomes activated by signaling through

262

the PDGF receptor pathway [64]. In these studies, binding of PDGF to the receptor activates Src, inducing c-Src to phosphorylate cytoplasmic c-Abl on tyrosine residues, which leads to the activation of c-Abl. The c-Abl tyrosine kinase is associated with an unknown inhibitor that down-regulates its tyrosine protein kinase activity [65]. Since SH2 domains are positive regulators of the non-receptor tyrosine kinases [58], we decided to determine whether this putative inhibitor would bind to an Abl SH2 protein construct and thus activate the c-Abl tyrosine kinase. We devised a lentivirus strategy for efficient transduction of the ABL SH2 DNA construct into various cell types. We used a lentivirus vector system [66] that we modified to increase expression of the HA-tagged ABL SH2 domain by use of an EF-1α promoter [67]. The lentivirus also encodes an internal ribosomal entry site (IRES) sequence to allow green fluorescent protein (GFP) synthesis from the same transcript (Fig. 9).

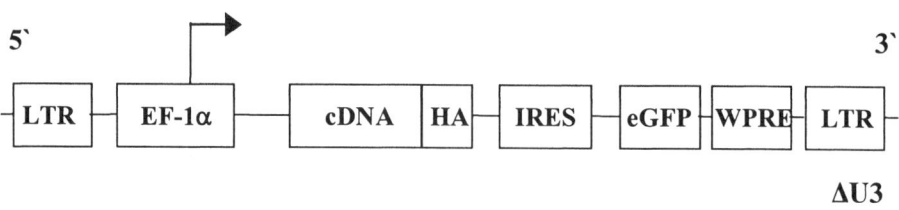

Figure 9. Diagram of a lentivirus gene transduction vector. EF-1 α is the elongation factor 1 α promoter; HA, refers to a hemaagglutin Tag for detection purposes; IRES is an internal ribosome entry site; eGFP is the enhanced green fluorescent protein gene; WPRE is the Woodchuck post transcription processing element; the 3' LTR has a deletion of the U3 sequence to reduce the risk of downstream promoter/enhancement activity. The integrated provirus will have a defective 5' and 3' promoter element as the direct result of this alteration.

This lentivirus was able to transduce efficient GFP expression in 55% to 85% of cells of various types (not shown). Expression of a 10 kDa Abl SH2 protein in COS-1 and Rat-1 cells activated the tyrosine kinase activity of p145 ABL, and induced both morphological transformation and foci formation in Rat-1 cells. Addition of the Abl kinase inhibitor STI-571 to ABL SH2 transformed Rat-1 cells inhibited tyrosine phosphorylation of p145 ABL. We have reported that the endogenous Bcr protein forms a complex with c-Abl in hematopoietic cells [31]; Pendergast et al. have shown that Bcr and c-Abl form a complex in insect cells [57]. We have shown that phosphoserine Bcr interacts with the Abl SH2 domain in a non-phosphotyrosine dependent manner [54], and that this interaction correlates with the Bcr's inhibition of Bcr-Abl oncogenic activity [54]. Importantly, expression of the Bcr protein in ABL SH2 transformed Rat-1 cells reversed the activation of the Abl kinase and prevented morphological transformation and foci formation [67]. Co-immunoprecipitation experiments showed that the 10 kDa Abl SH2 protein formed a complex with p160 BCR [67]. These results indicate that the cytoplasmic inhibitor of the c-Abl tyrosine kinase is Bcr, and that sequestration of the endogenous Bcr protein by the Abl SH2 protein activates both the kinase and oncogenic activities of the c-Abl tyrosine kinase. These findings suggest that the tyrosine kinase activity of the endogenous cytoplasmic form of c-Abl is persistently down-regulated by Bcr (Fig. 10). We propose that Bcr is bound to c-Abl through its SH2 domain by a process involving a serine-phosphorylated form of Bcr (see ref. 54). Our findings suggest that expression of the 10 kDa Abl SH2 domain protein competes with the c-Abl protein for binding to the endogenous phosphoserine Bcr protein. The end result is the activation of the c-Abl tyrosine kinase (probably induced by release of c-Abl from the complex with Bcr) that sets in motion a signaling process that induces oncogenic transformation. Importantly, the endogenous Bcr protein and the c-Abl protein are present in a complex [31,57]; our studies showed that endogenous Bcr is associated with c-Abl in SMS-SB cells [31], a pre-B leukemia cell line lacking the BCR-ABL fusion [68]. The c-Abl/Bcr complex is proposed to be formed by the binding of phosphoserine Bcr to the SH2 domain of c-Abl [54]. Thus, our studies indicate that expression of excess 10 kDa Abl SH2 protein would sequester the endogenous Bcr protein, thereby releasing c-Abl (in a form that can be activated) from the Bcr/c-Abl complex. The details of this process remain to be determined. Importantly then, transducing excess Bcr into the Abl SH2 transformed cell (Fig. 10) sequesters the excess Abl SH2 protein and therefore re-establishes the Bcr/c-Abl complex and c-Abl kinase inhibition. This reverses the oncogenic events, causing the transformed cells to revert to a normal phenotype (blocking foci formation and restoring the normal morphological pattern of growth [67]). It is known that PDGF is involved in activating the c-Abl tyrosine kinase in fibroblasts by a series of reactions involving the activation of the PDGF receptor, activation of c-Src and then activation of c-Abl by Src-induced tyrosine phosphorylation of key tyrosine residues within c-Abl [64]. How the Bcr protein fits into this scenario, if at all, is unknown. One possible scenario that could involve Bcr in the PDGF activation of c-Abl is as follows: inhibition of c-Abl within the Bcr/c-Abl complex would be over-ridden by activation of c-Abl by c-Src as a result of PDGF signaling [64]. Abl activation would then cause tyrosine phosphorylation of Bcr and thus inactivate the serine kinase activity of Bcr [24, 25], which in turn would lead to loss of Bcr serine auto-phosphorylation, loss of phosphate from serine residues by endogenous phosphatases and subsequent dissociation of c-Abl from the complex. The end result would be a transient activation of c-Abl. The Gag-Crk avian retrovirus provided the first evidence of the oncogenic

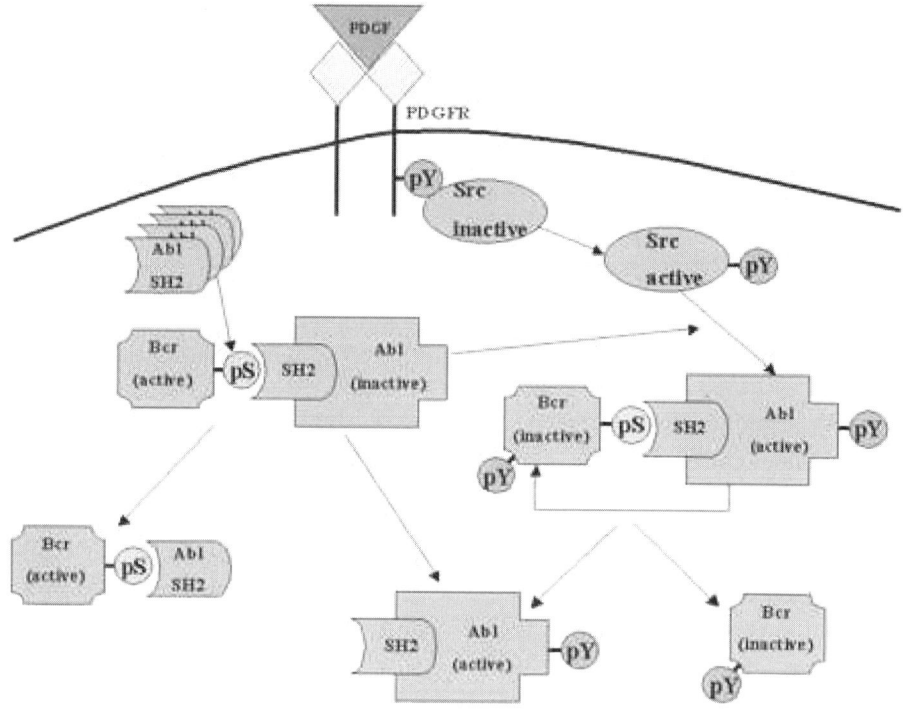

Figure 10. Regulation of the cytoplasmic Abl tyrosine kinase. This model is based on findings published by Ling et al. (67) and Plattner et al. (64). pS refers to phosphoserine; pY refers to phosphotyrosine. The upper-left part of the figure refers to overexpression of the ABL SH2 domain, which sequesters pSer Bcr, removing it from the Bcr/c-Abl complex, and thus activating c-Abl.

activity of SH2/SH3 motifs [69]. These viruses cause tumors and induce tyrosine phosphorylation of a variety of cellular proteins typically seen in cells infected with a virus that transduces a tyrosine kinase (e.g. avian sarcoma viruses encoding v-Src). The tyrosine kinases activated by Gag-Crk expression lead to development of tumors in birds [70]. To our knowledge, the findings presented by us [67] are the first report of the expression of an SH2 domain in the absence of the SH3 domain. It is of interest to test the effects of Bcr on other SH2 domains (e.g. c-Src SH2 domain) to determine whether similar anti-oncogenic effects are observed. However, we observed no effects of BCR expression on Src transformed cells [28]. The mechanism by which Bcr protein expression neutralizes the effects of the 10 kDa Abl SH2 protein probably relates to the ability of phosphoserine Bcr to interact with the Abl SH2 domain in a non-phosphotyrosine dependent manner (Fig. 10, ref. 54). Our studies indicate that a unique structural form of phosphoserine Bcr is required for its interaction with the Abl SH2 domain [54]. The structure of this altered phosphoserine Bcr protein is as yet unknown, but hyper-serine phosphorylation appears to be involved [54]. The binding or interaction of phosphoserine Bcr to the Abl SH2 is not affected by mutating the SH2 domain in a way that inhibits the binding of phosphotyrosine sequences, as the FLVRES mutant (R to L) of the Abl SH2 domain is not altered in its ability to bind a first exon deletion mutant of phosphoserine Bcr [Bcr(64-413)] [54]. The R to L FLVRES mutant ABL SH2 protein also retained the ability to activate the c-Abl kinase and induce foci formation in Rat-1 cells, consistent with the involvement of phosphoserine Bcr as the inhibitor of the c-Abl kinase. This experimental system will provide a useful system to further elucidate the molecular signals involved in oncogenic transformation of cells by the c-Abl protein and its mutant Gag-Abl and Bcr-Abl proteins. Our findings concerning Bcr and c-Abl interaction in Rat-1 cells provide useful information with regard to the Bcr-Abl fusion. For example, our recent experiments indicate that this phenomenon involving sequestering Bcr leads to activation of c-Abl in mouse and human hematopoietic cells [67]. It will be of interest to test the oncogenic potentials of such cells in mouse leukemia models. We have observed that mouse myeloid 32D cells that express an activated c-Abl tyrosine kinase as a result of Abl SH2 protein expression require less IL-3 for survival than control cells with down-regulated c-Abl. These cells were also more resistant to toxic drugs that normally induce apoptosis than control cells (X. Ling and R. Arlinghaus, unpublished). Will the c-Abl activated hematopoietic cells generate a lethal leukemia in mouse models like C3H/HeJ or NOD/scid? Will they have activated Jak2 similar to Bcr-Abl transformed hemjatopoietic cells that do not require IL-3? These studies are now underway.

7. CONCLUSIONS

The Bcr-Abl oncoprotein affects and perturbs several signal transduction pathways within the leukemic cell. Based on mouse leukemia model studies, the Ras pathway is critically required for the onset of leukemia in these mice. The IL-3/GM-CSF- receptor pathways are also critically involved. Stimulation of secretion of these cytokines provide an autocrine growth mechanism. However in parallel, independent activation of both the Jak2 and Stat5 pathways are also involved, although critical experiments in mouse models are needed to verify their importance to the leukemia syndrome. Other interesting alterations of signaling proteins also take place in Bcr-Abl positive leukemia cells. For example, reducing the activity of p53 occurs by a unique translational control mechanism that increases the synthesis of Mdm2, a negative regulator of p53 [40]. Another translational control affect is also mediated by Bcr-Abl. Thus, Bcr-Abl blocks expression of the transcription factor C/EBPα [41], which is required for initiating myeloid differentiation toward granulocytes. Bcr-Abl expression also increases the efficiency of DNA repair mechanisms [43], prolongs the cell cycle by blocking cells at the G2/M phase checkpoint, and stimulates the production of proapoptotic factors of the Bcl-2 family members, thereby providing an environment for drug/radiation resistance. In addition, the Bcr-Abl tyrosine kinase is aptly suited to functionally reduce its cellular inhibitor, Bcr, by a process of tyrosine phosphorylation [24,25]. Thus, the oncogenic effects of the Bcr-Abl oncoprotein can be strongly inhibited by modest over-expression of a cDNA of Bcr, either first exon sequences or the full length BCR sequence. Because of these findings discussed above, further investigations are warranted to develop new treatment strategies to enhance the effects of Gleevec with Bcr and other gene products for the therapy of CML [71].

Ralph B. Arlinghaus and Tong Sun
Department of Molecular Pathology, The University of Texas M.D. Anderson Cancer Center, Houston, Texas

8. REFERENCES

1. Kloetzer, W., Kurzrock, R., Smith, L., Talpaz, M., Spiller, M., Gutterman, J. and Arlinghaus, R.B. The human cellular abl gene product in the chronic myelogenous leukemia cell line K562 has an associated tyrosine protein kinase activity. Virology, 140:230-238, 1985.
2. Konopka J.B., Watanabe S.M., Witte O.N. An alteration of the human c-abl protein in K562 leukemia cells unmasks associated tyrosine kinase activity. Cell, 37, 1035-1042, 1984.
3. McWhirter J.R., Galasso D.L., Wang J.Y.J. A coiled-coil oligomerization domain of Bcr is essential for the transforming function of Bcr-Abl oncoproteins. Mol. Cell Biol., 13, 7587-7595,1993.
4. He, Y., Wertheim, J.A., Xu, L., Miller, J.P., Karnell, F. G., Choi, J. K., Ren, R., Pear, W.S. The coiled-coil domain and Tyr177 of bcr are required to induct a murine chronic myelogenous leukemia-like disease by bcr/abl. Blood, 99:2957-2968, 2002.
5. Zhang, X., Subrahmanyam, R., Wong, R., Gross, A.W., Ren, R. The NH (2) – terminal coiled-coil doman and tyrosine 177 play important roles in induction of a myeloproliferative disease mice by Bcr-Abl. Mol Cell Biol. 21:840-853, 2001.
6. Bedi, A., Zhenbauer, B.A., Barber, J.P., Sharkis, S.J. Inhibition of apoptosis by BCR-ABL in chronic myeloid leukemia. Blood, 83:2038-2044, 1994.
7. McWhiter, J.R. and Wang, J.Y.J. Activation of tyrosine kinase and microfilament-binding functions of c-Abl by Bcr sequences in bcr/abl fusion proteins. Mol. Cell Biol., 11:1553-1565, 1991.
8. Woodring, P.J., Hunter, T., Wang, J.Y.J. Regulation of F-actin-dependent processes by the Abl family of tyrosine kinases. Journal of Cell Science, 116:2613-2626, 2003..
9. Sattler, M., R. Salgia, K. Okuda, N., Uemura, M. A., Durstin, E., Pisick, G., Xu, J-L., Li, Prasad, K.V. Griffin, K.V. The protooncogene product p120CBL and the adaptor

proteins CRKL and c-CRK link c-ABL, p190BCR/ABL and p210BCR/ABL to the phosphatidylinositol-3' kinase pathway. Oncogene, 12:839-846, 1996.

10. Verfaillie, C.M., McCarthy, J. B., McGlave, P. B. Mechanisms underlying abnormal trafficking of malignant progenitors in chronic myelogenous leukemia: decreased adhesion to stroma and fibronectin but increased adhesion to the basement membrane components laminin and collagen type IV. J Clin. Invest., 90:1232-1241, 1992.

11. Van Etten, R.A., Jackson, P., Baltimore, D. The mouse type IV c-abl gene product is a nuclear protein, and activation of transforming ability is associated with cytoplasmic localization. Cell, 58:669-678, 1989.

12. Taagepera, S., McDonald, D., Loeb, J.E., Whitaker, L.L., McElroy, A.K., Wang, J.Y., Hope, T.J. Nuclear-cytoplasmic shuttling of C-ABL tyrosine kinase. Proc Natl Acad Sci USA, 95:7457-7462, 1998.

13. Vigneri, P., Wang, Jean, Y. J. Induction of apoptosis in chronic myelogenous leukemia cells through nuclear entrapment of BCR-ABL tyrosine kinase. Nature Medicine, 7:228-234, 2001.

14. Puil, L., Liu, J., Gish, G., Mbalamu, G., Arlinghaus, R., Pelicci, P.G., Pawson, T. BCR-ABL oncoproteins bind directly to activators of the Ras signalling pathway. EMBO, 13:764-773, 1994.

15. ten Hoeve, J., Morris, C., Heisterkamp, N., Groffen, J. Isolation and chromosomal localization of CRKL, a human crk-like gene. Oncogene, 8:2469-2474, 1993.

16. Gorre, M. E., Mohammed, M., Ellwood, K., Hsu, N., Paquette, R., Rao, P. N., Sawyers, C. L. Clinical resistance to STI-571 cancer therapy caused by BCR-ABL gene mutation or amplification. Science, 293:876-880, 2001.

17. ten Hoeve, J., Arlinghaus, R.B., Guo, J.Q., Heisterkamp, N., and Groffen, J. Tyrosine phosphorylation of CRKL in Ph-positive leukemia. Blood, 84:1731-1736, 1994.

18. Xie, S., Wang, Y., Liu, J., Sun, T., Wilson, M. B., Smithgall, T.E., Arlinghaus, R. B. Involvement of Jak2 tyrosine phosphorylation in Bcr-Abl Transformation. Oncogene, 20:6188-6195, 2001.

19. Wilson-Rawls, J., Xie, S.H., Liu, J., Laneuville, P., Arlinghaus, R.B. P210 Bcr-Abl interacts with the interleukin-3 bc subunit and constitutively induces its tyrosine phosphorylation. Cancer Res., 56:3426-3430, 1996.

20. Xie, S., Lin, H., Sun, T., Arlinghaus, R.B. Jak2 is involved in c-Myc induction by Bcr-Abl. Oncogene, 21:7137-7146, 2002.

21. Ilaria, R.L., Jr., Van Etten, R.A. P210 and P190(BCR/ABL) induce the tyrosine phosphorylation and DNA binding activity of multiple specific STAT family members. J. Biol. Chem., 271:31704-31710, 1996.

22. Klejman, A., Schreiner, S.J., Nieborowska-Skorska, M., Slupianek, A., Wilson, M., Smithgall, T.E., Skorski, T. The Src family kinase Hck couples BCR/ABL to STAT5 activation in myeloid leukemia cells. EMBL J. 21:5766-5774, 2002.

23. Liu J., Campbell, M., Guo, J.Q., Lu, D., Xian, Y.M., Andersson, B.S., Arlinghaus, R.B. BCR-ABL tyrosine kinase autophosphorylates itself or transphosphorylates P160 BCR on tyrosine predominantly within the first exon. Oncogene, 8:101-109, 1993.

24. Liu, J., Wu, Y., Lu, D., Haataja, L., Heisterkamp, N., Groffen, J., Arlinghaus, R.B. Inhibition of Bcr serine kinase by tyrosine phosphorylation. Mol. Cell Biol., 16:998-1005, 1996.

25. Wu, Y., Liu, J., Arlinghaus, R.B. Requirement of two specific tyrosine residues for the catalytic activity of Bcr serine/threonine kinase. Oncogene, 16:141-146, 1998.

26. Pendergast, A.M., Quilliam, L.D., Cripe, L.D., Bassing, C.H., Dai, Z., Li, N, Batzer A., Rabun, K.M., Der, C.J., Schlessinger, J., Gishizky, M.L. BCR-ABL-induced

268

Oncogenesis Is Mediated by Direct Interaction with the SH2 Domain of the GRB-2 Adaptor Protein. Cell, 75:175-185, 1993.

27. Pawson, T. Tyrosine kinase signaling pathways. Princess Takamatsu Symp., 24:303-322, 1994.

28. Wu, Y., Ma, G., Lu, D., Lin, F., Xu, H-J., Liu, J., Arlinghaus, R.B. Bcr: a negative regulator of the Bcr-Abl oncoprotein. Oncogene, 18:4416-4424, 1999.

29. Reuther, G. W., Fu, Cripe, L.D., Collier, R.J., Pendergast, A.M. Association of the Protein Kinases c-Bcr and Bcr-Abl with Proteins of the 14-3-3 Family. Science, 266:129-133, 1994.

30. Braselmann, S., McCormick F. BCR and RAF form a complex in vivo via 14-3-3 proteins. EMBO J., 14:4839-4848, 1995.

31. Ma, G., Lu, D., Wu, Y., Liu, J., and Arlinghaus, R.B. Bcr phosphorylated on tyrosine 177 binds Grb2. Oncogene, 14:2367-2372, 1997.

32. Heisterkamp, N., Jenster, G., ten Hoeve, J., Zovich., D., Pattengale, P.K., Groffen, J. Acute leukaemia in bcr/abl transgenic mice. Nature, 344:251-253, 1990.

33. Lugo, T.G., Witte, O.N. The BCR-ABL oncogene transforms Rat-1 cells and cooperates with v-myc. Mol. Cell Biol., 9:1263-1270, 1989.

34. Daley, G.Q., Baltimore, D. Transformation of an interleukin 3-dependent hematopoietic cell line by the chronic myelogenous leukemia-specific P210bcr/abl protein. Proc. Natl. Acad Sci U.S.A., 85:9312-9316, 1988.

35. Sirard, C., Laneuville, P., Dick, J.E. Expression of bcr-abl abrogates factor-dependent growth of human hematopoietic M07E cells by an autocrine mechanism. Blood, 83:1575-1585, 1994.

36. Jiang, X., Lopez, A., Holyoake, T., Eaves, A., Eaves, C. Autocrine production and action of IL-3 and granulocyte colony-stimulating factor in chronic myeloid leukemia. Proc. Natl. Acad. Sci. U.S.A., 96:12804-12809, 1999.

37. Rui, L., Carter-Su, C. Identification of SH2-bbeta as a potent cytoplasmic activator of the tyrosine kinase Janus kinase 2. Proc. Natl. Acad. Sci. U.S.A., 96:7172-7177, 1999.

38. Sawyers, C.L., Callahan, W., Witte, O.N. Dominant negative MYC blocks transformation by ABL oncogenes. Cell, 70:901-910, 1992.

39. Rui, L., Gunter, D. R., Herrington, J. Carter-Su, C. Differential binding to and regulation of JAK2 by the SH2 doman and N-termianl region of SH2-bbeta. Mol Cell Biol., 20:3168-3177, 2000.

40. Trotta, R., Vignudelli, T., Candini, O., Intine, R.V., Pecorari, L., Guerzoni, C., Santilli, G., Byrom, M. W., Goldoni, S., Ford, L. P., Caligiuri, M. A., Maraia, R. J., Perrotti, D., Calabretta, B. BCR/ABL activates mdm2 mRNA translation via the La antigen. Cancer Cell, 3:145-160, 2003.

41. Perrotti, D., Cesi, V., Trotta, R., Guerzoni, C., Santilli, G., Campbell, K., Iervolino, A., Condorelli, F., Gambacorti—Passerini, C., Caligiuri, M.A., Calabretta, B. BCR-ABL suppresses C/EBPalpha expression through inhibitory action of hnRNP E2. Nat. Genet., 30:48-58, 2002.

42. Feinstein, E. Cimino, G., Gale, R.P., Alimena, G., Berthier, R., Kishi, K., Goldman, J., Zaccaria, A., Berrebi, A., Canaani, E. p53 in chronic myelogenous leukemia in acute phase. Proc Natl Acad Sci U.S.A., 88:6293-6297, 1991.

43. Skorski, T. BCR/ABL regulates response to DNA damage: the role in resistance to genotoxic treatment and in genomic instability. Oncogene, 21:8591-8604, 2002.

44. Kurzrock, R., Shtalrid, M., Romero, P., Kloetzer, W.S., Talpas, M., Trujillo, J.M., Blick, M., Beran, M., Gutterman, J.U. A novel c-abl protein product in Philadelphia-positive acute lymphoblastic leukaemia. Nature, 325:631-635, 1987.

45. Li, W., Draezen, O., Kloetzer, W.S., Gale, R.P., and Arlinghaus, R.B. Characterization of bcr gene products in hematopoietic cells. Oncogene, 4:127-138, 1989.

46. Maru, Y., Witte, O.N. The BCR gene encodes a novel serine/threonine kinase activity within a single exon. Cell, 67:459-468, 1991.

47. Lu, D., Liu, J., Campbell, M., Guo, J.Q., Heisterkamp, N., Groffen, J., Canaani, E., and Arlinghaus, R.B. Tyrosine phosphorylation of P160 BCR by P210 BCR-ABL. Blood, 82:1257-1263, 1993.

48. Campbell, M., W. Li, and Arlinghaus, R.B. P210 BCR-ABL is complexed to P160 BCR and ph-P53 proteins in K562 cells. Oncogene, 5:773-776, 1990.

49. Liu, J., Wu, Y., Arlinghaus, R.B. Sequences within the first exon of BCR inhibit the activated tyrosine kinase of c-Abl and the Bcr-Abl oncoprotein. Cancer Res., 56:5120-5124, 1996.

50. Sun, T., Campbell, M., Gordon, W., Arlinghaus, R.B. Preparation and Application of Antibodies to Phosphoamino Acid Sequences. Biopolymers (Peptide Science), 60:61-75, 2001.

51. Arlinghaus, R.B. The involvement of Bcr in leukemias with the Philadelphia chromosome. Invited review, Journal of Critical Reviews in Oncogenesis, 9:1-18, 1998.

52. Annamalai, A. E., Colman, R. F. Reaction of the adenine nucleotide analogue 5'-p-fluorosulfonylbenzoyl adenosine at distinct tyrosine and cysteine residues of rabbit muscle pyruvate kinase. J. Biol. Chem., 256:10276-10283, 1981.

53. Tomich, J.M., Marti, C., Colman, R.F. Modification of two essential cysteines in rabbit muscle pyruvate kinase by the guanine nucleotide analogue 5' [p-(fluorosulfonyl) benzoyl] guanosine. Biochemistry, 20:6711-6720, 1981.

54. Hawk, N., Liu, J., Sun, T., Wang, Y., Wu, Y., Arlinghaus, R.B. Inhibition of the Bcr-Abl oncoprotein by Bcr requires phosphoserine 354. Cancer Res., 62:386-390, 2002.

55. Wang, Y., Liu, J., Wu, Y., Luo, W., Lin, S-H., Lin, H., Hawk, N., Sun, T., Guo, J.Q., Estrov, Z., Talpaz, M., Champlin, R., Arlinghaus, R.B. Expression of a truncated first BCR sequence in chronic myelogenous leukemia cells block cell growth and induces cell death. Cancer Res., 61:138-144, 2001.

56. Lin, F., Liu, J., Monaco, G., Sun, T., Liu, J., Lin, H., Stephens, C., Belmont, J., Arlinghaus, R.B. BCR gene expression blocks Bcr-Abl induced pathogenicity in a mouse model. Oncogene, 20:1873-1881, 2001.

57. Pendergast, A.M., Muller, A.J., Havlik, M.H., Maru Y., Witte O.N. BCR sequences essential for transformation by the BCR-ABL oncogene bind to the ABL SH2 regulatory domain in a nonphosphotyrosine-dependent manner. Cell, 66:161-171, 1991.

58. Songyang, Z., Carraway, K.L., III, Eck, M. J., Harrison, S.C., Feldman, R. A., Mohammadi, M. Schlessinger, J., Hubbard, S. R., Smith, D.P., Lorenzo, M. J., Ponder, B.A.J., Mayer, B. J. and Cantley, L.C. Cataytic specificity of protein-tyrosine kinases is critical for selective signaling. Nature, 373:536-539, 1995.

59. Mayer, B. J., Jackson, P.K., Van Etten, R. A., Baltimore, D. Point mutations in the abl SH2 domain coordinately impair phosphotyrosine binding in vitro and transforming activity in vivo. Mol. Cell Biol, 12:609-618, 1992.

60. Cleghon, V., Morrison, D.K. Raf-1 interacts with Fyn and SCR in a non-phosphotyrosine-dependent manner. Biol Chem., 269:17749-17755, 1994.

61. Park, I., Chung, J., Walsh, C.T., Yun, Y., Strominger, J.L., Shin, J. Phosphotyrosine-independent binding of a 62-kDa protein to the scr homology 2 (SH2) domain of p56lck and its regulation by phosphorylation of Ser-59 in the lck unique N-terminal region. Proc Natl Acad Sci USA, 92:12338-12342, 1995.

62. Muller, A.J., Pendergast, M., Havlik, H., Puil, L., Pawson, T., Witte, O.N. A limited set of SH2 domains binds Bcr through a high-affinity phosphotyrosine-independent interaction. Mol. Cell Biol., 12:5087-5083, 1992.

63. Pluk, H., Dorey, K., Superti-Furga, G. Autoinhibition of c-Abl. Cell, 108:247-259, 2002.

64. Plattner, R., Kadlec, L., DeMali, K. A., Kazlauskas, A., Pendergast, A.M. c-Abl is activated by growth factors and Src family kinases and has a role in the cellular response to PDGF. Genes & Dev., 13:2400-2411, 1999.

65. Pendergast, A.M., Muller, A.J., Havlik, M.H., Clark, R., McCormick, F., Witte, O.N. Evidence for regulation of the human ABL tyrosine kinase by a cellular inhibitor. Proc Natl Acad Sci USA, 88:5927-5931, 1991.

66. Naldini, L., Blomer, U., Gallay, P., Ory, D., Mulligan, R., Gage, F.H., Verma, I.M., Trono, D. In vivo gene delivery and stable transduction of non-dividing cells by a lentiviral vector. Science, 272:263-267, 1996.

67. Ling, X., Ma, G., Sun, T., Liu, J., Arlinghaus, R. B. Bcr and Abl Interaction: Oncogenic Activation of c-Abl by Sequestering Bcr. Advances in Brief, Cancer Research, 61:198-303, 2003.

68. Tsai LH, Nanu L, Smith RG, Ozanne B. Overexpression of c-fos in a human pre-B cell acute lymphocytic leukemia derived cell line, SMS-SB. EMBO J., 9:415-424, 1990.

69. Mayer, B.J., Hanafusa, H. Mutagenic Analysis of the v-crk Oncogene: Requirement for SH2 and SH3 Domains and Correlation between Increased Cellular Phosphotyrosine and Transformation. J. of Virology, 64:3581-3589, 1990.

70. Mayer, B.J., Hamaguchi, M., Hanafusa, H. A novel viral oncogene with structural similarity to phospholipase C. Nature, 332:272-275, 1988.

71. Druker, B.J. Inhibition of the Bcr-Abl tyrosine kinase as a therapeutic strategy for CM. Oncogene, 21:8541-8546, 2002.

72. Faderl, S., Talpaz, M., Estrov, Z., O'Brien, S., Kurzrock, R., Kantarjian, H. M. The biology of chronic myeloid leukemia. The New England J of Med., 341: 164--172, 1999.

73. Deininger, M. W. N., Goldman, J. M. Melo, J. V. The molecular biology of chronic myeloid leukemia. Blood, 96: 3343-3356, 2000.

ESTROGEN RECEPTORS AND
ANTI-ESTROGEN THERAPIES

LAKJAYA BULUWELA, DEMETRA CONSTANTINIDOU,
JOANNA PIKE AND SIMAK ALI

1. INTRODUCTION

The steroid hormone estrogen has wide ranging activities in the body and is most noted for the critical role it plays in the development and maintenance of the female and male reproductive systems and secondary sexual characteristics. Estrogens have also long been known to play a central role in promoting breast cancer growth. This understanding has led to the development of anti-estrogens as drugs for breast cancer treatment, including tamoxifen, the first target-directed anti-cancer drug. Tamoxifen has now become the first-line adjuvant treatment for estrogen responsive breast cancers and its widespread use has been a major factor in the significant improvement now seen in the survival of these patients. Estrogen action is achieved through the regulation of target genes by two estrogen receptors, ERα and ERß. Here, we describe the mechanisms by which these receptors act to regulate gene expression in an estrogen dependent and independent manner, the use of anti-estrogens in the treatment of breast cancer and the recent exciting findings arising from the use of anti-estrogens in breast cancer prevention.

2. ESTROGEN RECEPTORS ARE MEMBERS OF A SUPERFAMILY OF TRANSCRIPTION FACTORS

In the early 1960's Jensen and Jacobsen [1] put forward the hypothesis that the biological effects of estrogens were mediated by a receptor protein, termed the Estrogen Receptor, or ER. However, it was not until the cloning of the estrogen receptor, ERα, in the mid-1980's that a detailed understanding of this activity began to emerge [2]. Initial sequence comparisons showed that ERα is closely related to the glucocorticoid receptor [3]. Further sequence comparisons highlighted a homology between these two proteins and the v-erbA oncogene, encoded by the avian erythroblastosis virus. Subsequently, the cellular homologue of v-erbA was identified as the thyroid hormone receptor, a finding that suggested that the steroid hormone receptors were members of a larger gene family [4]. This gene family is now known as the nuclear receptor (NR) superfamily and defines one of the largest families of higher eukaryotic transcription factors. For example, analysis of genome sequences have identified 48 NR genes in man, including two distinct but closely related estrogen receptors, ERα and ERß, 68 NR genes in the pufferfish *Fugu rubripes* and 270 NR genes in the nematode *Caenorhabditis elegans* [5, 6].

In mammals, these receptors can be grouped into three subclasses on the basis of ligand binding. The first class are the so-called endocrine receptors, which include the steroid hormone receptors for androgens, estrogen, glucocorticoids and progesterone, as well as the retinoic acid, thyroid hormone and vitamin D3 receptors. These receptors bind with high affinity to ligands of endocrine origin. The second group comprise the so-called adopted orphan nuclear receptors, which bind with low affinity to physiological ligands that are generally dietary lipids. The final group, the so-called orphan nuclear receptors, have unknown ligand specificities, or have no requirement for ligands for activity. Collectively, NR have important functions in reproduction, development and in regulating metabolic processes [7, 8].

2.1. Estrogen Receptors

ERα and ERß share with other NR a common structure, being comprised of three domains that can act independently of each other, but which function together in the receptor (Figure 1). Ligand binding to ER triggers conformational changes that lead to binding to estrogen response elements (ERE) encoded in the promoters of estrogen-regulated genes. This results in the recruitment of transcription coregulators, the formation of a transcription preinitiation complex and increased transcription.

Region C, or the DNA binding domain (DBD), is the most highly conserved region amongst the NR family and is comprised of two zinc fingers. The N-terminal zinc finger encodes the P-box, which is required for specific contacts with DNA sequences in the response element. The second zinc finger participates in non-specific interactions with the DNA phosphate backbone, and is involved in receptor dimerization [9]. The DBDs of ERα and ERß are 96% homologous and are identical in the P-box, which explains their ability to bind to EREs with similar specificities and affinities.

Region E is also well conserved and encodes the ligand binding domain (LBD). The LBD mediates binding to estrogen and anti-estrogens, receptor dimerization and transactivation of target gene expression. Steroid receptors bind as homodimers to palindromic response elements, although ERα and ERß can also heterodimerise [10], potentially allowing differential expression of estrogen-regulated genes, depending on the relative levels of each ER isoform in a given cell type.

Ligand-regulated receptor dimerisation is mediated by the LBDs. Structural studies have shown that the LBD is a wedge-shaped structure generally composed of twelve α-helices, within which the ligand is accommodated [11]. Amino acid residues that line the surface of the ligand binding cavity, or that interact directly with bound ligands, span the LBD from helix 3 to helix 12. Structural studies also reveal that agonist binding reveals a hydrophobic groove comprising α-helices 3, 4, 5 and 12, which together form a surface for the recruitment of transcriptional coactivators. Coactivators interact with agonist-bound NR LBD via two-turn α-helical motifs containing the signature sequence containing leucine-X-X-leucine-leucine (LXXLL), where X represents any amino acid [12, 13].

In the ERα and ERß LBD structures with the anti-estrogen tamoxifen or raloxifene, helix 12 is displaced from its agonist position over the ligand-binding cavity and takes up position in the coactivator groove, thereby preventing coactivator recruitment [13-15]. Yet another mechanism of anti-estrogen action is indicated by the structure of the ERß LBD with the pure anti-estrogen ICI 164, 384, which is very similar in structure to Faslodex. Here it is the bulky side chain of ICI 164, 384 which protrudes into and occupies the coactivator binding pocket, thereby preventing helix 12 from assuming either the coactivator bound or the tamoxifen/raloxifene-bound conformation [15].

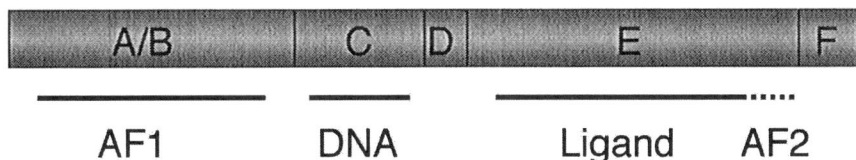

Figure 1. Schematic representation of the domain structure of nuclear receptors. The A/B domain contains AF1, which mediates transcriptional activation. The C domain contains the zinc fingers, which mediate binding to DNA and region E is the ligand binding domain. Region E is also involved in receptor dimerisation and contains AF2, which regulates the recruitment of transcriptional coactivators.

The A/B region, which encodes transcription activation function AF1, is required for transcriptional co-regulator recruitment and activation of the expression of estrogen-regulated genes. In the case of ERα, AF1 is highly active, its activity being dependent on cell and promoter context [16]. Moreover, the agonistic activity of partial estrogen antagonists, such as tamoxifen and raloxifene, is mediated by its inhibition of AF2, but activation of AF1. The ERß AF1, on the other hand, has negligible activity, which may explain why tamoxifen and raloxifene are pure antagonists for ERß [17, 18]. Moreover, it is likely that phosphorylation, particularly in AF1, can stimulate the activity of many NR [19, 20]. Notably, ERα is activated in a ligand-independent manner by stimulation of several protein kinase signalling cascades.

2.2. Transcription Regulation by Estrogen Receptors

In common with other steroid receptors, but unlike the non-steroid NR, the unliganded ERα is complexed with the heat shock protein, hsp90, which is believed to maintain the receptor in an inactive state [21]. It was originally thought that all steroid receptors were cytoplasmic in the absence of their cognate ligand. It is now clear, however, that this is only true for the androgen and glucocorticoid receptors, whereas other unliganded NR, including ER, localise mainly to the nucleus. Estrogen binding releases ERα from the hsp90 complex, thereby enabling dimerisation and sequence-specific binding to EREs.

ERα can interact with several components of the general transcription machinery to stimulate RNA polymerase II recruitment and transcription from naked DNA templates *in vitro* [22]. Using naked DNA templates similar findings have been obtained for other steroid receptors [23, 24]. However, this stimulation is ligand-independent and does not explain the ligand requirement for activation of gene expression by ERα *in vivo*. Moreover, in eukaryotes, genomic DNA is packaged around histone units to form nucleosomes, which are the repeating units of chromatin. Chromatin restricts access to transcription factors and the RNA polymerase complex and is thereby repressive for gene expression. Indeed, a critical role for chromatin in regulating gene expression by NR and obtaining ligand-regulated gene activation have been clearly demonstrated by *in vitro* studies using chromatinised templates [25]. It is now clear that transcription regulation by NR requires the recruitment of chromatin remodelling complexes, coactivators and corepressors.

Chromatin remodelling is an ATP-dependent process, involving alteration in chromatin structure to activate or repress gene expression. Many different ATP-dependent chromatin remodelling complexes have been identified, including the SWI/SNF and ISWI complexes [26, 27]. The SWI/SNF complexes can alter histone-DNA contacts, such that the path of DNA around the histone octamer is rearranged, thereby exposing specific DNA sequences, or through the transfer of histones from one section of DNA to another. The ISWI family of complexes loosen histone-DNA contacts to permit nucleosomes to "slide" in relation to each other.

Structural studies show that the N-terminal tails of the core histones protrude from the nucleosome structure and are involved in intranucleosomal interactions that influence chromatin compaction [28]. Post-translational modification of histone tails, including acetylation, methylation, and phosphorylation are all known to regulate gene activity [29, 30]. Acetylation of the ε-amino group of specific lysine residues within the N-terminal tails of core histones, is believed to result in reduced interaction with DNA and hence produce a "looser" chromatin conformation, allowing transcription factor access to the DNA. Histone acetylation is carried out by a family of enzymes known as histone acetyl transferases (HAT), many of which are known coactivators for NR. By contrast, histone deacetylases remove the acetyl group and are associated with transcriptional corepressor complexes. The dynamic reversibility of histone acetylation/deacetylation provides a mechanism for regulation of gene expression by NR and other transcription factors.

Methylation of arginine residues is also associated with gene activation, whilst lysine methylation is associated with transcriptional repression and activation, depending on the lysine that is methylated [30]. Phosphorylation, particularly at serine 10 of histone H3 has been correlated with initiation of transcription of immediate-early genes such as *c-jun*, *c-fos* and *c-myc* [31, 32]. Furthermore, modification at one residue can affect modification at other sites. For example, acetylation or methylation of lysine 9 of histone H3 are mutually exclusive. Methylation at this site inhibits phosphorylation of serine 10, whilst acetylation of lysine 9 and phosphorylation of serine 10 appear to be synergistic [29]. Additionally, histone acetylation enables the recruitment of bromo domain-containing proteins, including HATs such as CBP, GCN5 and P/CAF, whilst histone methylation allows the recruitment of chromo domain proteins, such as the heterochromatin protein HP1 and the lysine methyltransferase su(var)3-9. Together this diversity of histone tail modifications, acting synergistically or antagonistically in the recruitment of chromatin-associated proteins has been proposed as a "histone code" that provides epigenetic information for dynamic and long-term changes in gene expression [29].

2.3. Coactivators of the Estrogen Receptor

A large number of proteins that interact with ER and which potentially act as coactivators, have been described. Of these, the best characterised are the three members of the SRC1 family, namely SRC1/N-CoA1, TIF-2/GRIP1 and AIB1 [33, 34]. These coactivators are recruited to the LBD of liganded NR, where they predominantly function as platform proteins for the recruitment of other coactivator proteins, including the CREB Binding Protein (CBP) and its homologue p300, a factor that also interacts directly with NR. CBP/p300 in-turn recruits p300/CBP Associated Factor (P/CAF). P/CAF is also directly recruited to NR through interaction with the NR DBD [35]. CBP/p300 and P/CAF possess intrinsic HAT activity that can lead to chromatin modification and enhanced gene activation. SRC1 family members also recruit CARM1 and PRMT1, arginine methyltranserfeases that methylate arginine 17 of histone H3 and arginine 3 of histone H4, both modifications that result in a stimulation of gene expression [36].

NR also recruit the ISWI and SWI/SNF complexes. For example, the estrogen-bound ERα interacts with the BAF57 subunit of the SWI/SNF complex, to recruit SWI/SNF to promoters of estrogen-regulated genes [37]. Thus, ligand-induced coactivator recruitment to NR leads to chromatin modification and remodelling, which increases transcription factor accessibility to specific DNA sequences. In addition, the mammalian mediator complex, TRAP/DRIP/SMCC/Mediator, interacts with liganded NR and is required for thyroid hormone and vitamin D3 receptor, as well as ERα activities *in vitro* [38]. This complex purifies as a component of the RNA polymerase II holoenzyme and is required for mRNA synthesis and is likely to act as a bridging factor between NR and the basal transcription machinery. Hence, liganded NR activate gene expression by assembling an array of coactivator proteins, with diverse functions, on the promoters of target genes.

2.2.3. Corepressors and the Estrogen Receptor

Many nuclear receptors lower basal promoter activity in the absence of ligand through the recruitment of corepressor complexes. This is mediated by the recruitment of the NR corepressor (N-CoR) or the related factor SMRT (Silencing Mediator of Retinoid and Thyroid receptors) [39, 40]. N-CoR and SMRT are present in at least three distinct protein complexes. Several of these complexes contain histone deacetylases and their recruitment by unliganded receptors results in histone deacetylation and repression of the basal expression of regulated genes [41, 42]. However, N-CoR/SMRT have also been identified in association with the SWI/SNF ATP-dependent chromatin remodelling complex that also contains KAP-1, a corepressor that has been linked to heterochromatin silencing [43]. N-CoR/SMRT are also recruited by antagonist binding to some NR. Notably, the antagonist-bound ERα associates with N-CoR/SMRT and down-regulation of N-CoR protein levels has been shown to increase the agonist activity of tamoxifen [44, 45].

Several other proteins that interact with ERα to inhibit expression of estrogen-regulated reporter genes have been identified. Of particular interest are RIP140 [46] and L-CoR [47], which interact with the ERα LBD upon estrogen binding, to repress expression of estrogen-regulated reporter genes. These corepressors possess LXXLL motifs distinct from the corepressor motifs required for interaction of N-CoR and SMRT with NR [12, 48, 49]. Moreover, RIP140 and L-CoR repression appears to be mediated through complexes containing the transcriptional repressor CtBP and histone deacetylases [47, 50] and as such, they appear to be act in a manner quite distinct from N-CoR/SMRT. TIF1α, which also interacts with agonist-bound NR, similarly appears to be a corepressor, which interacts with HP1 [51, 52].

Collectively, these corepressor complexes may function to bring about gene repression in response to estrogen.

3. ESTROGENS AND BREAST CANCER

The mammary gland is an organ that exemplifies the activities of estrogen. In all mammalian species the structural development of the breast in females coincides with the acquisition of fertility, which is around the ages of 10 and 12 years in humans. With the approach of puberty, the rudimentary mammary gland begins to show growth both in the glandular tissue and in the surrounding stroma (differentiated mesenchyme). Mammary gland development during puberty starts before the first menstrual period and well before the ovary is fully functional. Hence, the growth of the breast is usually the first sign of the onset of puberty and precedes ovulation and secretion of the steroid hormone progesterone from the corpus luteum by several years. This suggests that mammary gland growth and development in early puberty is due to the effects of circulating estrogens, rather than progesterone, because the latter hormone is released only in the luteal phase of a fully established menstrual cycle [53].

During the estrogen-dominated follicular phase there is ductal elongation in the mammary gland. This growth appears to be induced by both estrogen and growth hormone. The full effect of growth hormone in mammary ductal morphogenesis may be mediated through insulin-like growth factor-1 (IGF-1) [54]. In contrast, growth stimulation by estrogen depends largely on the presence of ERα, as mammary development and function is phenotypically normal in ERß gene knockout mice [55], whereas ERα knockout mice do not respond to estrogenic signals and show lack of ductal morphogenesis [56]. ERα expression is confined to a sub-population of luminal epithelial cells and surrounding stromal fibroblasts. ERα expression in both of these cell types is required for normal mammary development in mice, although interestingly, mammary fat pad transplantation experiments using ERα knockout mice demonstrate that stromal ERα is sufficient for mammary development in adult mice [57]. In the resting human breast, only 15%-25% of the luminal epithelial cells are ERα-expressing. Interestingly, these cells do not proliferate, but appear to stimulate the proliferation of surrounding ERα-negative epithelial cells, presumably by stimulation of the synthesis and/or secretion of paracrine factors [58].

3.1. The Role of Estrogens in Breast Cancer

Breast cancer is the leading cancer in women in the Western World, with an incidence that continues to increase [59, 60]. Currently, in both the UK and USA, breast cancer affects around 140 women per 100, 000, with a lifetime risk as high as 1 in 8. However, breast cancer mortality has been falling significantly, primarily as a result of early diagnosis and screening programmes and the use of endocrine agents in an adjuvant setting [61].

The first indication of a connection between estrogen and breast cancer came as far back as 1896, when George Beatson demonstrated that surgical removal of ovaries (oophorectomy) of pre-menopausal women resulted in the regression of advanced breast tumours [62], although tumour regression was limited to a third of breast cancer patients [63]. The tumour regression obtained with oophorectomy is now known to be due to the elimination of ovarian produced estradiol, which promotes proliferation of breast cancer cells (for reviews and refs. see [64, 65]. In addition to its role in tumour progression, it is clear that the likelihood of breast cancer increases with the lifetime exposure to estrogens. Thus, increased risk of breast cancer has been associated with early onset of puberty, late first full-term pregnancy and, a late onset of menopause. Prolonged use of estrogen-containing oral contraceptives, as well as hormone replacement therapies, have also been linked to an increased risk of breast cancer. Finally, dietary and/or environmental estrogens have also been implicated in increased breast cancer risk, although the evidence is far from conclusive at present [66-68].

3.2. The Involvement of ERα and ERß in Breast Cancer

Studies of gene expression agree that the majority of breast cancers express ERß. However, different reports provide contradictory evidence regarding the importance of ERß in breast cancer progression and response to endocrine therapies [69]. The largest immunohistochemical study of ERß in breast cancer was indicative of this being a good prognostic marker [70]. However, current understanding of the importance of ERß in breast cancer is further complicated by the demonstration that alternative splice forms of ERß, some of which can act as dominant-negative receptors, are often co-expressed with ERα in breast cancer cells [71]. Further investigation is clearly required before a clear role for ERß in breast cancer processes can be established.

In contrast, the involvement of ERα is breast cancer is well documented. By immunohistochemical analysis 70-80% of primary breast cancers are seen to express ERα. The presence of ERα correlates with a better prognosis and the likelihood of response to endocrine therapies [65]. Further, high levels of ERα in benign breast epithelium correlate with increased risk of breast cancer development [72], indicating that ERα could be involved in breast cancer initiation, as well as in progression. Whilst the nature of the changes that result in high level ERα expression in breast cancer cells remains unclear, these findings suggest that altered ERα expression may be an early event in breast cancer.

4. ESTROGEN RECEPTOR AS A THERAPEUTIC TARGET IN BREAST CANCER

The realisation that breast cancer growth is regulated by estrogens has led to the development of strategies to interfere with estrogen action. This includes inhibition of estrogen biosynthesis and competitive inhibitors of estrogen. Each of these strategies is used clinically in the treatment and management of breast cancer and are described below.

4.1. Inhibition of Estrogen Biosynthesis as an Approach to Breast Cancer Treatment

In pre-menopausal women, the ovary is the most important site of estrogen biosynthesis and requires a cytochrome P450 enzyme complex, termed "aromatase" [73]. Estrogen synthesis in the ovary is subject to feedback regulation through hormonal signals from the pituitary, particularly luteinising hormone (LH), which controls the production of androstenedione by the theca cell compartment and follicle-stimulating hormone (FSH), which importantly acts to induce aromatase expression in granulosa cells [74]. Hence, by acting together, LH stimulates the production of the substrate for aromatase, while FSH increases aromatase enzyme levels, resulting in an overall increase in estrogen production, peaking at the time of ovulation. During the ovulation cycle, the initial release of LH and FSH is regulated by lutenising hormone releasing hormone (LHRH), making the activity of this molecule a key target for the pharmacological inhibition of estrogen biosysnthesis in the ovary [75]. This has led to the development of LHRH agonists, including goserelin (Zoladex), buserelin, leuprolide and triptorelin, all of which can achieve medical "ovarian ablation", resulting in circulating estrogen levels similar to those found in postmenopausal women [76]. Of the LHRH agonists, goserelin has been the most extensively studied and has been successfully used for the treatment of advanced breast cancer in pre-menopausal women, where it provides similar clinical benefit to that of oophorectomy in women with hormone-sensitive disease [77]. However, the use of LHRH agonists is limited, in that they do not affect estrogen synthesis in tissues outside the ovaries.

Aromatase inhibitors comprise a further and important class of drugs for inhibiting estrogen production in breast cancer. In post-menopausal women, the majority of circulating, biologically active estradiol is produced from estrone, which is made from androstenedione and testosterone secreted by the adrenal glands. This reaction is catalysed by the same cytochrome P450 aromatase enzyme complex found in the ovary [78] and occurs at peripheral sites, such as adipose tissue, liver, and muscle [79]. Breast cancer tissue is also an important target for aromatase inhibitors, since two-thirds of breast carcinomas contain aromatase, leading to the production of biologically significant amounts of estrogen locally in the tumour (Santen *et al*, 1994). Investigations into the development of aromatase inhibitors began in the 1970's with Aminoglutethimide being the first widely tested in post-menopausal women with metastatic hormone-dependent breast tumours (Santen *et al*, 1990; Brueggemeier, 2001). However, aminoglutethimide and other aromatase inhibitors proved to be inferior to treatment with tamoxifen. Recent advances have however, led to third generation aromatase inhibitors. In particular, clinical trials using anastrozole and letrozole show that these agents are more efficacious than tamoxifen as first-line therapy for post-menopausal women with advanced breast cancer [80, 81] and initial results from large scale trials using anastrozole (and comparing with tamoxifen) in the adjuvant setting for the treatment of early breast cancer show improved three-year disease-free survival and significantly reduced incidence of contra-lateral disease, compared to tamoxifen [82].

4.2. Anti-Estrogens in Breast Cancer Treatment

An important approach in breast cancer treatment has been the development of synthetic estrogen agonists. The first of these, tamoxifen, is a trans-geometric isomer of a substituted triphenylethylene that binds the estrogen receptors with an affinity approximately 2.5% that of estrogen, which was originally investigated as an antifertility drug [83]. Subsequently, ten-year survival studies have shown that for women with ERα-positive tumours, adjuvant tamoxifen treatment for as little as one year reduces recurrence and death, with maximal reductions being obtained with treatment for five years. Five years of adjuvant tamoxifen treatment provides a 50% reduction in recurrence and 26% reduction in mortality. The survival benefits are similar for pre- and post-menopausal women. Tamoxifen also reduces the incidence of contra-lateral breast cancer [84]. Collectively, these findings have resulted in tamoxifen becoming the first-line adjuvant treatment for ERα-positive breast cancer in pre- and post-menopausal women.

Tamoxifen competes with estrogens for receptor binding, to inhibit receptor activation through the ligand-dependent AF2 domain of ERα and subsequently preventing the induction of estrogen-regulated genes. Such genes include estrogen-regulated growth factors and angiogenic factors that act to promote the establishment and growth of the tumour [85, 86]. Additionally, tamoxifen causes a G1 cell cycle arrest [87, 88], a response that is likely to involve c-Myc [89] and cyclin D1 [90] as estrogen-regulated targets. Finally, tamoxifen is known to induce apoptosis in breast cancer [91], although it is not clear if this important anti-tumour activity results from inhibition of estrogen receptor activity [92].

As described in section 1.1, tamoxifen and its metabolic derivative, 4-hydroxytamoxifen (OHT), possess mixed estrogenic/anti-estrogenic actions. The estrogenic activity of tamoxifen is due to its ability to stimulate AF1, whilst inhibiting AF2 [16]. This agonist activity has been proposed to underlie the estrogen-like actions of tamoxifen in maintaining bone density and in reducing levels of circulating cholesterol, thereby protecting against coronary heart disease and atherosclerosis. Indeed, clinical trials have shown that tamoxifen can increase bone density in post-menopausal women, with a reduction in fractures [93]. Similarly, women with breast cancer who received adjuvant tamoxifen for 5 years have a reduced incidence of fatal myocardial infarction [94].

The clinical success of tamoxifen and the benefits arising from its estrogenic activity in bone and the cardiovascular system are tempered by the fact that a proportion of patients with ERα-positive disease do not respond to tamoxifen. Moreover, a large proportion of patients who initially present with primary breast cancer and all of the patients that present with metastatic disease eventually relapse. Further, in women with breast cancer receiving tamoxifen for 1 or 2, years the incidence of endometrial cancer was doubled and was quadrupled with 5 years of treatment [84]. Other side effects have been attributed to the estrogenic action of tamoxifen and include hot flushes, retinopathy, thromboembolism and ovarian cysts [84, 95]. This has spurred the development of new compounds with anti-estrogenic activity in the breast, but which show estrogenic activity in other tissues. Of note is raloxifene, another non-steroidal anti-estrogen that is less potent than tamoxifen as an anti-cancer agent, but has low estrogen-like actions in the rat uterus [96]. Anti-estrogens like tamoxifen and raloxifene, which are anti-estrogenic in some tissues, but are estrogen-like in other tissues have now been named selective estrogen receptor modulators (SERMs).

A second class of antiestrogens are the Type II, or pure-antiestrogens, which possess no detectable estrogen-like properties *in vivo* or *in vitro* and inhibit activation of both AF-1 and AF-2 of ERα. The most potent of these, ICI 182,780, a steroidal anti-estrogen also known as faslodex or fulvestrant, binds to the estrogen receptor with an affinity similar to that of estrogen. In addition to displaying no SERM-like activity, faslodex inhibits ERα dimerisation, DNA binding in vitro and reduced shuttling of ERα from the cytoplasm to the nucleus [97-99]. In addition, or in combination with these effects, the faslodex reduces ERα half-life, resulting in a rapid loss of cellular ERα [100]. In clinical trials faslodex gave responses in a proportion of patients with metastatic disease who had relapsed on prior tamoxifen treatment, with a mean duration of response that was better than that obtained with anastrazole [101]. The concern with faslodex, as with the potent aromatase inhibitors such as anastrazole is the lack of the beneficial effects in bone and on the cardiovascular system observed with SERMs.

4.3. Anti-Estrogens in Breast Cancer Prevention

The idea that agents that antagonise the effects of estrogen could be used to prevent breast cancer was advanced more than 65 years ago [102]. The success of tamoxifen in reducing breast cancer mortality, as well as a reduction in incidence of second breast cancers, together with the estrogen protective effects in other tissues, observed in patients treated with tamoxifen has resulted in clinical trials to determine whether tamoxifen can be used for breast cancer chemoprevention in women in high risk groups. The largest of these trials reported a reduction in the early incidence of breast cancer by 69%. However, other trials using more stringent selection criteria observed only small reductions in breast cancer incidence. Finally, a trial using raloxifene to evaluate its efficacy at preventing fractures saw a 76% reduction in the incidence of breast cancer [103, 104]. The estrogenic action of tamoxifen in some tissues remains a concern regarding its use in breast cancer prevention. But these findings open the way for the development of other SERMs such as raloxifene, that can be used for breast cancer chemoprevention, and which are estrogenic in some tissues, but anti-estrogenic in other tissues.

5. RESISTANCE TO ENDOCRINE THERAPIES

Following initial response to tamoxifen, a proportion of patients who present with primary breast cancer and all of the patients who first present with metastatic disease, relapse. The continued involvement of estrogen receptor in the majority of resistant tumours is demonstrated by the response obtained in these patients with faslodex or anastrozole [101]. These and other findings are indicative of altered ERα function as a mechanism underlying endocrine resistance. To date, there is little to support a role for ERß in endocrine response and/or resistance (see section 2.2). In this respect, mechanisms that involve ERα would include mutations leading to altered ERα activity and post-translational modification of ERα.

5.1. ERα Mutations and Breast Cancer

In principle, it would be possible to obtain endocrine resistance as a result of mutations in ERα whereby the protein becomes constitutively active, acquires a broader range of ligand-specificity or, features mutations that result in an increased sensitivity to the agonist activity of tamoxifen. Surprisingly, it has been difficult to identify mutations in ERα in breast cancer. A screen of 30 tumours from metastatic breast cancer patients identified three point mutations, two of which had no measurable effect. The third resulted in a substitution of tyrosine at amino acid position 537 by asparagine. Tyrosine

537 is located within the crucial helix 12 of the LBD and its replacement has indeed been shown to result in transcriptional activity and coactivator recruitment in the absence of ligand [105-107], although these mutants receptors are not activated by anti-estrogens. In a different study, mutation of ERα has been seen in 30% of pre-malignant hyperplastic lesions, resulting in the substitution of lysine at position 303 by arginine. Lysine 303 is a substrate for acetylation by CBP and its mutation to arginine has been shown to increase sensitivity of ERα for estrogen [108, 109]. However, the clinical significance of these mutations remains unclear, given that only 1% of breast cancers carry ERα mutations [110]. Moreover, no such mutations have been associated with endocrine resistance.

ERα gene amplification may be another mechanism underlying increased ERα expression in some breast tumours. However, although a 1.6 to 3-fold ERα gene amplification is observed in 50% of ERα-positive breast tumours, there is no correlation with levels of ERα protein present in the tumours [111]. Thus, rearrangements, amplification and/or mutations in the ERα gene do not appear to play a significant role in breast cancer progression or in resistance.

5.2. Stimulation of Estrogen Receptor Activity by Phosphorylation

In addition to its activation by binding ligand, ER activity is regulated through crosstalk with growth factor-regulated signalling cascades. *In vivo*, epidermal growth factor (EGF), transforming growth factor α (TGFα) and IGF-1 can mimic the stimulatory effect of estrogen on DNA synthesis and proliferation in the uterus. The EGF-mediated stimulation of uterine proliferation is ERα-mediated, since it can be blocked by ICI 164, 384 (a derivative of faslodex) and is not observed in ERα-deficient mice [112-114]. In cell culture, IGF-1, EGF, TGF-α and dopamine, as well as inducers of protein kinase A and protein kinase C can stimulate ERα activity in the absence of ligand and/or in synergism with tamoxifen [19].

ERα is phosphorylated in AF1, the DBD and the LBD upon stimulation by these signals. AF1 is phosphorylated by mitogen activated protein kinases (MAPK) and the TFIIH kinase at serine 118 Phosphorylation of serine residues at positions 104 and/or 106 by cdk2 also stimulates ERα activity, whilst phosphorylation of serine 167 by p90RSK and AKT activates ERα [65]. Much interest has been engendered by these findings, particularly because phosphorylation at these sites results in ERα activation in the absence of ligand and/or at low concentrations of estrogen, as well as increased agonist activated by the tamoxifen-bound receptor. Moreover, ERK1/2 MAPK activity and/or levels are significantly increased in a large proportion of breast cancers [115], as are the PI3 kinase-regulated protein kinases known as AKT or PKB, whilst the negative regulator of AKT, the tumour suppressor gene PTEN is frequently mutated in breast cancer [116]. Thus, increased activity of these protein kinases could result in increased AF1 phosphorylation and subsequent resistance to endocrine therapies. Indeed, expression of constitutively active AKT3 resulted in estrogen-dependent MCF7 breast cancer cells forming tumours in nude mice in the absence of estrogen, which were stimulated by tamoxifen [117]. As AKT is known to activate ERα upon serine 167 phosphorylation, it is possible that AKT overexpression is important in endocrine resistance. This suggests that AKT overexpression and the specific phosphorylation of serine 167 may both be features of endocrine resistance. Additionally, increased MAPK activity in breast cancer has been correlated with poor response to tamoxifen [118].

Another interesting finding is the recent demonstration that expression of constitutively active p21 activated protein kinase (Pak1) promotes murine mammary gland hyperplasia in transgenic mice, an effect that may be mediated by its phosphorylation of ERα at serine 305 in the LBD and consequent stimulation of ERα activity [119]. Collectively, these findings indicate that ERα phosphorylation may play an important role in breast cancer progression and in the response to endocrine therapies.

5.3. Coactivators and Corepressors in Endocrine Resistance

The identification of nuclear receptor coactivators and corepressors has suggested the possibility that their mutation and/or altered expression could contribute to endocrine resistance. Inactivating mutations in the p300, CBP and p/CAF genes have also been identified in breast cancer cell lines, but appear to be rare in breast tumours [120, 121]. However, support for the latter possibility has been forthcoming. The p160 family member AIB1 was identified as a result of gene identification studies of a region of chromosome 20q that is frequently amplified in breast cancer [122]. Also, overexpression of the related p160 protein SRC1 increases the agonist activity of tamoxifen [44, 123], whilst decreased expression of the corepressor N-CoR, correlates with resistance to tamoxifen in a mouse model for human breast cancer [45]. Moreover, the action of tamoxifen as an estrogen antagonist in breast cancer cells and as an agonist in endometrial cells was recently shown to correlate with the recruitment of corepressors by the tamoxifen-bound receptor in breast cancer cells, but recruitment of the coactivator SRC1 in endometrial cells, apparently due to the high levels of SRC1 in endometrial cells [124]. These and other findings indicate that the balance between coactivators and corepressors could be key in determining response to anti-estrogens. However, no differences in expression of SMRT or RIP140 co-repressors were observed in tamoxifen-resistant MCF-7 cells, nor in tamoxifen resistant breast tumours [125], although low level expression of N-CoR has recently been found correlate with short relapse-free period for patients receiving tamoxifen [126] and levels of the corepressor known as REA are also reduced in advanced breast cancer [127]. Thus the balance between coactivators and corepressors appears to be central to the agonist/antagonist properties of SERMs and a change in the ratios of these coregulators could contribute to the development of endocrine resistance.

6. CONCLUSION

The realisation that breast cancer growth is regulated by the hormone estrogen led to the development of tamoxifen as the first target-directed cancer drug. Clearly, there is considerable headway to be made in further elucidating estrogen receptor function, in particular in identifying the mechanisms underlying endocrine resistance. It is hoped that these studies will lead to the development of novel therapies aimed at overcoming endocrine resistance. Nevertheless, the success of anti-estrogen therapies in the treatment of breast cancer have paved the way for another first in cancer treatment, that of breast cancer prevention.

7. Acknowledgments

The authors would like to thank Cancer Research UK, the Association for International Cancer Research and the Wellcome Trust for supporting work in our laboratory.

Lakjaya Buluwela, Demetra Constantinidou, Joanna Pike and Simak Ali
Department of Cancer Medicine, Imperial College London, London,
United Kingdom

8. REFERENCES

1. Jensen E.V.Jacobsen H.I. Basic guides to the mechanism of estrogen action. Rec. Prog. Horm. Res. 1962; 18:387-414.
2. Green S., Walter P., Kumar V., Krust A., Bornert J.M., Argos P.Chambon P. Human oestrogen receptor cDNA: sequence, expression and homology to v-erb-A. Nature 1986; 320:134-139.
3. Hollenberg S.M., Weinberger C., Ong E.S., Cerelli G., Oro A., Lebo R., Thompson E.B., Rosenfeld M.G.Evans R.M. Primary structure and expression of a functional human glucocorticoid receptor cDNA. Nature 1985; 318:635-41.
4. Mangelsdorf D.J., Thummel C., Beato M., Herrlich P., Schutz G., Umesono K., Blumberg B., Kastner P., Mark M., Chambon P.et al. The nuclear receptor superfamily: the second decade. Cell 1995; 83:835-839.
5. Maglich J.M., Sluder A., Guan X., Shi Y., McKee D.D., Carrick K., Kamdar K., Willson T.M.Moore J.T. Comparison of complete nuclear receptor sets from the human,

286

Caenorhabditis elegans and Drosophila genomes. Genome Biol 2001; 2:research0029.1-research0029.7.

6. Maglich J.M., Caravella J.A., Lambert M.H., Willson T.M., Moore J.T.Ramamurthy L. The first completed genome sequence from a teleost fish (Fugu rubripes) adds significant diversity to the nuclear receptor superfamily. Nucleic Acids Res 2003; 31:4051-8.

7. Mangelsdorf D.J., Thummel C., Beato M., Herrlich P., Schutz G., Umesono K., Blumberg B., Kastner P., Mark M., Chambon P.et al. The nuclear receptor superfamily: the second decade. Cell 1995; 83:835-9.

8. Chawla A., Repa J.J., Evans R.M.Mangelsdorf D.J. Nuclear receptors and lipid physiology: opening the X-files. Science 2001; 294:1866-70.

9. Schwabe J.W., Chapman L., Finch J.T.Rhodes D. The crystal structure of the estrogen receptor DNA-binding domain bound to DNA: how receptors discriminate between their response elements. Cell 1993; 75:567-78.

10. Pace P., Taylor J., Suntharalingam S., Coombes R.C.Ali S. Human estrogen receptor beta binds DNA in a manner similar to and dimerizes with estrogen receptor alpha. J Biol Chem 1997; 272:25832-25838.

11. Moras D.Gronemeyer H. The nuclear receptor ligand-binding domain: structure and function. Curr Opin Cell Biol 1998; 10:384-391.

12. Heery D.M., Kalkhoven E., Hoare S.Parker M.G. A signature motif in transcriptional co-activators mediates binding to nuclear receptors. Nature 1997; 387:733-736.

13. Shiau A.K., Barstad D., Loria P.M., Cheng L., Kushner P.J., Agard D.A.Greene G.L. The structural basis of estrogen receptor/coactivator recognition and the antagonism of this interaction by tamoxifen. Cell 1998; 95:927-937.

14. Brzozowski A.M., Pike A.C., Dauter Z., Hubbard R.E., Bonn T., Engstrom O., Ohman L., Greene G.L., Gustafsson J.A.Carlquist M. Molecular basis of agonism and antagonism in the oestrogen receptor. Nature 1997; 389:753-8.

15. Pike A.C., Brzozowski A.M., Walton J., Hubbard R.E., Thorsell A.G., Li Y.L., Gustafsson J.A.Carlquist M. Structural insights into the mode of action of a pure antiestrogen. Structure (Camb) 2001; 9:145-53.

16. Gronemeyer H. Transcription activation by estrogen and progesterone receptors. Annu Rev Genet 1991; 25:89-123.

17. McInerney E.M., Weis K.E., Sun J., Mosselman S.Katzenellenbogen B.S. Transcription activation by the human estrogen receptor subtype beta (ER beta) studied with ER beta and ER alpha receptor chimeras. Endocrinology 1998; 139:4513-22.

18. Barkhem T., Carlsson B., Nilsson Y., Enmark E., Gustafsson J.Nilsson S. Differential response of estrogen receptor alpha and estrogen receptor beta to partial estrogen agonists/antagonists. Mol Pharmacol 1998; 54:105-12.

19. Smith C.L. Cross-talk between peptide growth factor and estrogen receptor signaling pathways. Biol Reprod 1998; 58:627-632.

20. Rochette-Egly C. Nuclear receptors: integration of multiple signalling pathways through phosphorylation. Cell Signal 2003; 15:355-66.

21. Ylikomi T., Wurtz J.M., Syvala H., Passinen S., Pekki A., Haverinen M., Blauer M., Tuohimaa P.Gronemeyer H. Reappraisal of the role of heat shock proteins as regulators of steroid receptor activity. Crit Rev Biochem Mol Biol 1998; 33:437-66.

22. Elliston J.F., Fawell S.E., Klein-Hitpass L., Tsai S.Y., Tsai M.J., Parker M.G.O'Malley B.W. Mechanism of estrogen receptor-dependent transcription in a cell-free system. Mol Cell Biol 1990; 10:6607-12.

23. Klein-Hitpass L., Tsai S.Y., Weigel N.L., Allan G.F., Riley D., Rodriguez R., Schrader W.T., Tsai M.J.O'Malley B.W. The progesterone receptor stimulates cell-free transcription by enhancing the formation of a stable preinitiation complex. Cell 1990; 60:247-57.

24. Tsai S.Y., Srinivasan G., Allan G.F., Thompson E.B., O'Malley B.W.Tsai M.J. Recombinant human glucocorticoid receptor induces transcription of hormone response genes in vitro. J Biol Chem 1990; 265:17055-61.

25. Kraus W.L.Kadonaga J.T. p300 and estrogen receptor cooperatively activate transcription via differential enhancement of initiation and reinitiation. Genes Dev 1998; 12:331-42.

26. Neely K.E.Workman J.L. The complexity of chromatin remodeling and its links to cancer. Biochim Biophys Acta 2002; 1603:19-29.
27. Narlikar G.J., Fan H.Y.Kingston R.E. Cooperation between complexes that regulate chromatin structure and transcription. Cell 2002; 108:475-87.
28. Luger K., Mader A.W., Richmond R.K., Sargent D.F.Richmond T.J. Crystal structure of the nucleosome core particle at 2.8 A resolution. Nature 1997; 389:251-60.
29. Jenuwein T.Allis C.D. Translating the histone code. Science 2001; 293:1074-80.
30. Kouzarides T. Histone methylation in transcriptional control. Curr Opin Genet Dev 2002; 12:198-209.
31. Thomson S., Clayton A.L., Hazzalin C.A., Rose S., Barratt M.J.Mahadevan L.C. The nucleosomal response associated with immediate-early gene induction is mediated via alternative MAP kinase cascades: MSK1 as a potential histone H3/HMG-14 kinase. Embo J 1999; 18:4779-93.
32. Chadee D.N., Hendzel M.J., Tylipski C.P., Allis C.D., Bazett-Jones D.P., Wright J.A.Davie J.R. Increased Ser-10 phosphorylation of histone H3 in mitogen-stimulated and oncogene-transformed mouse fibroblasts. J Biol Chem 1999; 274:24914-20.
33. McKenna N.J., Lanz R.B.O'Malley B.W. Nuclear receptor coregulators: cellular and molecular biology. Endocr Rev 1999; 20:321-44.
34. Glass C.K.Rosenfeld M.G. The coregulator exchange in transcriptional functions of nuclear receptors. Genes Dev 2000; 14:121-41.
35. Blanco J.C., Minucci S., Lu J., Yang X.J., Walker K.K., Chen H., Evans R.M., Nakatani Y.Ozato K. The histone acetylase PCAF is a nuclear receptor coactivator. Genes Dev 1998; 12:1638-51.
36. Stallcup M.R. Role of protein methylation in chromatin remodeling and transcriptional regulation. Oncogene 2001; 20:3014-3020.
37. Belandia B., Orford R.L., Hurst H.C.Parker M.G. Targeting of SWI/SNF chromatin remodelling complexes to estrogen-responsive genes. Embo J 2002; 21:4094-103.
38. Kang Y.K., Guermah M., Yuan C.X.Roeder R.G. The TRAP/Mediator coactivator complex interacts directly with estrogen receptors alpha and beta through the TRAP220 subunit and directly enhances estrogen receptor function in vitro. Proc Natl Acad Sci U S A 2002; 99:2642-2647.
39. Chen J.D.Evans R.M. A transcriptional co-repressor that interacts with nuclear hormone receptors. Nature 1995; 377:454-7.
40. Horlein A.J., Naar A.M., Heinzel T., Torchia J., Gloss B., Kurokawa R., Ryan A., Kamei Y., Soderstrom M., Glass C.K.et al. Ligand-independent repression by the thyroid hormone receptor mediated by a nuclear receptor co-repressor. Nature 1995; 377:397-404.
41. Heinzel T., Lavinsky R.M., Mullen T.M., Soderstrom M., Laherty C.D., Torchia J., Yang W.M., Brard G., Ngo S.D., Davie J.R., Seto E., Eisenman R.N., Rose D.W., Glass C.K.Rosenfeld M.G. A complex containing N-CoR, mSin3 and histone deacetylase mediates transcriptional repression. Nature 1997; 387:43-8.
42. Huang E.Y., Zhang J., Miska E.A., Guenther M.G., Kouzarides T.Lazar M.A. Nuclear receptor corepressors partner with class II histone deacetylases in a Sin3-independent repression pathway. Genes Dev 2000; 14:45-54.
43. 43. Underhill C., Qutob M.S., Yee S.P.Torchia J. A novel nuclear receptor corepressor complex, N-CoR, contains components of the mammalian SWI/SNF complex and the corepressor KAP-1. J Biol Chem 2000; 275:40463-70.
44. Smith C.L., Nawaz Z.O'Malley B.W. Coactivator and corepressor regulation of the agonist/antagonist activity of the mixed antiestrogen, 4-hydroxytamoxifen. Mol Endocrinol 1997; 11:657-666.
45. Lavinsky R.M., Jepsen K., Heinzel T., Torchia J., Mullen T.M., Schiff R., Del-Rio A.L., Ricote M., Ngo S., Gemsch J., Hilsenbeck S.G., Osborne C.K., Glass C.K., Rosenfeld M.G.Rose D.W. Diverse signaling pathways modulate nuclear receptor recruitment of N-CoR and SMRT complexes. Proc Natl Acad Sci U S A 1998; 95:2920-2925.

288

46. Cavailles V., Dauvois S., L'Horset F., Lopez G., Hoare S., Kushner P.J.Parker M.G. Nuclear factor RIP140 modulates transcriptional activation by the estrogen receptor. EMBO J 1995; 14:3741-3751.

47. Fernandes I., Bastien Y., Wai T., Nygard K., Lin R., Cormier O., Lee H.S., Eng F., Bertos N.R., Pelletier N., Mader S., Han V.K., Yang X.J.White J.H. Ligand-dependent nuclear receptor corepressor LCoR functions by histone deacetylase-dependent and -independent mechanisms. Mol Cell 2003; 11:139-50.

48. Hu X.Lazar M.A. The CoRNR motif controls the recruitment of corepressors by nuclear hormone receptors. Nature 1999; 402:93-96.

49. Nagy L., Kao H.Y., Love J.D., Li C., Banayo E., Gooch J.T., Krishna V., Chatterjee K., Evans R.M.Schwabe J.W. Mechanism of corepressor binding and release from nuclear hormone receptors. Genes Dev 1999; 13:3209-3216.

50. Vo N., Fjeld C.Goodman R.H. Acetylation of nuclear hormone receptor-interacting protein RIP140 regulates binding of the transcriptional corepressor CtBP. Mol Cell Biol 2001; 21:6181-8.

51. Le Douarin B., You J., Nielsen A.L., Chambon P.Losson R. TIF1alpha: a possible link between KRAB zinc finger proteins and nuclear receptors. J Steroid Biochem Mol Biol 1998; 65:43-50.

52. Nielsen A.L., Ortiz J.A., You J., Oulad-Abdelghani M., Khechumian R., Gansmuller A., Chambon P.Losson R. Interaction with members of the heterochromatin protein 1 (HP1) family and histone deacetylation are differentially involved in transcriptional silencing by members of the TIF1 family. Embo J 1999; 18:6385-95.

53. Russo J.Russo I.H., Development of the human mammary gland. *In The Mammary Gland. Development, Regulation and Function*, M. Neville and C.W. Daniel ed. New York: Plenum, 1987.

54. Ruan W.Kleinberg D.L. Insulin-like growth factor I is essential for terminal end bud formation and ductal morphogenesis during mammary development. Endocrinology 1999; 140:5075-81.

55. Krege J.H., Hodgin J.B., Couse J.F., Enmark E., Warner M., Mahler J.F., Sar M., Korach K.S., Gustafsson J.A.Smithies O. Generation and reproductive phenotypes of mice lacking estrogen receptor beta. Proc Natl Acad Sci U S A 1998; 95:15677-82.

56. Bocchinfuso W.P.Korach K.S. Mammary gland development and tumorigenesis in estrogen receptor knockout mice. J Mammary Gland Biol Neoplasia 1997; 2:323-34.

57. Mueller S.O., Clark J.A., Myers P.H.Korach K.S. Mammary gland development in adult mice requires epithelial and stromal estrogen receptor alpha. Endocrinology 2002; 143:2357-65.

58. Anderson E., Clarke R.B.Howell A. Estrogen responsiveness and control of normal human breast proliferation. J Mammary Gland Biol Neoplasia 1998; 3:23-35.

59. Botha J.L., Bray F., Sankila R.Parkin D.M. Breast cancer incidence and mortality trends in 16 European countries. Eur J Cancer 2003; 39:1718-29.

60. Howe H.L., Wingo P.A., Thun M.J., Ries L.A., Rosenberg H.M., Feigal E.G.Edwards B.K. Annual report to the nation on the status of cancer (1973 through 1998), featuring cancers with recent increasing trends. J Natl Cancer Inst 2001; 93:824-42.

61. Peto R., Boreham J., Clarke M., Davies C.Beral V. UK and USA breast cancer deaths down 25% in year 2000 at ages 20-69 years. Lancet 2000; 355:1822.

62. Beatson G.T. On the treatment of inoperable cases of carcinoma of the mamma: suggestions for a new method of treatment with illustrative cases. Lancet 1896; 2:104-107.

63. Boyd S. On oophorectomy in cancer of the breast. Br Med J 1900; 2:1161-1167.

64. MacGregor J.I.Jordan V.C. Basic guide to the mechanisms of antiestrogen action. Pharmacol Rev 1998; 50:151-96.

65. Ali S.Coombes R.C. Endocrine-responsive breast cancer and strategies for combatting resistance. Nature Rev. Cancer 2002; 2:101-112.

66. Key T.J., Verkasalo P.K.Banks E. Epidemiology of breast cancer. Lancet Oncol 2001; 2:133-40.

67. Clemons M.Goss P. Estrogen and the risk of breast cancer. N Engl J Med 2001; 344:276-85.
68. Safe S.H. Interactions between hormones and chemicals in breast cancer. Annu Rev Pharmacol Toxicol 1998; 38:121-58.
69. Speirs V. Oestrogen receptor beta in breast cancer: good, bad or still too early to tell? J Pathol 2002; 197:143-7.
70. Fuqua S.A.W., Schiff R., Parra I., Moore J.T., Mohsin S.K., Osborne C.K., Clark G.M.Allred D.C. Estrogen Receptor ß Protein in Human Breast Cancer: Correlation with Clinical Tumor Parameters. Cancer Res 2003; 63:2434-2439.
71. Palmieri C., Cheng G.J., Saji S., Zelada-Hedman M., Warri A., Weihua Z., Van Noorden S., Wahlstrom T., Coombes R.C., Warner M.Gustafsson J.A. Estrogen receptor beta in breast cancer. Endocr Relat Cancer 2002; 9:1-13.
72. Khan S.A., Rogers M.A., Khurana K.K., Meguid M.M.Numann P.J. Estrogen receptor expression in benign breast epithelium and breast cancer risk. J Natl Cancer Inst 1998; 90:37-42.
73. Osawa Y., Higashiyama T., Fronckowiak M., Yoshida N.Yarborough C. Aromatase. J Steroid Biochem 1987; 27:781-9.
74. Steinkampf M.P., Mendelson C.R.Simpson E.R. Regulation by follicle-stimulating hormone of the synthesis of aromatase cytochrome P-450 in human granulosa cells. Mol Endocrinol 1987; 1:465-71.
75. Schally A.V. Luteinizing hormone-releasing hormone analogs: their impact on the control of tumorigenesis. Peptides 1999; 20:1247-62.
76. Kaufmann M., Jonat W., Schachner-Wunschmann E., Bastert G.Maass H. The depot GnRH analogue goserelin in the treatment of premenopausal patients with metastatic breast cancer--a 5-year experience and further endocrine therapies. Cooperative German Zoladex Study Group. Onkologie 1991; 14:22-4, 26-8, 30.
77. Taylor C.W., Green S., Dalton W.S., Martino S., Rector D., Ingle J.N., Robert N.J., Budd G.T., Paradelo J.C., Natale R.B., Bearden J.D., Mailliard J.A.Osborne C.K. Multicenter randomized clinical trial of goserelin versus surgical ovariectomy in premenopausal patients with receptor-positive metastatic breast cancer: an intergroup study. J Clin Oncol 1998; 16:994-9.
78. Santen R.J.Harvey H.A. Use of aromatase inhibitors in breast carcinoma. Endocr Relat Cancer 1999; 6:75-92.
79. Buzdar A.Howell A. Advances in aromatase inhibition: clinical efficacy and tolerability in the treatment of breast cancer. Clin Cancer Res 2001; 7:2620-35.
80. 80. Bonneterre J., Buzdar A., Nabholtz J.M., Robertson J.F., Thurlimann B., von Euler M., Sahmoud T., Webster A.Steinberg M. Anastrozole is superior to tamoxifen as first-line therapy in hormone receptor positive advanced breast carcinoma. Cancer 2001; 92:2247-58.
81. Mouridsen H., Gershanovich M., Sun Y., Perez-Carrion R., Boni C., Monnier A., Apffelstaedt J., Smith R., Sleeboom H.P., Janicke F., Pluzanska A., Dank M., Becquart D., Bapsy P.P., Salminen E., Snyder R., Lassus M., Verbeek J.A., Staffler B., Chaudri-Ross H.A.Dugan M. Superior efficacy of letrozole versus tamoxifen as first-line therapy for postmenopausal women with advanced breast cancer: results of a phase III study of the International Letrozole Breast Cancer Group. J Clin Oncol 2001; 19:2596-606.
82. Baum M., Budzar A.U., Cuzick J., Forbes J., Houghton J.H., Klijn J.G.Sahmoud T. Anastrozole alone or in combination with tamoxifen versus tamoxifen alone for adjuvant treatment of postmenopausal women with early breast cancer: first results of the ATAC randomised trial. Lancet 2002; 359:2131-9.
83. Harper M.J.Walpole A.L. Contrasting endocrine activities of cis and trans isomers in a series of substituted triphenylethylenes. Nature 1966; 212:87.
84. Group E.B.C.T.C. Tamoxifen for early breast cancer: an overview of the randomised trials. Lancet 1998; 351:1451-67.
85. Murphy L.C. Antiestrogen action and growth factor regulation. Breast Cancer Res Treat 1994; 31:61-71.

290

86. Arteaga C.L.Osborne C.K. Growth factors as mediators of estrogen/antiestrogen action in human breast cancer cells. Cancer Treat Res 1991; 53:289-304.

87. Foster J.S., Henley D.C., Ahamed S.Wimalasena J. Estrogens and cell-cycle regulation in breast cancer. Trends Endocrinol Metab 2001; 12:320-7.

88. Doisneau-Sixou S.F., Sergio C.M., Carroll J.S., Hui R., Musgrove E.A.Sutherland R.L. Estrogen and antiestrogen regulation of cell cycle progression in breast cancer cells. Endocr Relat Cancer 2003; 10:179-86.

89. Prall O.W., Rogan E.M., Musgrove E.A., Watts C.K.Sutherland R.L. c-Myc or cyclin D1 mimics estrogen effects on cyclin E-Cdk2 activation and cell cycle reentry. Mol Cell Biol 1998; 18:4499-508.

90. Watts C.K., Brady A., Sarcevic B., deFazio A., Musgrove E.A.Sutherland R.L. Antiestrogen inhibition of cell cycle progression in breast cancer cells in associated with inhibition of cyclin-dependent kinase activity and decreased retinoblastoma protein phosphorylation. Mol Endocrinol 1995; 9:1804-13.

91. Ellis P.A., Saccani-Jotti G., Clarke R., Johnston S.R., Anderson E., Howell A., A'Hern R., Salter J., Detre S., Nicholson R., Robertson J., Smith I.E.Dowsett M. Induction of apoptosis by tamoxifen and ICI 182780 in primary breast cancer. Int J Cancer 1997; 72:608-13.

92. Mandlekar S.Kong A.N. Mechanisms of tamoxifen-induced apoptosis. Apoptosis 2001; 6:469-77.

93. Love R.R., Mazess R.B., Barden H.S., Epstein S., Newcomb P.A., Jordan V.C., Carbone P.P.DeMets D.L. Effects of tamoxifen on bone mineral density in postmenopausal women with breast cancer. N Engl J Med 1992; 326:852-6.

94. McDonald C.C., Alexander F.E., Whyte B.W., Forrest A.P.Stewart H.J. Cardiac and vascular morbidity in women receiving adjuvant tamoxifen for breast cancer in a randomised trial. The Scottish Cancer Trials Breast Group. Bmj 1995; 311:977-80.

95. Osborne C.K. Tamoxifen in the treatment of breast cancer. N Engl J Med 1998; 339:1609-18.

96. Jordan V.C.Morrow M. Tamoxifen, raloxifene, and the prevention of breast cancer. Endocr Rev 1999; 20:253-78.

97. Dauvois S., White R.Parker M.G. The antiestrogen ICI 182780 disrupts estrogen receptor nucleocytoplasmic shuttling. J Cell Sci 1993; 106 (Pt 4):1377-88.

98. Fawell S.E., White R., Hoare S., Sydenham M., Page M.Parker M.G. Inhibition of estrogen receptor-DNA binding by the "pure" antiestrogen ICI 164,384 appears to be mediated by impaired receptor dimerization. Proc Natl Acad Sci U S A 1990; 87:6883-7.

99. Metzger D., Berry M., Ali S.Chambon P. Effect of antagonists on DNA binding properties of the human estrogen receptor in vitro and in vivo. Mol Endocrinol 1995; 9:579-91.

100. Dauvois S., Danielian P.S., White R.Parker M.G. Antiestrogen ICI 164,384 reduces cellular estrogen receptor content by increasing its turnover. Proc Natl Acad Sci U S A 1992; 89:4037-41.

101. Morris C.Wakeling A. Fulvestrant ('Faslodex')-a new treatment option for patients progressing on prior endocrine therapy. Endocr Relat Cancer 2002; 9:267-76.

102. Lacassagne A. Hormonal pathogenesis of adenocarcinoma of the breast. Am. J. Cancer 1936; 27:217-25.

103. Jordan V.C., Gapstur S.Morrow M. Selective estrogen receptor modulation and reduction in risk of breast cancer, osteoporosis, and coronary heart disease. J Natl Cancer Inst 2001; 93:1449-57.

104. Cuzick J., Powles T., Veronesi U., Forbes J., Edwards R., Ashley S.Boyle P. Overview of the main outcomes in breast-cancer prevention trials. Lancet 2003; 361:296-300.

105. Zhang Q.X., Borg A., Wolf D.M., Oesterreich S.Fuqua S.A. An estrogen receptor mutant with strong hormone-independent activity from a metastatic breast cancer. Cancer Res 1997; 57:1244-9.

106. Weis K.E., Ekena K., Thomas J.A., Lazennec G.Katzenellenbogen B.S. Constitutively active human estrogen receptors containing amino acid substitutions for tyrosine 537 in the receptor protein. Mol Endocrinol 1996; 10:1388-98.

107. White R., Sjoberg M., Kalkhoven E.Parker M.G. Ligand-independent activation of the oestrogen receptor by mutation of a conserved tyrosine. Embo J 1997; 16:1427-35.
108. Fuqua S.A., Wiltschke C., Zhang Q.X., Borg A., Castles C.G., Friedrichs W.E., Hopp T., Hilsenbeck S., Mohsin S., O'Connell P.Allred D.C. A hypersensitive estrogen receptor-alpha mutation in premalignant breast lesions. Cancer Res 2000; 60:4026-9.
109. Wang C., Fu M., Angeletti R.H., Siconolfi-Baez L., Reutens A.T., Albanese C., Lisanti M.P., Katzenellenbogen B.S., Kato S., Hopp T., Fuqua S.A., Lopez G.N., Kushner P.J.Pestell R.G. Direct acetylation of the estrogen receptor alpha hinge region by p300 regulates transactivation and hormone sensitivity. J Biol Chem 2001; 276:18375-83.
110. Roodi N., Bailey L.R., Kao W.Y., Verrier C.S., Yee C.J., Dupont W.D.Parl F.F. Estrogen receptor gene analysis in estrogen receptor-positive and receptor-negative primary breast cancer. J Natl Cancer Inst 1995; 87:446-51.
111. Nembrot M., Quintana B.Mordoh J. Estrogen receptor gene amplification is found in some estrogen receptor-positive human breast tumors. Biochem Biophys Res Commun 1990; 166:601-7.
112. Ignar-Trowbridge D.M., Nelson K.G., Bidwell M.C., Curtis S.W., Washburn T.F., McLachlan J.A.Korach K.S. Coupling of dual signaling pathways: epidermal growth factor action involves the estrogen receptor. Proc Natl Acad Sci U S A 1992; 89:4658-62.
113. Curtis S.W., Washburn T., Sewall C., DiAugustine R., Lindzey J., Couse J.F.Korach K.S. Physiological coupling of growth factor and steroid receptor signaling pathways: estrogen receptor knockout mice lack estrogen-like response to epidermal growth factor. Proc Natl Acad Sci U S A 1996; 93:12626-30.
114. Klotz D.M., Hewitt S.C., Ciana P., Raviscioni M., Lindzey J.K., Foley J., Maggi A., DiAugustine R.P.Korach K.S. Requirement of estrogen receptor-alpha in insulin-like growth factor-1 (IGF-1)-induced uterine responses and in vivo evidence for IGF-1/estrogen receptor cross-talk. J Biol Chem 2002; 277:8531-7.
115. Sivaraman V.S., Wang H., Nuovo G.J.Malbon C.C. Hyperexpression of mitogen-activated protein kinase in human breast cancer. J Clin Invest 1997; 99:1478-83.
116. Testa J.R.Bellacosa A. AKT plays a central role in tumorigenesis. Proc Natl Acad Sci U S A 2001; 98:10983-5.
117. Faridi J., Wang L., Endemann G.Roth R.A. Expression of constitutively active Akt-3 in MCF-7 breast cancer cells reverses the estrogen and tamoxifen responsivity of these cells in vivo. Clin Cancer Res 2003; 9:2933-9.
118. Gee J.M., Robertson J.F., Ellis I.O.Nicholson R.I. Phosphorylation of ERK1/2 mitogen-activated protein kinase is associated with poor response to anti-hormonal therapy and decreased patient survival in clinical breast cancer. Int J Cancer 2001; 95:247-54.
119. Wang R.A., Mazumdar A., Vadlamudi R.K.Kumar R. P21-activated kinase-1 phosphorylates and transactivates estrogen receptor-alpha and promotes hyperplasia in mammary epithelium. Embo J 2002; 21:5437-47.
120. Gayther S.A., Batley S.J., Linger L., Bannister A., Thorpe K., Chin S.F., Daigo Y., Russell P., Wilson A., Sowter H.M., Delhanty J.D., Ponder B.A., Kouzarides T.Caldas C. Mutations truncating the EP300 acetylase in human cancers. Nat Genet 2000; 24:300-3.
121. Ozdag H., Batley S.J., Forsti A., Iyer N.G., Daigo Y., Boutell J., Arends M.J., Ponder B.A., Kouzarides T.Caldas C. Mutation analysis of CBP and PCAF reveals rare inactivating mutations in cancer cell lines but not in primary tumours. Br J Cancer 2002; 87:1162-5.
122. Anzick S.L., Kononen J., Walker R.L., Azorsa D.O., Tanner M.M., Guan X.Y., Sauter G., Kallioniemi O.P., Trent J.M.Meltzer P.S. AIB1, a steroid receptor coactivator amplified in breast and ovarian cancer. Science 1997; 277:965-968.
123. Webb P., Nguyen P., Shinsako J., Anderson C., Feng W., Nguyen M.P., Chen D., Huang S.M., Subramanian S., McKinerney E., Katzenellenbogen B.S., Stallcup M.R.Kushner P.J. Estrogen receptor activation function 1 works by binding p160 coactivator proteins. Mol Endocrinol 1998; 12:1605-18.
124. Shang Y.Brown M. Molecular determinants for the tissue specificity of SERMs. Science 2002; 295:2465-8.

125. Chan C.M., Lykkesfeldt A.E., Parker M.G.Dowsett M. Expression of nuclear receptor interacting proteins TIF-1, SUG-1, receptor interacting protein 140, and corepressor SMRT in tamoxifen-resistant breast cancer. Clin Cancer Res 1999; 5:3460-7.
126. Girault I., Lerebours F., Amarir S., Tozlu S., Tubiana-Hulin M., Lidereau R.Bieche I. Expression analysis of estrogen receptor alpha coregulators in breast carcinoma: evidence that NCOR1 expression is predictive of the response to tamoxifen. Clin Cancer Res 2003; 9:1259-66.
127. Simon S.L., Parkes A., Leygue E., Dotzlaw H., Snell L., Troup S., Adeyinka A., Watson P.H.Murphy L.C. Expression of a repressor of estrogen receptor activity in human breast tumors: relationship to some known prognostic markers. Cancer Res 2000; 60:2796-9.

TARGETING ENDOTHELIN AXIS IN CANCER

ANNA BAGNATO AND PIER GIORGIO NATALI

1. INTRODUCTION

The endotelin (ET) axis, that includes a family of 3 small (21-amino acid) peptides ET-1, ET-2, and ET-3, and the ET receptors, plays an important physiological role in normal tissue, acting as a modulator of vasomotor tone, tissue differentiation and development, cell proliferation, and hormone production. Endothelins exert their effects by binding to cell surface ET receptors, of which there are 2 types, ET_A and ET_B. The ET_B receptor (ET_BR) binds the three peptide isotypes with equal affinity. In contrast, ET_AR binds ET-1 with higher affinity than the other isoforms. Both receptors belong to the G protein-coupled receptor (GPCR) system and mediate biological responses from a variety of stimuli, including growth factors, vasoactive polypeptides, neurotransmitters, hormones, and phospholipids [1-3] (Figure 1).

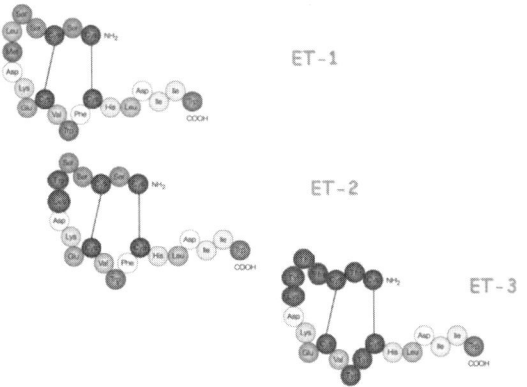

Figure 1. Amino-acid structure of ET-1, ET-2 and ET-3.

ET-1 is produced by a variety of normal cells, including endothelial cells, vascular smooth muscle cells, and various epithelial tissues (eg, bronchial, endometrial, mammary, and prostatic) and is mitogenic for many cell types including endothelial cells, vascular and bronchial smooth muscle cells, fibroblasts, glomerular mesangial cells, osteoblasts, melanocytes, and endometrial stromal cells. The mitogenic activity of ET-1 can be amplified by synergistic interactions with other growth factors including epidermal growth factor (EGF), basic fibroblast growth factor (bFGF), insulin, insulin-like growth factor (IGF), platelet-derived growth factor (PDGF), transforming growth factor (TGF), and interleukin-6 (IL-6) [4].

ET-1 is a relevant growth factor in several tumor types including carcinoma of the prostate, ovary, colon, cervix, breast, kidney, lung, colon, central nervous system (CNS) as well as melanoma, Kaposi's sarcoma (KS) and bone metastasis [5-34]. ETs and their receptors have been implicated in cancer progression through autocrine and paracrine pathways [35]. There is increasing evidence that ET-1 participates to a range of disease processes involved in neoplasia, such as cell proliferation, inhibition of apoptosis, matrix remodeling, bone deposition, and metastases. The demonstration of ET-1 as an important mediator in the progression of many tumors clearly identifies the ET axis as a potential therapeutic target [36]. This has propelled the development of several potent and selective ET-1 receptor antagonists. These small molecules have contributed to our understanding of the physiopathological relevance of the ET axis and the beginning of translation of this information into clinical trials [37-39].

2. ROLE OF THE ENDOTHELIN AXIS IN OVARIAN CANCER

2.1. Signaling Pathways Activated by ET-1

ET-1, which is the most common circulating form of ETs, is produced by many epithelial tumors. The peptide signals through two GPCR, ET_A and ET_B, that have different affinities for ETs.

ET-1 and the ET_AR are overexpressed in primary and metastatic ovarian carcinomas, as compared to normal ovaries [9]. In ovarian tumor cells, ET-1 acts as an autocrine growth factor selectively through the ET_AR. Ligand binding to the receptor results in activation of a pertussis toxin-insensitive G protein that stimulates phospholipase C activity and increases intracellular Ca^{2+} levels, activation of protein kinase C, mitogen activated protein kinase (MAPK) and p125 focal adhesion kinase (FAK) phosphorylation [40]. Among downstream events after ET_AR activation in ovarian carcinoma, ET-1 causes EGF receptor transactivation, which is partly responsible for MAPK activation, suggesting that the coexistence of ET-1 and EGF autocrine circuits in these tumor cells could enhance their growth potential [41] (Figure 2).

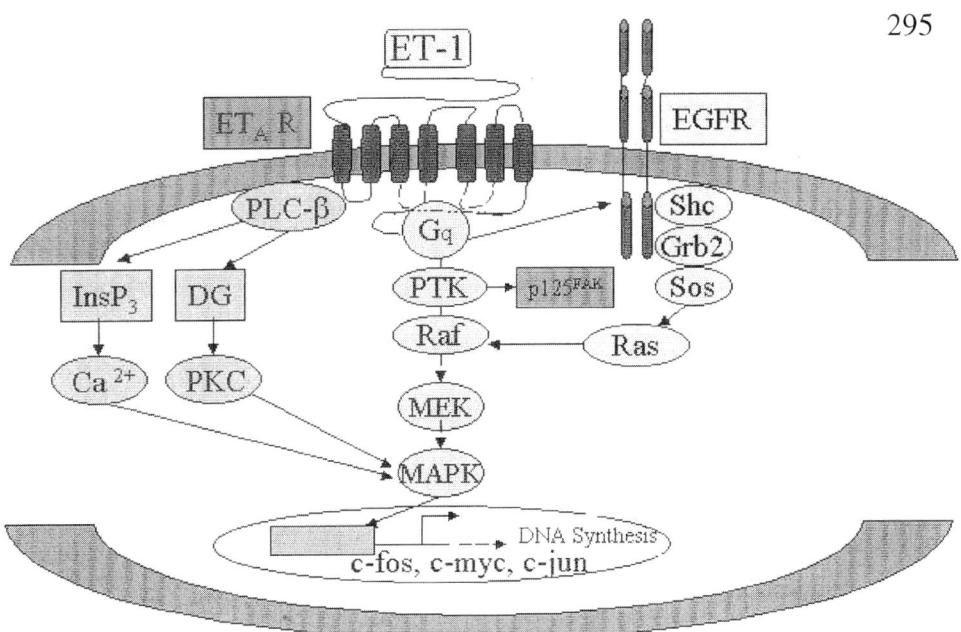

Figure 2. Endothelin-induced signal transduction pathways. Binding of ET-1 to the ET_AR in ovarian carcinoma cells triggers signal transduction pathways through G_9, a pertussis toxin-insensitive G protein. Activation of phospholipase C (PLC), protein tyrosine kinases (PTKs; such as FAK) results in activation of the RAF/MEK/MAPK pathway. Mobilization of intracellular calcium (Ca^{2+}), activation of protein kinase C (PKC), and activated MAPK induce nuclear transcription of protooncogenes (such as c-fos, c-myc, and c-jun) leading DNA synthesis and cell growth. Further analysis of the signaling pathway shoved that ET-1 stimulated epidermal growth factor receptor (EGFR) transactivation, followed by phosphorylation of Shc and its recruitment complexed with Grb2, suggesting that this pathway may contribute to the activation of MAPK induced by ET-1.

2.2. ET-1 and Tumor Neovascularization

ETs, which function as mitogens for endothelial cells, vascular smooth muscle, fibroblasts, and pericytes, are angiogenic factors. Endothelial cell mitogenesis is thought to be mediated by ET_BR, while vascular smooth muscle cells and pericyte mitogenesis appear to be mediated predominantly or solely by the ET_AR. ET-1 modulates various stages of neovascularization, including endothelial cell proliferation, migration, invasion, protease production, tube formation and stimulates neovascularization *in vivo* [42]. Neovascularization is an early and critical event in ovarian cancer progression. In this regard, the elevated expression of ET-1 and its cognate receptor is significantly associated with microvessel density (MVD) and vascular endothelial growth factor (VEGF) expression [13]. Furthermore, it has been found that the elevated level of ET-1 released by ovarian carcinoma cells in ascitic fluid is primarily responsible for *in vitro* endothelial cell migration, acting through ET_BR, as demonstrated by its inhibition by ET_BR antagonists.

The significant inhibition obtained by co-incubating HUVECs with the ET_BR antagonist, BQ 788, and with anti-VEGF antibody, strongly suggests that ET-1 and VEGF might have a complementary and coordinated role during neovascularization in ovarian carcinoma [11].

In OVCA 433 and HEY ovarian carcinoma cell lines, ET-1 increases VEGF expression, at mRNA and protein level, in a time- and dose-dependent fashion, and does so to a greater extent during hypoxia. It was also found that these actions of ET-1 were mediated through the ET_AR, because the specific antagonist, BQ 123, reversed the stimulation of VEGF production.

Transcriptional upregulation plays an important role in the induction of VEGF expression. The transcriptional mechanism has been linked to a critical mediator of hypoxia signaling, the hypoxia inducible factor-1α (HIF-1α). HIF-1 comprises HIF-1α and HIF-1β, which bind as a dimer to HRE. HIF-1 controls the expression of several pro-angiogenic factors, such as VEGF, erythropoietin and ET-1 in response to hypoxia in different tumor cell lines. Although the HIF-1β protein is present in all cells, HIF-1α is virtually undetectable under normal oxygen conditions and is strongly and rapidly induced by hypoxia. In the presence of oxygen, HIF-1α binds to the VHL tumor suppressor protein, leading to its ubiquitination and rapid degradation. Although hypoxia has been shown to be the major inducer of HIF-1α in all cells tested, other stimuli, such as growth factors, hormones, nitric oxide, transition metals and iron chelators, can also induce VEGF expression in an HIF-1-dependent manner in normoxic cells. It has been demonstrated that ET-1 promotes VEGF production through HIF-1α and that this mechanism might be responsible for increasing tumor angiogenesis. Steady-state levels of HIF-1α protein are regulated at the levels of synthesis and stability. Analysis of HIF-1α protein stability in cycloheximide-treated cells showed that degradation of HIF-1α was reduced in ET-1-treated ovarian carcinoma cells compared with controls under both hypoxic and normoxic conditions, indicating that the induction of HIF-1α protein level by ET-1 is due to enhanced HIF-1α stability. After ET-1 stimulation, HIF-1α protein levels increase in the cells, and the HIF-1 transcription complex is formed and binds to the HRE binding site. Therefore, ET-1-induced HIF-1 accumulation in ovarian carcinoma cell lines activates all the signals necessary for a complete HIF-1 response [43]. These findings indicate that ET-1 is able, under normoxic conditions, to activate the hypoxia response pathway, which is the strongest physiological regulator of VEGF expression. The HIF-1α-mediated transcription of VEGF by ET-1 under normoxic conditions points to a general mechanism through which oncogenes and growth factors might upregulate VEGF, and could synergize with hypoxia during tumor growth. These findings which demonstrate that ET_AR activation by ET-1 participates in ovarian carcinoma pathogenesis, driving tumor cell growth and angiogenesis in a paracrine fashion might be clinically relevant. Addition of a specific ET_AR antagonist blocked the ET-1-induced upregulation of VEGF expression and secretion as well as the ET-1-induced activation of HIF-1 transcription complex, indicating the direct involvement of ET_AR in the control of angiogenesis. Because the regulation of VEGF production is a crucial event in tumor neovascularization, one can envisage that, in pathological conditions such as cancer ET-1 might be upregulated by hypoxia and could promote angiogenesis by increasing VEGF production through an HIF-1α-dependent mechanism. Thus, under hypoxic conditions, ET-1 potentiates hypoxia stimulus by amplifying HIF-1α stability and VEGF production [42]. These results define a novel mechanism that could act with a hypoxic-independent pathway and/or could enhance the hypoxic-dependent molecular machinery, resulting in tumor angiogenesis.

All together these results indicate that ET-1 could modulate tumor angiogenesis through direct angiogenic effects on endothelial cells and in part through VEGF stimulation (Figure 3).

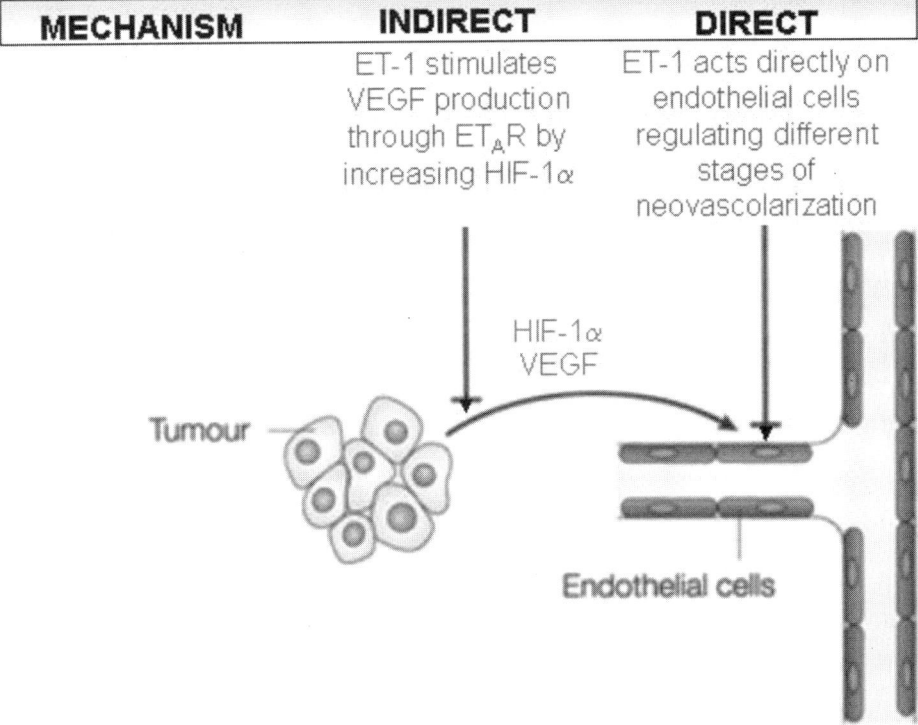

Figure 3. Role of ET-1 in tumor angiogenesis throught direct effects on endothelial cells stimulating proliferation, migration, protease secretion, invasion. ET-1 indirectly modulates angiogenesis stimulating VEGF production by increasing the accumulation of the transcription factor HIF-1α.

2.3. ET-1 and Apoptosis

ET-1 is an antiapoptotic factor in different cell types, suggesting that the peptide may also modulate cell survival pathways. Therefore the role of ET-1 in modulating the sensitivity of two ovarian carcinoma cell lines (OVCA 433 and HEY) to apoptosis induced by two different stimuli was examined. The addition of ET-1 markedly inhibited serum-withdrawal and paclitaxel-induced apoptosis in a concentration-dependent manner, as demonstrated by sub-G_1 peak in DNA content histograms, internucleosomal DNA fragmentation and TdT-mediated dUTP biotin nick-end labeling (TUNEL) method. Paclitaxel-induced apoptosis resulted in the phosphorylation of Bcl-2 that was suppressed by the addition of ET-1. Further analysis of the survival pathway in OVCA 433 cells demonstrated that ET-1 stimulated Akt activation. The phosphatidylinositol 3-kinase (PI3-K) inhibitor, wortmannin, blocked ET-1-induced stimulation of Akt, indicating that ET-1-induced Akt phosphorylation is dependent on PI3-K. Interestingly, the addition of a specific ET_AR antagonist blocked the ET-1-induced resistance to paclitaxel-mediated apoptosis. Furthermore, BQ 123 blocked the ET-1-induced activation of Akt, indicating that ET-1 contribute to trigger resistance to paclitaxel through ET_AR binding *via* activation of anti-apoptotic signaling pathways such as Akt.

The pharmacologic use of specific ET_AR antagonist may therefore provide an additional approach to the treatment of ovarian carcinoma in which ET_AR blockade could result in the tumor growth inhibition by reducing tumor growth as well as by inducing massive apoptosis. Furthermore when combined with the conventional chemotherapy the ET_AR antagonists would more effectively induce apoptosis by contributing to the reversal of paclitaxel resistance [43].

2.4. ET-1 and Tumor Invasion

As mentioned, high levels of ET-1 are detected in the majority of ascitic fluids of ovarian cancer patients and are significantly correlated with VEGF ascitic concentrations, suggesting that ET-1 may enhance the secretion of extracellular matrix-degrading proteinases [11]. Thus, ET-1 acting through the ET_AR consistently induced the activity of two families of metastasis-related proteinases, the matrix metalloproteinases (MMPs) and the urokinase type plasminogen activator system at several levels: mRNA transcription, zymogen secretion and pro-enzymes activation. ET-1 in fact induces the activation of MMP-2, MMP-9, MMP-3, MMP-7 and MMP-13. In addition to soluble MMPs, ET-1 enhanced the activation of MT1-MMP. The upregulation of this protein was concomitant with the overproduction and activation of MMP-2 and MMP-13 indicating that in ovarian cancer cells ET-

1 could participate in the co-ordinate secretion and activation of different MMPs and that the combination of these active enzymes could result in rapid degradation of extracellular matrix (ECM). Furthermore the demonstration of the concomitant production of TIMP-1 and -2 associated with the upregulation of MMPs induced by ET-1 in ovarian carcinoma cell lines strongly supports the pivotal role of ET-1 to regulate the net MMP activity. In the ovarian carcinoma cells, co-induction of uPA system, by the concomitant stimulation of production and secretion of uPA and uPAR, and MMPs by ET-1 resulted into the highest invasive potential of tumor cells through the Matrigel. Interestingly, the addition of BQ 123 blocked the ET-1-induced activation of MMPs and uPA system and ovarian carcinoma cell migration and invasion, indicating that ET-1 contributes to metastatic progression via activation of tumor proteases and subsequent increase in cell migration and invasion through ET_AR binding. In ovarian carcinoma cells, ET_AR activation by ET-1 has been shown to stimulate FAK and paxillin phosphorylation [40]. These effect directly correlate with tumor cell migration and invasion induced by ET-1 and suggest that ET_AR antagonist can inhibit cell migration and possibly other FAK-associated processes which also contributes to invasion and metastasis in ovarian carcinoma cells [45] (Figure 4).

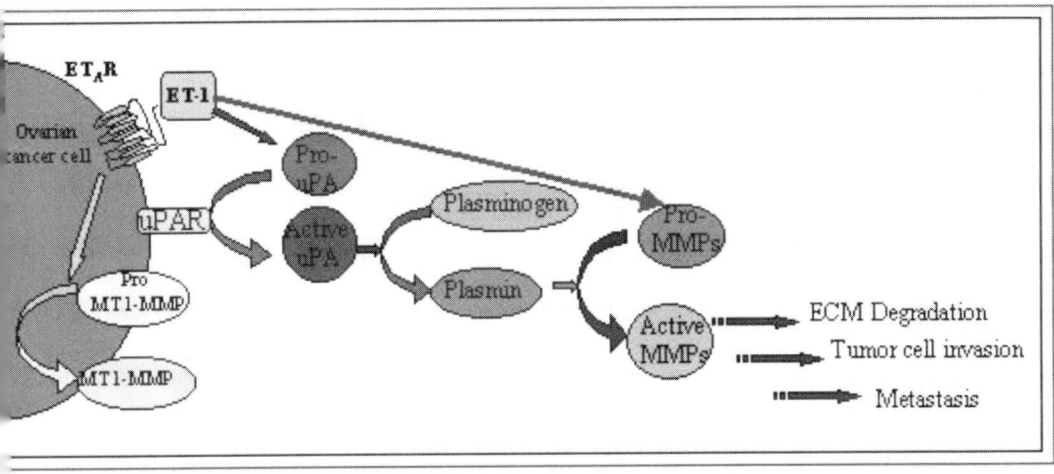

Figure 4. ET-1 enhances proteolytic activity of ovarian carcinoma cells. This peptide modulates the cooperation between MMPs and serine proteinases for enzymatic activation involved in ECM degradation, tumor cell invasion and metastasis.

Following malignant transformation, stepwise changes in intercellular communications provide tumor cells with the ability to overcome microenvironmental control from the normal surrounding tissue, thus allowing local invasiveness and metastatization. Various tumor-promoting agents and different growth factors decrease gap junction intercellular communication (GJIC), either by suppressing connexin (Cx) expression, or by inducing their post-translational modifications such as phosphorylation, a process that is closely related to cellular processes such as trafficking, assembly/disassembly, gating of gap junction channels and altered susceptibility to degradation "gating" and turnover. Several studies have shown that the turnover of Cx is exceptionally rapid, and that degradation of Cx43 involves both the lysosome and the proteasome pathways. Phosphorylation, in most cases, is a prerequisite for ubiquitination that marks the protein for proteasomal destruction. The COOH-terminal tail of Cx43 contains several serine and tyrosine phosphorylation sites suggesting that this molecular domain contains a complex array of potential regulatory sites. Human ovarian surface epithelial cells exhibit extensive GJIC and expression of different types of Cx (e.g. Cx26, Cx32 and Cx43). Defects in intercellular communication, including reduced or inappropriate expression of Cx43, the main gap junction protein in normal human ovarian surface epithelium, has emerged as key factors in ovarian carcinoma progression. Because phosphorylation of Cx43 is believed to be causally linked with disruption of GJIC, the phosphorylation of Cx43 by ET-1 was analyzed. In ovarian carcinoma cells, ET-1 via ET_AR induces a transient and a dose-dependent reduction of GJIC (50-75%). Western blot and immunolocalization analysis clearly showed that Cx43 becomes more phosphorylated and less gap junction plaques were apparent suggesting that ET-1 promotes the cellular uncoupling at the level of connexin maturation and subsequent degradation. Cx43-tyrosine phosphorylation was mainly responsible for ET-1-induced loss of cell-cell communication in these tumor cells as demonstrated by experiments performed with specific tyrosine kinase inhibitor that prevented the ET-1-induced GJIC reduction, and with an inhibitor of tyrosine phosphatase that mimicked the ET-1 action on GJIC (Figure 5).

Furthermore, the increased Cx43 tyrosine phosphorylation was correlated with ET-1-induced increase of c-Src activity. The selective c-Src inhibitor, PP2, suppressed the ET-1-induced Cx43 tyrosine phosphorylation indicating that inhibition of Cx43-based GJIC is mainly mediated by Src tyrosine kinase pathway [46]. From these experiments it could be established that the signaling mechanisms involved in GJIC disruption on ovarian carcinoma cells depend on the Cx43 tyrosine phosphorylation mediated by c-Src. Interestingly, the addition of BQ 123 blocked the ET-1-induced loss of GJIC and Cx43 phosphorylation demonstrating that ET_AR activation by ET-1 contributes to loss of growth control via a Cx43-mediated disruption of GJIC. Because the dynamic behavior of cell interactions and communication

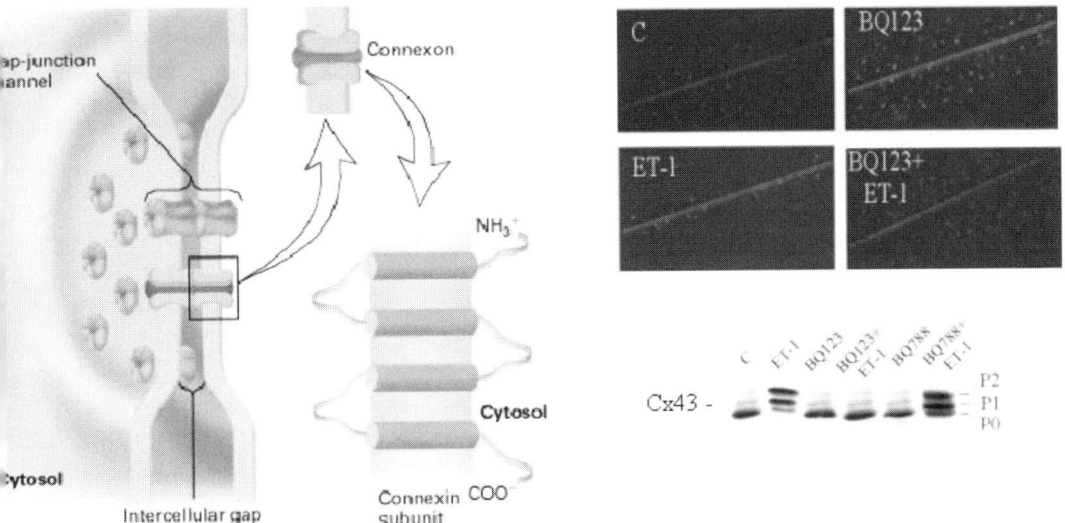

Figure 5. ET-1 decreases gap-juctional intercellular communication by inducing phosphorylation of connexin 43 (Cx 43) in human ovarian carcinoma cells

is affected in migrating cells that lack cell-cell contacts, one can envision that a GPCR, such as ET_AR, which mediates inhibition of Cx43-based junctional communication, might alter intercellular interactions which are responsible for contact mediated regulatory control. The capacity of ET-1 to disrupt gap junctions through phosphorylation of Cx43 via c-Src could serve as a basis to further evaluate the cell-cell metabolic uncoupling and cell detachment that occurs during tumor progression and add further information on the overall relevance of ET_AR in regulating the complex assay of cell-cell or cell-matrix interactions promoting ovarian carcinoma growth [46].

3. TARGETING ET_A RECEPTOR AS NOVEL APPROACH IN CANCER TREATMENT

3.1. Ovarian Carcinoma

The ET_AR autocrine pathway contributes to ovarian cancer progression by inducing cell proliferation, survival, angiogenesis and invasiveness.

In view of the above findings, the ET_AR has been proposed as a potential target for anticancer therapy. The recent identification of low molecular weight compounds that inhibit ligand-induced activation of the ET_AR now offers the possibility of testing this therapeutic approach in a clinical setting. Among various ET_AR antagonists, ABT-627 (atrasentan,

Abbott Laboratories, Abbott Park, IL) is an orally bioavailable endothelin antagonist that potently ($K_i = 34$ pM) and selectively binds to the ET_AR, blocking signal transduction pathways implicated in cancer cell proliferation and other host-dependent processes promoting cancer growth (Figure 6).

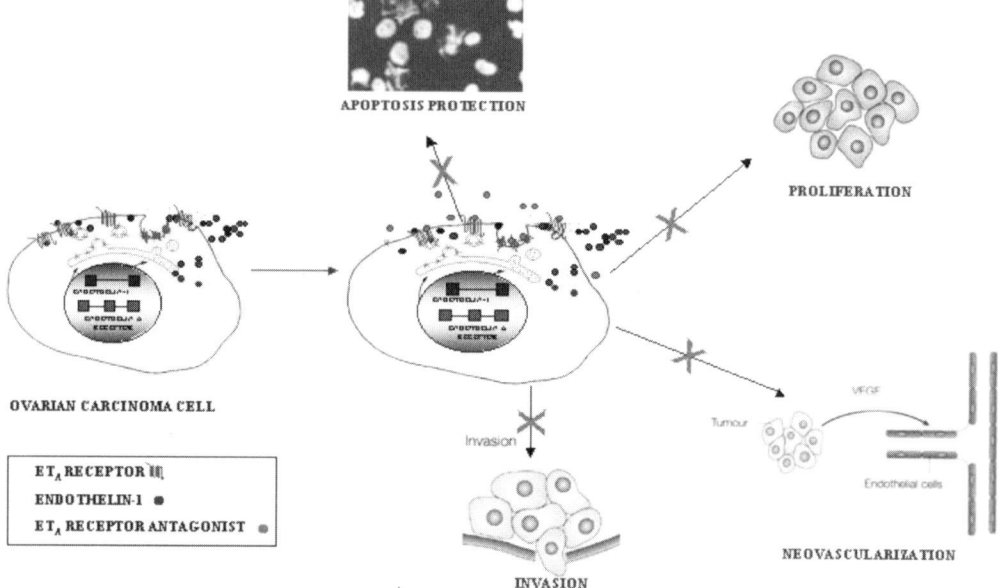

Figure 6. Cancer-promoting functions of endothelin axis: proliferation, survival, angiogenesis, invasion

To evaluate the effect of ABT-627 on the proliferation of various ovarian carcinoma cells, two primary cultures (PMOV1 and PMOV2) and two established cell lines (HEY and OVCA 433) were employed. All these cells express functional ET_AR and secrete high levels of ET-1. Cells were incubated for up to 5 days in the absence or presence of ABT-627 and ET_BR antagonists. In all ovarian carcinoma cells, spontaneous growth was significantly inhibited in the presence of ABT-627. Addition of the ET_BR antagonist did not affect the basal growth rate of the cells demonstrating that endogenous ET-1 acts as an autocrine modulator of ovarian carcinoma cell proliferation only through ET_AR, and can be selectively inhibited by ABT-627.

To determine whether the antiproliferative effect of ABT-627 resulted in the induction of programmed cell death, the percentage of dying cells in ABT-627-treated and control cultures was measured. ABT-627 treatment increased the percentage of apoptotic HEY and OVCA 433 ovarian cancer cells after 48 hours of treatment. In these cells, activation of ET_AR by ET-1 prevents paclitaxel-induced apoptosis. The potential combined proapoptotic effect of treatment with ET$_A$R antagonist and paclitaxel was therefore determined. HEY and OVCA 433 cells were incubated with paclitaxel alone or in combination with ABT-627. As expected, the addition of ABT-627 significantly increased paclitaxel-induced apoptosis in both cell lines establishing that ET$_A$R-activated survival pathways can be blocked by treatment with ABT-627. Therefore ET$_A$R overexpression can promote tumor development through stimulation of cancer cell growth and survival. However, ET$_A$R may also regulate angiogenesis by stimulating tumor production of VEGF. Treatment with either ABT-627 or paclitaxel alone caused ~ 45% inhibition of VEGF secretion. Combination of ABT-627 with paclitaxel exerted a marked inhibitory effect, reaching almost 60% reduction of VEGF secretion.

The potential antitumor effect of ABT-627 was then assessed *in vivo*. Human ovarian carcinoma cells HEY, which overexpress ET$_A$R and secrete high levels of ET-1, were grown as subcutaneous tumors in nude mice. Seven days later, when well established HEY xenografts were palpable with a tumor size of ~ 0.25 cm^3, mice were randomized into treatment and vehicle control groups of 10 animals each. The treated mice were injected i.p. for 21 days with two different concentrations of ABT-627, 2 mg/kg/day and 10 mg/kg/day. Treatment with ABT-627 produced a 65% inhibition of HEY tumor growth on day 40 after tumor injection with either low (2 mg/kg/day) or high (10 mg/kg/day) doses. ABT-627 treatment was generally well tolerated, with no detectable signs of acute or delayed toxicity even at the highest ABT-627 dose. Tumor growth suppression by treatment with 2 mg/kg/day ABT-627 was comparable to that achieved by treatment with paclitaxel (20 mg/kg i.v. for three times each four days). Furthermore, the tumor growth inhibition obtained with ABT-627 persisted up to 4 weeks following termination of treatment. The cooperative proapoptotic effect of ABT-627 and paclitaxel that was observed *in vitro* was then assayed *in vivo*. More marked tumor growth inhibition (90% of controls) was obtained by combined treatment with ABT-627 and paclitaxel, with no histological evidence of HEY tumors in 4 of 10 mice. The dual treatment at the dose and schedule tested were well tolerated, as judged by the absence of weight loss or other signs of acute or delayed toxicity. As compared with control tumor xenografts, the growth delay in established tumors persisted for up to 4 weeks following termination of treatment with ABT-627 combined with paclitaxel .

Because HEY cells express ET$_A$R and various autocrine and paracrine angiogenesis-related factors including ET-1, VEGF and MMP-2, the expression of these factors *in vivo* after ABT-627 treatment at a lower dosage (2mg/kg/day) was investigated. Immunohistochemical analysis of HEY tumors on day 40 after tumor cell injection, revealed a marked reduction in the percentage of VEGF and MMP-2 positive HEY cells in ABT-627-treated mice. Tumor-induced vascularization, which was quantified as microvessel density (MVD) using antibody against CD31, was directly proportional to the expression of VEGF. There was a parallel reduction (45%) in MVD in tumors after treatment with ABT-627. A significant increase in the percentage of TUNEL-positive cells was found in HEY tumors treated with ABT-627 [47]. The inhibition of human ovarian tumor growth in nude mice induced by the potent ET_AR antagonist, ABT-627, was also associated with a reduction of Cx43 phosphorylation and to increase Cx43-based intercellular communication, suggesting that ET_AR blockade may contribute to the control of ovarian carcinoma growth and progression also by preventing the loss of GJIC [46].

Almost complete inhibition of VEGF, MMP-2 expression, and tumor neovascularization, and an increase in apoptosis, were observed following combined treatment of ABT-627 with paclitaxel.

These findings demonstrate the antitumor activity of ABT-627 *in vivo* and provide a rationale for the clinical evaluation of this molecule, alone and in combination with cytotoxic drugs, in patients with ovarian tumors and potentially in other epithelial tumors that overexpress functional ET_AR.

3.2. Prostate Carcinoma

The ET axis has recently been identified as contributing to the pathophysiology of prostate cancer [5-8]. In the normal prostate gland, ET-1 is produced by epithelial cells; the highest concentrations of ET-1 in the body are found in seminal fluid. In prostate cancer, key components of the ET-1 clearance pathway, ET_BR and neutral endopeptidease (NEP), are diminished, resulting in an increase in local ET-1 concentrations. Increased ET_AR expression is also seen with advancing tumor stage and grade in both primary and metastatic prostate cancer. There are multiple pathways by which the ET-1/ET_AR axis may promote prostate cancer progression. ET-1 is a mitogen for prostate cancer cell lines in vitro and acts synergistically with other peptide growth factors. ET-1 modulates apoptosis by affecting cell survival [5-8]. Selective ET_AR antagonists may block the proliferative effects of exogenous ET-1 in both prostate cancer cells and osteoblasts. In a phase I clinical trial, patients with advanced prostate cancer were treated with atrasentan (ABT-627). In this phase I trial, stabilization and decline of PSA occurred in 66% of patients. In a phase II study, atrasentan suppressed markers of biochemical and clinical prostate cancer progression in bone [48-49]. These data substantiate the role of the ET-1/ET_AR axis as a growth and survival pathway and as a therapeutic target in hormone-refractory prostate cancer (HPCA). Atrasentan may inhibit tumor growth in bone both by direct effects on the tumor cells and by disrupting important bone/tumor interactions. In a randomized double-blind phase II study, 288 patients with HPCA were enrolled and were treated with atrasentan 2.5 mg and 10 mg administered orally once daily. In the evaluable patients (n=244), atrasentan significantly delayed time to clinical and biochemical (PSA) progression. The most common side effects included rhinitis, peripheral edema, and headache. Atrasentan also maintained total and bone alkaline phosphatase concentrations at baseline values compared to the placebo-treated group [48-50]. These findings suggest that atrasentan may inhibit progression of hormone refractory prostate cancer in men with metastatic disease.

3.3. Osteoblastic Bone Metastases

Osteoblastic bone metastases occur in advanced cases of prostate cancer and frequently in breast cancer, but the mechanism by which tumor cells stimulate new bone formation are unclear [51]. Many tumor-linked factors have been implicated in the development and progression of skeletal metastases, including IGFI and II, TGFβ, PSA, uPA, FGF1 and 2, bone morphogenic proteins, and ET-1 [51-52]. ET-1 stimulates mitogenesis in osteoblasts and decreases osteoclastic bone resorption and osteoclast motility.

Accumulating data have identified ET-1 as a factor that stimulates bone formation associated with metastatic tumors. ET-1 is increased in the circulation of patients with prostate cancer with osteoblastic metastases, and is also expressed by breast cancer cell lines that cause osteoblastic metastases. Substantial date implicate a causal role for ET-1 in the pathogenesis of osteoblastic bone metastases. Recently three human breast cancer cell lines that produce ET-1 and cause osteoblastic bone metastases in a nude mice model have been identified. Treatment with atrasentan dramatically decreased bone metastases and tumor burden in mice inoculated with ZR-75-1 cells [53]. The growing list of potential clinical applications for ET_AR blockade should now include treatment and eventually prevention of osteoblastic bone metastases.

3.4. Cervical Carcinoma

Human papillomavirus (HPV)-positive human cervical carcinoma cell lines overexpress ET-1 and ET_AR mRNA and secrete ET-1 protein. In contrast, HPV-negative cervical carcinoma cells do not secrete detectable ET-1. Binding studies show that HPV-infected cells express increased numbers of functional ET_AR, and a specific antagonist of ET_A inhibits ET-1-induced proliferation. An ET_BR selective antagonist had no effect. These results indicate that the ET-1/ET_AR axis participates in the progression of HPV-associated cervical carcinoma, in which ET_AR is overexpressed and could be targeted for antitumor therapy. Thus, atrasentan inhibited the growth of cervical carcinoma cell xenografts. Two cycles of treatment completely reverted tumor growth. As reported for ovarian cancer, this small molecule displayed additive effects when administrated in combination with the cytotoxic drug paclitaxel supporting its clinical use either in mono- or combination regimens [28-29].

3.5. Targeting ET_A and ET_B Receptors in Kaposi's Sarcoma Treatment

A different approach in targeting ET-1 receptor in cancer treatment is represented by Kaposi's sarcoma (KS) in which ET-1 acts as an autocrine growth factor through both ET_AR and ET_BR. Binding of ET-1 and ET-3 to both receptors increased the proliferation, migration and invasiveness of the KS derived cell line, KS IMM cells by stimulating secretion and activation of multiple tumor proteases. Therefore we tested the ET_A/ET_BR antagonist A182086 on KS tumor growth in nude mice. Treatment of mice bearing KS xenografts with small molecules A182086 resulted in tumor growth inhibition most likely related to antiproliferative effect on tumor cells and to the antiangiogenic effect on endothelial cells expressing ET_BR. Thus ET-1 receptor antagonists may be effective for treatment of this malignancy because capable of interfering simultaneously with cell proliferation, invasiveness and angiogenesis [30-31].

3.6. Targeting ET_B Receptors in Cutaneous Melanoma Treatment

Melanoma cells express both ET_AR and ET_BR [54]. Gene expression profiling [55] and immunophenotyping of human cutaneous melanoma [56] have recently identified ET_BR as critical in the progression of this malignancy. Through the same receptor, ET-1 acts as antiapoptotic factor for melanoma cells and melanocytes [57]. Thus, ET_BR blockade by the ET_BR peptide antagonist BQ788 resulted in growth inhibition and death of melanoma cells in vivo and in vitro [58]. While these studies define a relevant role of the ET-1/ET_BR pathway in the biology of melanocytic tumors, the molecular events underlying this activity have not been investigated. Early melanoma growth is the result of disrupted intercellular homeostatic regulation [59]. Once this balance is lost and malignant transformation has occurred, microenviromental factors such as cell adherence to extracellular matrix, host-tumor interactions, degradation of matrix components, migration and invasion became essential for the tumor progression to the metastatic phenotype [58]. Changes in cadherins, gap junctions and matrix metalloproteinases expression have emerged as key factors in melanoma progression [60-61]. Therefore the role played by ETs and ET_BR in melanoma cell proliferation, integrin expression, cell-cell adhesion and communication, migration, tumor proteinase activation and invasion was investigated. The selective blockade of this receptor results into inhibition of FAK and MAPK phosphorylation and cell proliferation. Loss of melanoma-microenvironment interactions are required at the early stages of tumor progression [61]. In this regards ET-1 and ET-3 by ET_BR signalling induce the inactivation of the gap junctions through the phosphorylation of the Cx43, which are responsible for contact mediated regulatory control of keratinocytes [60]. Additionally, activation of ET_BR pathway by ET-1 and ET-3 contributes to disruption of normal host-tumor interactions by downregulating the expression of E-cadherin and associated β-catenin adhesion proteins. Significant mRNA expression of the transcription factor *Snail*, which has recently been identify a potent inhibitor of E-cadherin expression in melanoma, closely correlates with downregulation of E-cadherin. ETs also cause a tyrosine phosphorylation of β-catenin which contributes to loss of E-cadherin function, with a concomitant upregulation of N-cadherin. This latter change can mediate homotypic adhesive interactions as well as heterotypic (i.e. fibroblasts, endothelial cells) melanoma cell-cell interactions. Concurrently ETs increase $\alpha_v\beta_3$ and $\alpha_2\beta_1$ integrin expression and matrix metalloproteinase (MMP)-2, -9, and membrane type-1-MMP activation. These effects were associated with ET_BR-mediated enhancement of cell adhesion, migration and invasiveness. Due to the resistance of melanoma to current therapies, the identification of molecular mechanisms underlying local and metastatic growth is mandatory for the development of novel treatments. In view of the availability of orally bioavailable non peptide ET_BR antagonists [53] we utilize these compounds to assess their anti-tumor activity *in vivo*. The small molecule A-192621, an orally biovailable non peptide ET_BR antagonist, significantly inhibited melanoma growth in nude mice. In conclusion, multiple molecular pathways elicited by ET-1 and ET-3 which regulated melanoma local and metastatic growth have been identified. Because all the molecular effectors involved in melanoma cell migration mechanisms, including integrins, tumor proteases, cell-cell adhesion and cell-cell communication molecules, are triggered by the ET_BR activity, blockade of this receptor by small molecules results into inhibition of melanoma growth in vitro and in vivo, thus offering an unprecedented opportunity of targeted therapy in this malignancy [62].

4. CONCLUSIONS

A large body of *in vitro* and *in* vivo studies indicate the ET axis as a potential novel therapeutic target for cancer. ET-1 is overexpressed in many malignancies, acting as an autocrine growth factor. Engagement of the ET_AR by ET-1 triggers tumor proliferation, VEGF-induced angiogenesis, metastatic potential, antiapoptotic action and is synergistic with other growth factor. Direct mechanistic evidence of the role of ET_ARs in prostate, ovarian, cervical, colorectal, and other cancers supports the concept that ET_AR antagonists may have a strong impact on the malignant phenotype [36]. The antitumor activity of selective ET_AR antagonists observed in preclinical studies has been strongly supported in early clinical trials in which ET_AR antagonists have demonstrated a promising therapeutic potential (Figure 7).

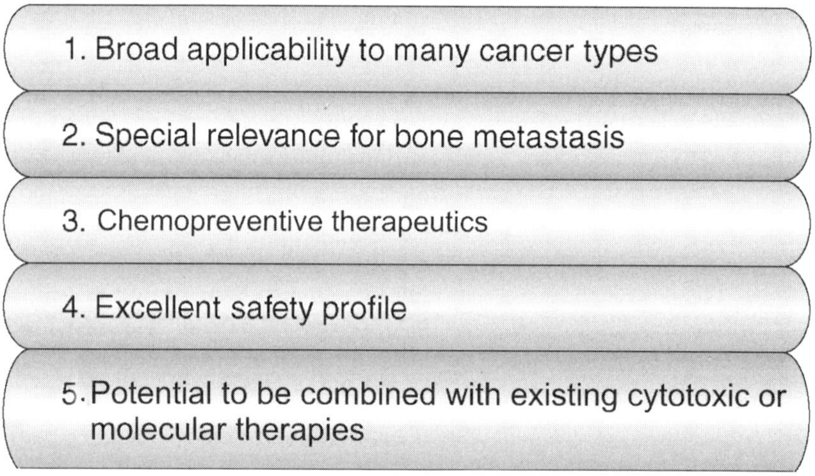

1. Broad applicability to many cancer types

2. Special relevance for bone metastasis

3. Chemopreventive therapeutics

4. Excellent safety profile

5. Potential to be combined with existing cytotoxic or molecular therapies

Figure 7. Characteristics of ET_AR antagonist as candidates for cancer therapy

Additional studies are warranted in combination with chemotherapy in ovarian cancer and other solid tumors expressing ET_AR. Moreover, these findings demonstrate that ET-1 and ET-3 through ET_BR activation trigger signalling pathways involved in uncontrolled proliferation and progression of cutaneous melanoma and that pharmacological interruption of ET_BR signalling may open new possibilities in the treatment of this malignancy. Development of ET-1 receptor antagonists has provided a better understanding of the ET axis in cancer pathogenesis and suggests an important role for these molecules exploring new therapeutic interventions in tumors such as prostatic, ovarian, renal, pulmonary, colorectal, cervical, mammary carcinoma, Kaposi's sarcoma, central nervous system (CNS) tumors, melanoma, and bone metastases (Table 1) [37-39].

Table 1. Role of ET-1 and its receptors in different malignancie*s*

	Action of ET-1	Endothelin receptors	Receptor antagonists and their effects
Prostate cancer	ET-1 promotes prostate cancer growth, inhibits apoptosis through the ET_AR	High ET_AR expression, decreased or absent ET_BR expression, frequent methylation of ET_AR gene	Atrasentan relieved pain, and delayed time to clinical and biochemical progression
Ovarian cancer	ET-1 promotes cell proliferation survival, invasion and VEGF-dependent angiogenesis through ET_AR	ET_AR mRNA was detected in 84% carcinoma examined, ET_BR in only 40% ET_AR mediated all ET-1 induced tumor promoting effects	BQ123 showed the growth rate of tumor cells. Atrasentan inhibited growth of ovarian carcinoma xenografts and displayed additive effects in combination with taxanes
Melanoma	ET-1 and ET-3 promotes melanoma cell proliferation and invasion	ET_AR are downregulated in melanoma cells. ET_BR expression seems to be increased in melanoma cells in comparison to benign nevi	ET_BR antagonist inhibited growth of 7 melanoma cells lines, and reduced human melanoma tumor growth in nude mice.
Bone malignancies	ET-1 increase in the expression of osteocalcin protein and new bone formation	Both ET_AR and ET_BR are expressed	ET_AR antagonist blocked ET-induced osteocalcin expression; decreased ET-1-induced new bone formation and inhibited progression of skeletal metastases in prostate cancer patients
Breast cancer	Increased ET-1 expression inversely correlated with the degree of tumor cell differentiation	Elevated expression of ET_AR have been detected in breast cancer tissue in comparison to normal	
Renal cancer	ET-1 opposed the paclitaxel-induced apoptosis in renal carcinoma cell lines	All cell lines expressed ET_AR	
Lung cancer	ET-1 was detected in most squamous cell and adenocarcinomas	Both ET_AR and ET_BR are expressed; ET_AR is downregulated in comparison to normal bronchial tissue.	
Colon cancer	ET-1 protected colon carcinoma cells from FasL-induced apoptosis	Increased expression of ET_AR and ET_BR in neoplastic tissue	$ET_{A,B}R$ antagonist, potentiated FasL-induced apoptosis of tumor cells
Cervical cancer	ET-1 induced proliferation of HPV-positive cervical cell lines	Express both ET_AR and ET_BR. Increased expression of ET_AR on HPV-positive cells	Atrasentan inhibited cell proliferation and growth of cervical cancer xenografts and displayed additive effects in combination with taxane
Kaposi's	ET-1 and ET-3 induced cell proliferation, migration and invasion	Both ET_AR and ET_BR are expressed	$ET_{A,B}R$ antagonist blocked ET-1 induced cell proliferation and invasion and inhibited growth in nude mice
CNS tumors	ET-1 promoted meningioma cell proliferation	Both ET_AR and ET_BR are expressed	BQ123 blocked ET-1-induced effects; ET_AR antagonist had no effect

Among the ET_AR antagonists, atrasentan deserves at the present particular attention for clinical testing, in view of its oral bioavailability, suitable pharmacokinetic and toxicity profile. Preliminary results from clinical trials investigating atrasentan in patients with prostate cancer are encouraging. The role of the ET axis and the therapeutic relevance of ET-1 receptor antagonists in a range of malignancies requires future investigation that may lead to a new generation of molecularly targeted therapies for cancer.

5. ACKNOWLEDGEMENTS

We thank L. Rosanò. F. Spinella, V. Di Castro for their contributions to studies on ovarian cancer cells, M.V. Sarcone for secretarial assistance and P. Nisen (Abbott Laboratories, Global Oncology Develepment, Abbott Park, IL) for kindly provided atrasentan. This work was supported by the Associazione Italiana Ricerca sul Cancro, Ministero della Salute, and CNR-MIUR.

Anna Bagnato
Molecular Pathology and Ultrastructure Laboratory, Regina Elena Cancer Institute,Rome, Italy

Pier Giorgio Natali
Immunology Laboratory, Regina Elena Cancer Institute, Rome, Italy

6. REFERENCES

1. Levin E.R. Endothelins. N Engl J Med 1995; 333:356-63
2. Goldie R.G. Endothelins in health and disease: an overview. Clin Exp Pharmacol Physiol 1999; 26:145-48
3. Masaki T. The endothelin family: an overview. J Cardiovasc Pharmacol. 2000; 35:S3-5.
4. Battistini B., Chailler P., D'Orleans-Juste P., Briere N., Sirois P. Growth regulatory properties of endothelins. Peptides 1993; 14:385-99
5. Nelson J.B., Chan-Tack K., Hedican S.P., Magnuson S.R., Opgenorth T.J., Bova G.S., Simons J.W. Endothelin-1 production and decreased endothelin B receptor expression in advanced prostate cancer. Cancer Res 1996; 56:663-68
6. Nelson J.B., Hedican S.P., George D.J., Reddi A.H., Piantadosi S., Eisenberger M.A., Simons J.W. Identification of endothelin-1 in the pathophysiology of metastatic adenocarcinoma of the prostate. Nat Med 1995; 1:944-49
7. Gohji K., Kitazawa S., Tamada H., Katsuoka Y., Nakajima M. Expression of endothelin receptor a associated with prostate cancer progression. J Urol 2001; 165:1033-36
8. Nelson J.B., Lee W.H., Nguyen S.H., Jarrard D.F., Brooks J.D., Magnuson S.R., Opgenorth T.J., Nelson W.G., Bova G.S. Methylation of the 5' CpG island of the

endothelin B receptor gene is common in human prostate cancer. Cancer Res. 1997; 57:35-7

9. Bagnato A., Salani D., Di Castro V., Wu-Wong J.R., Tecce R., Nicotra M.R., Venuti A., Natali P.G. Expression of endothelin 1 and endothelin A receptor in ovarian carcinoma: evidence for an autocrine role in tumor growth. Cancer Res 1999; 59:720-27

10. Moraitis S., Langdon S.P., Miller W.R. Endothelin expression and responsiveness in human ovarian carcinoma cell lines. Eur J Cancer 1997; 33:661-8

11. Salani D., Di Castro V., Nicotra M.R., Rosano L., Tecce R., Venuti A., Natali P.G., Bagnato A. Role of endothelin-1 in neovascularization of ovarian carcinoma. Am J Pathol 2000; 157:1537-47

12. Moraitis S., Miller W.R., Smyth J.F., Langdon S.P. Paracrine regulation of ovarian cancer by endothelin. Eur J Cancer 1999 ; 35:1381-7

13. Bagnato A., Tecce R., Moretti C., Di Castro V., Spergel D., Catt K. Autocrine actions of endothelin-1 as a growth factor in human ovarian carcinoma cells Clin Cancer Res 1995; 1:1059-66

14. Yohn J.J., Smith C., Stevens T., Hoffman T.A., Morelli J.G., Hurt D.L., Yanagisawa M., Kane M.A., Zamora M.R. Human melanoma cells express functional endothelin-1 receptors. Biochem Biophys Res Commun 1994; 201:449-57

15. Kikuchi K., Nakagawa H., Kadono T., Etoh T., Byers H.R., Mihm M.C., Tamaki K. Decreased ET(B) receptor expression in human metastatic melanoma cells. Biochem Biophys Res Commun 1996; 219:734-9

16. Eberle J., Weitmann S., Thieck O., Pech H., Paul M., Orfanos C.E. Downregulation of endothelin B receptor in human melanoma cell lines parallel to differentiation genes. J Invest Dermatol 1999; 112:925-32

17. Demunter A., De Wolf-Peeters C., Degreef H., Stas M., van den Oord J.J. Expression of the endothelin-B receptor in pigment cell lesions of the skin. Evidence for its role as tumor progression marker in malignant melanoma. Virchows Arch 2001; 438:485-91

18. Lahav R., Heffner G., Patterson P.H. An endothelin receptor B antagonist inhibits growth and induces cell death in human melanoma cells in vitro and in vivo. Proc Natl Acad Sci U S A 1999; 96:11496-500

19. Nambi P., Wu H.L., Lipshutz D., Prabhakar U. Identification and characterization of endothelin receptors on rat osteoblastic osteosarcoma cells: down-regulation by 1,25-dihydroxy-vitamin D_3. Mol Pharmacol. 1995; 47:266-71

20. Yamashita J., Ogawa M., Inada K., Yamashita S., Matsuo S., Takano S. A large amount of endothelin-1 is present in human breast cancer tissues. Res Commun Chem Pathol Pharmacol 1991; 74:363-9.

21. Alanen K., Deng D.X., Chakrabarti S. Augmented expression of endothelin-1, endothelin-3 and the endothelin-B receptor in breast carcinoma. Histopathology 2000; 36:161-7.

22. Giaid A., Hamid Q.A., Springall D.R., Yanagisawa M., Shinmi O., Sawamura T., Masaki T., Kimura S., Corrin B., Polak J.M. Detection of endothelin immunoreactivity and mRNA in pulmonary tumours. J Pathol 1990; 162:15-22

23. Thevarajah S., Udan M.S., Zheng H., Pfluyg B.R., Nelson J.B. Endothelin axis expression in renal cell carcinoma. J Urol 1999; 161:137-43

24. Ahmed S.I., Thompson J., Coulson J.M., Woll P.J. Studies on the expression of endothelin, its receptor subtypes, and converting enzymes in lung cancer and in human bronchial epithelium. Am J Respir Cell Mol Biol 2000; 22:422-31

25. Egidy G., Juillerat-Jeanneret L., Jeannin J.F., Korth P., Bosman F.T., Pinet F. Modulation of human colon tumor-stromal interactions by the endothelin system. Am J Pathol 2000; 157:1863-74

26. Eberl L.P., Egidy G., Pinet F., Juillerat-Jeanneret L. Endothelin receptor blockade potentiates FasL-induced apoptosis in colon carcinoma cells via the protein kinase C-pathway. J Cardiovasc Pharmacol 2000; 36:S354-6

27. Eberl L.P., Valdenaire O., Saintgiorgio V., Jeannin J.F., Juillerat-Jeanneret L. Endothelin receptor blockade potentiates FasL-induced apoptosis in rat colon carcinoma cells. Int J Cancer 2000; 86:182-7

28. Venuti A., Salani D., Manni V., Poggiali F., Bagnato A. Expression of endothelin-1 and endothelin A receptor in HPV-associated cervical carcinoma: new potential targets for anticancer therapy. FASEB J 2000; 14:2277-83

29. Bagnato A., Cirilli A., Salani D., Simeone P., Muller A., Nicotra M.R., Natali P.G., Venuti A. Growth inhibition of cervix carcinoma cells in vivo by endothelin A receptor blockade. Cancer Res 2002; 62:6381-4

30. Bagnato A., Rosano L., Di Castro V., Albini A., Salani D., Varmi M., Nicotra M.R., Natali P.G. Endothelin receptor blockade inhibits proliferation of Kaposi's sarcoma cells. Am J Pathol 2001; 158:841-7

31. Rosanò L., Spinella F., Di Castro V., Nicotra M.R., Albini A., Natali P.G., Bagnato A. Endothelin receptor blockade inhibits molecular effectors of tumor invasion in Kaposi's sarcoma. Am J Pathol 2003;163:753-62

32. Pagotto U., Arzberger T., Hopfner U., Sauer J., Renner U., Newton C.J., Lange M., Uhl E., Weindl A., Stalla G.K. Expression and localization of endothelin-1 and endothelin receptors in human meningiomas. Evidence for a role in tumoral growth. J Clin Invest 1995; 96:2017-25

33. Pagotto U., Arzberger T., Hopfner U., Weindl A., Stalla G.K. Cellular localization of endothelin receptor mRNAs (ET_A and ET_B) in brain tumors and normal human brain. J Cardiovasc Pharmacol 1995; 26(Suppl 3):S104-6

34. Harland S.P., Kuc R.E., Pickard J.D., Davenport A.P. Expression of endothelin$_A$ receptors in human gliomas and meningiomas, with high affinity for the selective antagonist PD156707. Neurosurgery 1998; 43:890-8

35. Bagnato A., Catt K.J. Endothelin as autocrine regulators of tumor cell growth. Trends Endocrinol Metab 1998; 9:378-83

36. Nelson J.B., Bagnato A., Battistini B., Nisen P. The endothelin axis: emerging role in cancer. Nature Rev Cancer 2003; 3:110-16

37. Opgenorth T.J. Endothelin receptor antagonism. Adv Pharmacol 1995; 33:1-65

38. Wu C., Radford Decker E., Holland G.W., Brown P.M., Stavros F.D., Brock T.A., Dixon R.A.F. Nonpeptide endothelin antagonist in clinical development. Drugs of Today 2001; 37:441-53

39. Remuzzi G., Perico N., Benigni A. New therapeutics that antagonize endothelin: promises and frustations. Nature Rev Drug Disc 2002; 1: 986-1000

40. Bagnato A., Tecce R., Di Castro V., Catt K.J. Activation of mitogenic signaling by endothelin-1 in ovarian carcinoma cells. Cancer Res 1997; 57: 1306-11

41. Vacca F., Bagnato A., Catt K.J., Tecce R. Transactivation of the epidermal growth factor receptor in endothelin-1-induced mitogenic signaling in human ovarian carcinoma cells. Cancer Res 2000; 60:5310-17

42. Salani D., Taraboletti G., Rosanò L., Di Castro V., Borsotti P., Giavazzi R., Bagnato A. Endothelin-1 induces an angiogenic phenotype in cultured endothelial cells and stimulates neovascularization in vivo. Am J Pathol 2000; 157:1703-11

43. Spinella F., Rosanò L., Di Castro V., Natali P.G., Bagnato A. Endothelin-1 induces vascular endothelial growth factor by increasing hypoxia-inducible factor 1α in ovarian carcinoma cells. J Biol Chem 2002; 277:27850-5

44. Del Bufalo D., Di Castro V., Biroccio A., Varmi M., Salani D., Rosanò L., Trisciuoglio D., Spinella S., Bagnato A. Endothelin-1 protects ovarian carcinoma

cells against paclitaxel-induced apoptosis: requirement for Akt activation. Mol Pharmacol 2002; 61:524-32

45. Rosanò L., Varmi M., Salani D., Di Castro V., Spinella F., Natali P.G., Bagnato A. Endothelin-1 induces tumor proteinase activation and invasiveness of ovarian carcinoma cells. Cancer Res 2001; 61:8340-6
46. Spinella F., Rosanò L., Di Castro V., Nicotra M.R., Natali P.G., Bagnato A. Endothelin-1 decreases gap-junctional intercellular communication by inducing phosphorylation of connexin 43 in human ovarian carcinoma cells. J Biol Chem 2003; 278: 41294-301
47. Rosanò L., Spinella F., Salani D., Di Castro V., Venuti A., Nicotra M.R., Natali P.G., Bagnato A. Therapeutic targeting of endothelin A receptor in human ovarian carcinoma. Cancer Res 2003; 63:2447-53
48. Carducci M.A., Nelson J.B., Bowling M.K., Rogers T., Eisenberger M.A., Sinibaldi V., Donehower R., Leahy T.L., Carr R.A., Isaacson J.D., Janus T.J., Andre A., Hosmane B.S., Padley R.J. Atrasentan, an endothelin-receptor antagonist for refractory adenocarcinomas: safety and pharmacokinetics. J Clin Oncol 2002; 20:2171-218
49. Carducci M.A., Padley R.J., Breul J., Vogelzang N.J., Zonnenberg B.A., Daliani D.D., Schulman C.C., Nabulsi A.A., Humerickhouse R.A., Weinberg M.A., Schmitt J.L., Nelson J.B. Effect of endothelin-A receptor blockade with atrasentan on tumor progression in men with hormone-refractory prostate cancer: a randomized, phase II, placebo-controlled trial. J Clin Oncol 2003; 21:679-89
50. van der Boon J. New drug slows prostate-cancer progression. Lancet Oncol 2002; 3:201
51. Guise T.A. Molecular mechanisms of osteolytic bone metastases. Cancer 2000; 88:2892-8
52. Chiao J.W., Moonga B.S., Yang Y.M., Kancherla R., Mittelman A., Wu-Wong J.R., Ahmed T. Endothelin-1 from prostate cancer cells is enhanced by bone contact which blocks osteoclastic bone resorption. Br J Cancer 2000; 83:360-5
53. Yin, J.J, Mohammed, K.S., Kakonen, S.M., Harris S., Wu-Wong J.R., Wessale J.L., Padley R.J., Garrett I.R., Chirgwin J.M., Guise T.A. A causal role for endothelin-1 in the pathogenesis of osteoblastic bone metastasis. Proc Natl Acad Sci USA 2003; 100:10954-9
54. Yohn, J.J., Smith, C., Stevens, T., Hoffman, T.A., Morelli, J.G., Hurt, D.L., Yanagisawa, M., Kane, M.A., Zamora, M.R. Human melanoma cells express functional endothelin-1 receptors. Biochem. Biophys. Res Commun 1994; 201:449-57
55. Bittner, M., Meltzer, P., Chen, Y., Jiang, Y., Seftor, E., Hendrix, M., Radmacher, M., Simon, R., Yakhini, Z., Ben-Dor, A., Sampas N., Dougherty E., Wang E., Marincola F., Gooden C., Lueders J., Glatfelter A., Pollock P., Carpten J., Gillanders E., Leja D., Dietrich K., Beaudry C., Berens M., Alberts D., Sondak V. Molecular classification of cutaneous malignant melanoma by gene expression profiling. Nature 2000; 406:536-40
56. Demunter, A., De Wolf-Peeters, C., Degreef, H., Stas, M., van den Oord, J.J. Expression of the endothelin-B receptor in pigment cell lesions of the skin. Evidence for its role as tumor progression marker in malignant melanoma. Virchows Arch 2001; 438:485-91
57. Eberle, J., Fecker, L.F., Orfanos, C.E., Geilen, C.C. Endothelin-1 decreases basic apoptotic rates in human melanoma cell lines. J Invest Dermatol 2002; 119:549-55
58. Lahav, R., Heffner, G., Patterson, P.H. An endothelin receptor B antagonist inhibits growth and induces cell death in human melanoma cells in vitro and in vivo. Proc Natl Acad Sci USA 1999; 96:11496-500

314

59. Li, G., Satyamoorthy, K., Meier, F., Berking, C., Bogenrieder T., Herlyn, M. Function and regulation of melanoma-stromal fibroblast interactions: when seeds meet soil. Oncogene 2003; 22:3162-71

60. Hsu, M., Andl, T., Li, G., Meinkoth, J.L., Herlyn, M. Cadherin repertoire determines partner-specific gap junctional communication during melanoma progression. J Cell Science 2000; 113:1535-42

61. Ruiter, D., Bogenrieder, T., Elder, D., Herlyn, M. Melanoma-stroma interactions: structural and functional aspects. Lancet Oncol 2002; 3:35-43

62. Bagnato A., Rosanò L. Spinella F., Di Castro V. Tecce R., Natali P.G. Endothelin B receptor blockade inhibits dynamic of cell interactions and communications in melanoma cell progression. Cancer Res 2004; 64:

km23: A NOVEL TGFβ SIGNALING TARGET ALTERED IN OVARIAN CANCER

WEI DING AND KATHLEEN M. MULDER

1. INTRODUCTION

Transforming growth factor β superfamily members play a pivotal role in almost every aspect of cellular and tissue activities, including cell cycle control, regulation of early embryonic development, cell differentiation, cell motility, extracellular matrix formation, angiogenesis, and induction of apoptosis [1, 2].

TGFβ has been shown to initiate at least two prominent signaling cascades, the Smad and the Ras/ MAPK pathways, to regulate cellular and tissue activities [2]. In the Smad pathway, TGFβ binds to TβRII and recruits TβRI into a heterotetrameric complex, resulting in transphosphorylation of TβRI in the GS-domain by TβRII [3, 4]. Phosphorylated Smad2 and/or Smad3 by activated TβRI form(s) a heteromeric complex with Smad4 and enter(s) the nucleus to regulate gene transcription [1, 2, 5]. As we first demonstrated, TGFβ can activate the Ras/MAPK pathways in TGFβ-sensitive epithelial cells [6, 7, 8, 9]. Three members of this family, including the extracellular signal-regulated kinases (ERKs), c-Jun N-terminal kinases (JNKs)/stress-activated protein kinases (SAPKs), and p38, can be activated by TGFβ in a wide variety of cell types [2]. These pathways have been shown to be required for several of the major TGFβ responses, including growth inhibition, production of TGFβ$_1$, and induction of fibronectin [1, 2, 10].

In addition to the Smad and Ras/MAPK pathways, various proteins have been identified based upon their interaction with the TβRs and/or Smads. Smad anchor for receptor activation (SARA) and hepatic growth factor-regulated tyrosine kinase substrate (Hgs/Hrs) have been shown to facilitate Smad2/3 recruitment to the activated TβRI/activin receptor I (ActRI) [11, 12]. Disabled-2 (Dab-2) is an adaptor protein that can bind TβRs and Smad2/Smad3 to bridge the TβR complex to the Smad pathway [13]. Serine-threonine kinase receptor-associated protein (STRAP) recruits Smad7 to activated TβRI and forms a complex. The complex stabilizes the association between Smad7 and the activated receptor, thereby blocking Smad2/3-mediated transcriptional activation [14]. The immunophilin FK506 binding protein 12 (FKBP12) functions as an inhibitor of TβRI by binding to the ligand-free form. FKBP12 is released upon ligand-induced, TβRII-mediated phosphorylation of TβRI [15]. TGFβ-receptor interacting protein-1 (TRIP-1) associates with TβRII in a kinase-dependent manner, but does not interact with the type II activin receptor or TβRI. It does, however, associate with the heteromeric TGFβ receptor complex (16). Death-associated protein (Daxx) interacts with the cytoplasmic domain of TβRII and also associates with the Fas receptor, which mediates activation of JNK and programmed cell death [17]. TGFβ receptor I-associated protein-1 (TRAP-1) only interacts with TβRI that has been activated through mutation or ligand binding. In the absence of TGFβ, TRAP-1 will not interact with wild-type TβRI [18]. TRAP-1-like protein (TLP) has been identified to interact with both active and kinase-deficient TβRs and activin type II receptors, but interacts with the common-mediator Smad4 only in the presence of TGFβ□activin signaling [19].

Herein, we describe a novel TGFβ receptor-interacting protein, termed km23. This protein is the mammalian homologue of the LC7 family of DLCs [20]. TGFβ stimulates not only the phosphorylation of km23, but also the recruitment of km23 to the DIC. Kinase-active TβRs are required for km23 phosphorylation and interaction with DIC [20]. We also demonstrate that km23 interacts with Smad2 both *in vitro* and *in vivo*, and it co-localizes with Smad2 in the cytoplasm at very early time points after TβR activation. Moreover, km23 can mediate specific TGFβ responses, including JNK activation, c-Jun phosphorylation, and growth inhibition [20]. Furthermore, we have identified altered forms of km23 both in TGFβ-resistant human ovarian cancer cell lines and in cancer tissues from ovarian cancer patients. Our data suggest that these alterations in km23 may modify km23 functions in TGFβ signaling and tumorgenesis.

2. CLONING OF km23 IN A HIGHLY TGFβ-RESPONSIVE RAT EPITHELIAL CELL LINE

We have developed a novel method for the identification of TGFβ receptor-interacting proteins [20]. The intracellular portions both of TβRI and TβRII were expressed in E. coli, purified by affinity chromatography, and phosphorylated by in vitro kinase assays. These activated cytoplasmic domains of the TβRs were used as probes to screen an expression protein library prepared from a highly TGFβ-responsive rat intestinal epithelial IEC4-

1 cell line [21]. Several positive clones were isolated using this method. Among these clones, km23 was the most interesting because it is also a homologue of the regulatory region of the Drosophila's Ultrabithorax (Ubx) unit, through which transcription can be activated by one of the TGFβ superfamily members, Dpp. It was conceivable that km23 might be an important intermediate in a TGFβ signaling pathway.

3. GENE STRUCTURE AND HOMOLOGY OF HUMAN km23

Human km23 is a 96-amino acid protein encoded by a 291-base pair open reading frame. km23 is localized on human chromosome 20 (20 q11.21) and is composed of 4 exons. Northern blots have shown that km23 is ubiquitously expressed, with higher levels of message detected in brain, kidney, and placenta tissues. Western blots have indicated km23 is an 11-kDa cytoplasmic protein. The human and rat km23 amino acid sequences differ by only three amino acids and show 98% similarity. Additional alignments of human km23 with sequences in the National Center for Biotechnology Information database indicate that km23 is also the mammalian homologue of the Chlamydomonas LC7 class of DLCs. It is relatively conserved across different species, including Chlamydomonas (ch/LC7, 74% similarity), Drosophila (robl gene, 82% similarity), Caenorhabditis elegans (T24H10.6 gene, 76% similarity), and Danio Rerio (ZFIN gene, 93% similarity). Another mammalian form of km23 has been identified, termed km23-2. Human km23-2 localizes to chromosome 16 (16 q23.3), can be translated into a 96-amino acid protein, and displays 70% homology and 91% similarity with human km23. The rat and mouse forms of km23-2 display 71% homology and 91% similarity with human km23.

4. km23 DIRECTLY INTERACTS WITH TβII, BUT NOT WITH TβRI ALONE

Since km23 was isolated using activated intracellular regions of TβRI and TβRII, it was of interest to verify this interaction *in vivo*. ^{125}I-TGFβ cross-linking assays indicated that both TβRI and TβRII were present in km23 immunocomplexes. Also, TGFβ induced the interaction of TβRII with km23 in two epithelial cell lines at early times after TGFβ addition. The results of immunoprecipitation/blot (IP/blot) assays in either direction indicated that km23 can interact with TβRII and with the TGFβ receptor complex, but that TβRI may not be a direct binding partner of km23 [20].

5. km23 PHOSPHORYLATION

km23 can be phosphorylated when both TβRI and TβRII are expressed in 293T cells, but it is not constitutively phosphorylated when expressed alone. km23 could not be phosphorylated by kinase-deficient (KN) RII when it was co-expressed with wild-type (wt) RI, suggesting that the kinase activity of TβRII is required for km23 phosphorylation. In contrast, km23 could still be phosphorylated after co-expression of wt RII and KN-RI, suggesting that the RI kinase does not directly phosphorylate km23. In support of this finding, km23 was also not phosphorylated when a kinase active RI mutant (T204D) was expressed alone. Thus, the mechanism for the phosphorylation of km23 differs from that of the R-Smads [22]. Collectively, our results strongly suggest that although both RI and RII are present in the complex with km23, the RII kinase is the primary mediator of km23 phosphorylation. The kinase activity of RI does appear to play some role in km23 phosphorylation, however, since expression of KN-RI with wt RII reduced the level of phosphorylation relative to that observed after expression of RII alone.

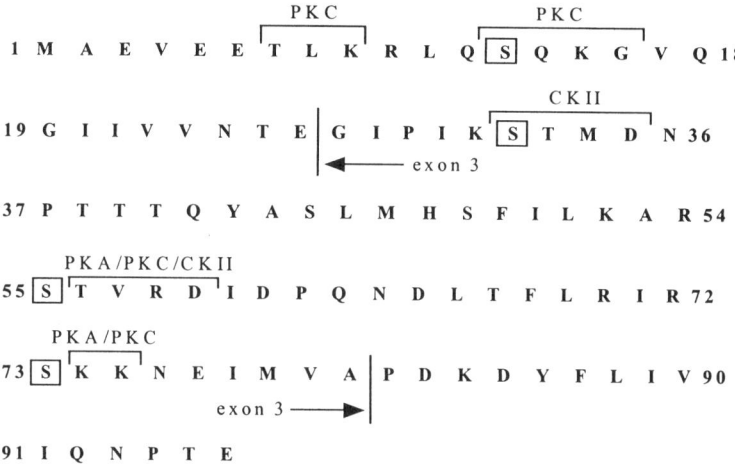

Figure 1. Human km23 sequence depicting expected phosphorylation sites. Potential phosphorylation sites for PKA, PKC, and CKII were predicted using PhosphoBase v. 2.0 and NetPhos 2.0 [24]. For these sites, prediction scores for serine and threonine were all over 0.91. Four serine residues that are conserved among the mammalian km23 forms are boxed. Exon 3 of km23 contains three potential phosphorylation sites for PKA, PKC, and/or CKII.

Phosphoamino acid analysis indicated that km23 is phosphorylated primarily on serine residues in response to TGFβ receptor activation [20]. This finding is in keeping with the known serine-threonine kinase specificity of the TβRs [1, 2, 22]. In contrast, epidermal growth factor (EGF), known to phosphorylate many downstream targets on tyrosine residues after receptor activation [23], could not phosphorylate km23.

km23 also contains several potential phosphorylation sites for a variety of other kinases, including protein kinase A (PKA), protein kinase C (PKC), and/or casein kinase II (CKII) predicted by PhosphoBase v. 2.0 and NetPhos 2.0 [24]. Among the serine residues, four are conserved between the human and rat forms (Fig. 1).

6. ACTIVATION OF JNK AND PHOSPHORYLATION OF c-Jun

Since our data indicate that km23 may function as a signaling intermediate for TGFβ, it was of interest to investigate whether km23 could mediate any of the known TGFβ signaling events. We have previously reported that TGFβ rapidly activates the Jun N-terminal kinase (JNK) family of MAPKs [9]. Further, JNK activation by TGFβ appears to play a role in TGFβ-mediated growth inhibition, either through the amplification of $TGF\beta_1$ production, via cross-talk with the Smads, and/or by regulation of cell cycle inhibitors [2, 25]. Thus, we examined whether stable expression of km23 could activate JNK as TGFβ does. We found that JNK was super-activated in the absence of TGFβ upon expression of km23 in Mv1Lu cells [20]. JNK activity was about 15 times greater in the km23-expressing cells than in the empty vector-expressing cells during an early period after TGFβ addition. Furthermore, phospho-c-Jun, a downstream effector of JNK, was also super-activated in the absence of TGFβ in the same Mv1Lu cells, compared with cells only stably expressing the empty vector [20]. These findings suggest that km23 may function as a signaling intermediate for TGFβ, in a pathway that involves JNK.

7. RECRUITMENT OF km23 TO DIC REQUIRESTGFβ RECEPTOR ACTIVATION

DLCs such as km23 are known to interact with the dynein motor complex through the DIC [20, 26]. For the DIC termed IC74-1A, the km23/roadblock-binding domain has been localized to a region spanning amino acids 243 to

282 [26]. In contrast, two other DLC families, Tctex-1 and DLC8, have been shown to bind to the same DIC through different binding domains. DLCs are also known to interact with a diverse array of cargo to exert their diverse functions. For example, Tctex-1 was found to interact with Doc2 [27] and the N-terminus of the p59^{fyn} Src family tyrosine protein kinase [28] to exert diverse functions. Tctex-1 also can interact directly with the C-terminal tail of rhodopsin to mediate the transport of the visual pigment to the base of the connecting cilium within the photoreceptor [29]. Similarly, DLC8 binds to IC74 and a diverse set of other proteins, such as neuronal nitric-oxide synthase (nNOS) [30, 31], the proapoptotic Bcl-2 family member Bim [32], the Drosophila mRNA localization protein Swallow [33], and the rabies virus P protein [34]. The binding of DLC8 to these diverse proteins occurs at a conserved GIQVD motif or (K/R)XTQT motif [34, 35, 36].

With regard to the role of the DLC km23 in TGFβ signaling, we have shown that TGFβ could rapidly stimulate the recruitment of km23 to DIC, suggesting a connection between TGFβ signaling and DLC recruitment. Moreover, the association between km23 and DIC required the kinase activity of TβRII, since KN-RII blocked the recruitment of km23 to the DIC [20]. It is likely that the binding of km23 to the DIC after TGFβ receptor activation is important for specifying the nature of the cargo that will be transported along the MTs. Thus, km23 may behave similar to the Tctex-1 and DLC8.

8. CO-LOCALIZATION AND INTERACTION WITH Smad 2 IN VITRO AND IN VIVO

km23 associates with the TGFβ signal transducer Smad2 both *in vitro* and *in vivo*. In addition, immunofluorescence studies indicate that km23 and Smad2 are co-localized at very early times after TGFβ addition to cells. However, once Smad2 has translocated to the nucleus (within 15 min), the two proteins are no longer co-localized. Since km23 may function as a motor receptor to recruit TGFβ signaling components to the dynein motor for intracellular transport along microtubules (MTs) toward the nucleus, our results suggest that Smad2 may be one of the cargoes that km23 links to the dynein motor. It was reported that Smad2 must be translocated to the nucleus prior to its transcriptional regulation of target genes [37]. Thus, the interaction and co-localization of Smad2 with km23 may be an early step in Smad2 signaling of TGFβ responses. That is, km23 may transport Smad2 in the cytoplasm along MTs, prior to Smad2 nuclear translocation and transcriptional regulation.

9. km23 ALTERATION IN OVARIAN CANCER TISSUES

Ovarian carcinoma is often diagnosed at an advanced stage and is the leading cause of death from gynecological neoplasia, accounting for more than 14,000 deaths per year [38]. Overall, the molecular changes that underlie the initiation and development of this tumor are poorly understood. It has been reported that more than 75% of ovarian carcinomas are resistant to TGFβ, particularly recurrent ones [39, 40]. As such, the loss of TGFβ responsiveness may play an important role in the pathogenesis and/or progression of ovarian cancer.

It has been reported that TGFβ1, the TGFβ receptors, and several TGFβ signaling components (ie, Smad2, Smad4, and Dab-2) are altered in different types of cancers [1, 2, 13, 25]. Alterations in TβRII have been identified in 25% of ovarian carcinomas [41], while mutations in TβRI were reported in 33% of such cancers [42]. Loss of function mutations in TGFβ1, TβRI, and TβRII can lead to disruption of TGFβ signaling pathways and subsequent loss of cell cycle control [41, 42, 43, 44]. However, these alterations only account for a minority of TGFβ-resistant ovarian carcinomas, suggesting that other alterations in TGFβ signaling components may be involved in the pathogenesis of this type of cancer.

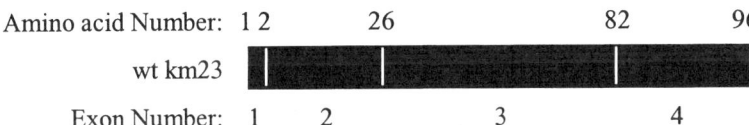

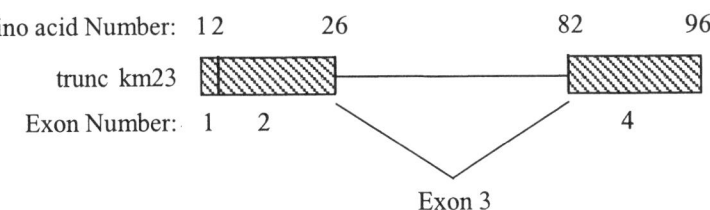

Figure 2. Amino acid sequence alignment of the truncated form of km23 with the wild-type form, indicating exon and amino acid sequence numbers. The 27th-81st amino acid residues have been deleted in the truncated form of km23. This form was detected both in ovarian tumor cell lines and ovarian cancer patients, resulting in a smaller 41-amino acid peptide.

In order to investigate whether this novel TGFβ receptor-interacting protein was altered in human ovarian cancer cells, we analyzed 19 ovarian carcinoma samples from patients, as well as six ovarian cancer cell lines, using laser capture micro-dissection (LCM) and nested reverse transcription-polymerase chain reaction (RT-PCR) strategies. Our data revealed a truncated form of km23, missing exon 3 of km23, in two ovarian cancer cell lines (SK-OV-3 and IGROV-1), both of which are resistant to TGFβ [45, 46]. The same truncated form of km23 was also detected in two out of nineteen ovarian cancer patients (Fig. 2). As mentioned above, km23 consists of 4 exons, and among them, the third exon is the longest, encoding amino acid residues 27-81 of km23. In addition, exon 3 contains three potential phosphorylation sites for PKA, PKC, and/or CKII (Fig. 1). The loss of these potential phosphorylation sites in the truncated protein may result in a disruption of signal transduction.

In *Drosophila*, a deletion mutant of km23/robl, which lacks portions of intron 2 and exon 3, displays mitotic defects [47]. This alteration is similar to the truncated form of km23 we identified in the ovarian tumor cell lines and cancer patients. The mitotic index of the mutant Drosophila hemizygotes was about five times that of the wt form [47]. In addition, the km23/robl mutant homozygotes cannot be fully rescued by the genomic or cDNA rescue constructs, suggesting that this mutation can act in a dominant fashion to inhibit the action of wt km23/robl [47]. Here we have shown that the wt form of km23 was still present in the tumor cell lines and in the tumor tissues from patients, despite the presence of the truncated form. Thus, the alterations in km23 we have identified here may also act in a dominant fashion in ovarian tumors. Increases in mitotic index and mitotic defects are commonly observed in the majority of cancers, including ovarian carcinomas. Thus, the truncated form of km23 may play an important role in the formation or progression of ovarian and other types of tumors.

In addition to the truncated form of km23 described above, several missense mutations were identified in the ovarian cancer patients. In two patients, the stop codon of km23 was altered from TAA to CAC. This alteration resulted in a larger protein, encoding 107 amino-acid residues, instead of the wt 96-amino acid form of km23. In addition to this missense mutation, five other missense mutations, including T38I, S55G, T56S, I89V and V90A, were detected in four other patients.

of the five missense mutations identified, T56S and S55G appeared to be the most interesting. As mentioned above, there are several potential PKA, PKC, and/or CKII phosphorylated sites in km23 (Fig. 1). Among them, the threonine at the 56th amino-acid residue spans a predicted phosphorylation site for PKA, PKC, and CKII. Mutation of this site may result in disruption of signals transduced by these kinases. Further, we have reported that km23 is phosphorylated primarily on serines after activation of TGFβ receptors [20]. There are 4 serine residues in km23 that are conserved among the mammalian forms (boxed in Fig. 1). Among these, three are located in exon 3, including Ser55, which was found to be mutated in one of the ovarian carcinomas. Thus, this site and/or the other two in exon 3 may be potential phosphorylation sites modified by TGFβ.

10. CONCLUSIONS

km23 is a novel motor receptor found to be linked to a TGFβ signaling pathway. Based upon our observations, we have constructed the following model (Fig. 3). Upon phosphorylation of km23 by TβRs, DIC is recruited to form a motor complex. This complex transports membrane vesicles containing TGFβ signaling components such as Smad2 along the microtubules to a new vesicular compartment. After reaching the new compartment, the Smad complex may be translocated to the nucleus for transcriptional regulation of target genes. At this time point, km23 is no longer co-localized with Smad2. Future studies are still required to answer several questions about this novel TGFβ signaling intermediate. For example, we would like to identify other cargoes that km23 carries along the MTs. In addition, it would be of interest to know precisely which serine residues are phosphorylated by TGFβ, as well as what the consequences of dephosphorylation of these sites may be. Overall, we would like to develop km23-based cancer diagnostics and therapeutics, since km23 is altered at such a high rate in ovarian cancer. km23 most likely plays an important role in tumorgenesis or tumor progression in at least this type of cancer. Future studies will reveal whether this is the case for other tumor types as well.

Wei Ding and Kathleen M. Mulder
Department of Pharmacology, Penn State University College of Medicine, Hershey, Pennsylvania

324

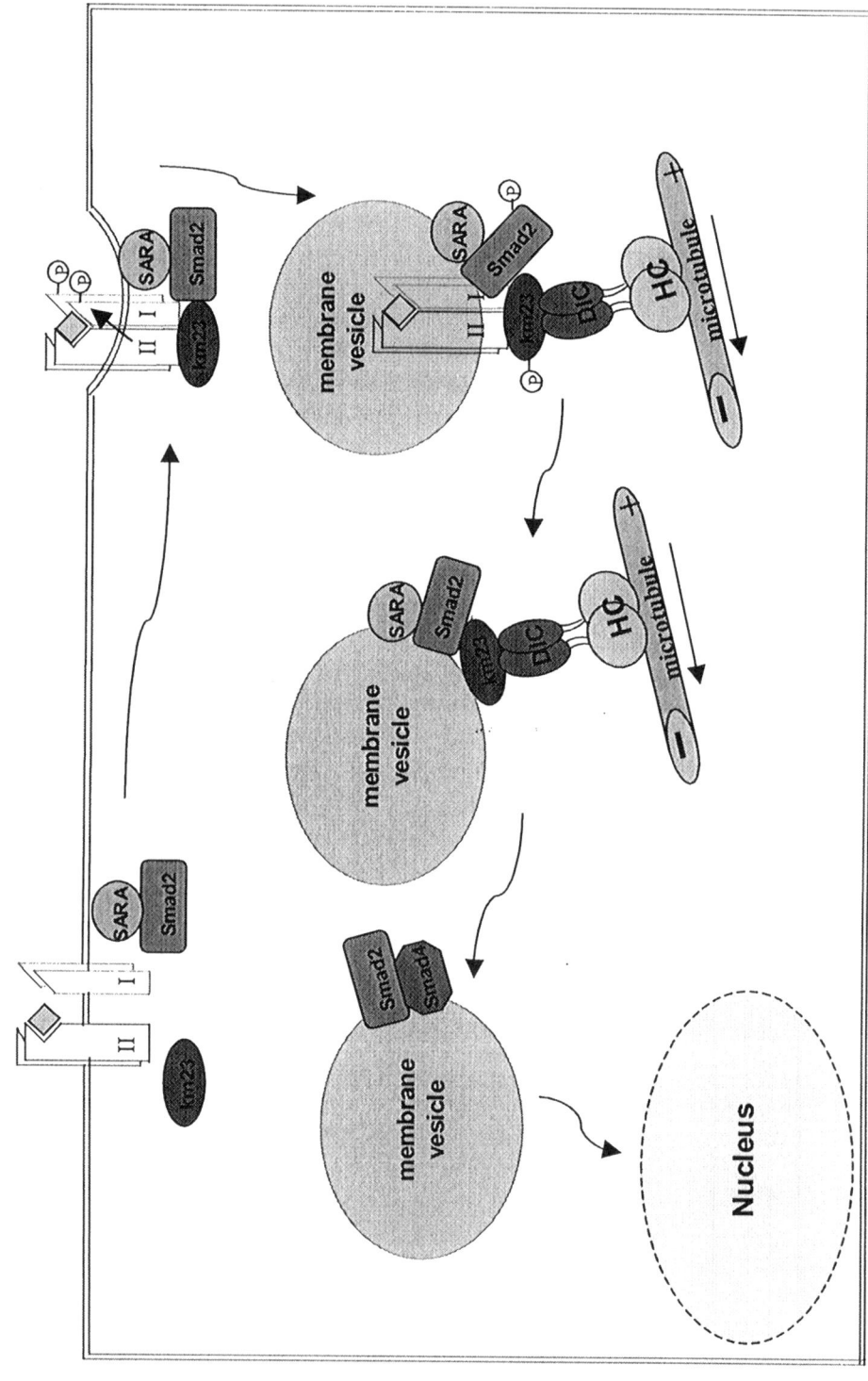

Figure 3. km23 transports membrane vesicles containing TGFβ signaling components along the MTs toward the nucleus. Phosphorylated by ligand binding, the activated TβRII recruits and transphosphorylates TβRI. TβRII also phosphorylates km23, which then recruits DIC to form a motor complex. This complex transports the vesicles containing Smad2 along the microtubules toward the nucleus. Upon reaching the next compartment, the Smad complex may translocate from here to the nucleus for transcriptional regulation of target genes.

11. REFERENCES

1. Massague J., Blain S.W., and Lo R.S. TGFβ signaling in growth control, cancer, and heritable disorders. Cell 2000; 103: 295-309.
2. Yue J. and Mulder K.M. Transforming growth factor-β signaling transduction in epithelial cells. Pharmacol Ther 2001; 91: 1-34
3. Heldin C.H., Miyazono K., and ten Dijke P. TGF-β signalling from cell membrane to nucleus through SMAD proteins. Nature 1997; 390: 465-471.
4. Massague J. TGF-β signal transduction. Annu Rev Biochem 1998; 67: 753-791.
5. Abdollah S., Macias-Silva M., Tsukazaki T., Hayashi H., Attisano L., and Wrana J.L. TβRI phosphorylation of Smad2 on Ser465 and Ser467 is required for Smad2-Smad4 complex formation and signaling. J Biol Chem 1997; 272: 27678-27685.
6. Mulder K.M. and Morris S.L. Activation of p21ras by transforming growth factor β in epithelial cells. J Biol Chem 1992; 267: 5029-5031.
7. Hartsough M.T. and Mulder K.M. Transforming growth factor β activation of p44mapk in proliferating cultures of epithelial cells. J Biol Chem 1995; 270: 7117-7124.
8. Hartsough M.T., Frey R.S., Zipfel P.A., Buard A., Cook S.J., McCormick F., and Mulder K.M. Altered transforming growth factor signaling in epithelial cells when ras activation is blocked. J Biol Chem 1996; 271: 22368-22375.
9. Frey R.S. and Mulder K.M. Involvement of extracellular signal-regulated kinase 2 and stress-activated protein kinase/Jun N-terminal kinase activation by transforming growth factor β in the negative growth control of breast cancer cells. Cancer Res 1997; 57: 628-633.
10. Hocevar B.A., Brown T.L., and Howe P.H. TGF-β induces fibronectin synthesis through a c-Jun N-terminal kinase-dependent, Smad4-independent pathway. EMBO J 1999; 18: 1345-1356.
11. Tsukazaki T., Chiang T.A., Davison A.F., Attisano L., and Wrana J.L. SARA, a FYVE domain protein that recruits Smad2 to the TGFβ receptor. Cell 1998; 95: 779-791.
12. Miura S., Takeshita T., Asao H., Kimura Y., Murata K., Sasaki Y., Hanai J.I., Beppu H., Tsukazaki T., Wrana J.L., Miyazono K., and Sugamura K. Hgs (Hrs), a FYVE domain protein, is involved in Smad signaling through cooperation with SARA. Mol Cell Biol 2000; 20: 9346-9355.
13. Hocevar B.A., Smine A., Xu X.X., and Howe P.H. The adaptor molecule disabled-2 links the transforming growth factor β receptors to the Smad pathway. EMBO J 2001; 20: 2789-2801.
14. Datta P.K. and Moses H.L. STRAP and Smad7 synergize in the inhibition of transforming growth factor β signaling. Mol Cell Biol 2000; 20: 3157-3167.
15. Wang T., Li B.Y., Danielson P.D., Shah P.C., Rockwell S., Lechleider R.J., Martin J., Manganaro T., and Donahoe P.K. The immunophilin FKBP12 functions as a common inhibitor of the TGFβ family type I receptors. Cell 1996; 86: 435-444.
16. Chen R.H., Miettinen P.J., Maruoka E.M., Choy L., and Derynck R. A WD-domain

protein that is associated with and phosphorylated by the type II TGF-β receptor. Nature 1995; 377: 548-552.

17. Perlman R., Schiemann W.P., Brooks M.W., Lodish H.F., and Weinberg R.A. TGFβ-induced apoptosis is mediated by the adapter protein Daxx that facilitates JNK activation. Nat Cell Biol 2001; 3: 708-714.

18. Charng M.J., Zhang D., Kinnunen P., and Schneider M.D. A novel protein distinguishes between quiescent and activated forms of the type I transforming growth factor β receptor. J Biol Chem 1998; 273: 9365-9368.

19. Felici A., Wurthner J.U., Parks W.T., Giam L.R., Reiss M., Karpova T.S., McNally J.G., and Roberts A.B. TLP, a novel modulator of TGF-β signaling, has opposite effects on Smad2- and Smad3-dependent signaling. EMBO J 2003; 22 : 4465-4477.

20. Tang Q., Staub C.M., Gao G., Jin Q., Wang Z., Ding W., Aurigemma R.E., and Mulder K.M. A novel TGFβ receptor-interacting protein that is also a light chain of the motor protein dynein. Mol Biol Cell 2002; 13: 4484-4496.

21. Mulder K.M., Segarini P.R., Morris S.L., Ziman J.M., and Choi H.G. Role of receptor complexes in resistance or sensitivity to growth inhibition by TGFβ in intestinal epithelial cell clones. J Cell Physiol 1993; 54: 162-174.

22. Shi Y. and Massague J. Mechanisms of TGF-β signaling from cell membrane to the nucleus. Cell 2003; 113: 685-700.

23. Jorissen R.N., Walker F., Pouliot N., Garrett T.P., Ward C.W., and Burgess A.W. Epidermal growth factor receptor: mechanisms of activation and signalling. Exp Cell Res 2003; 284: 31-53.

24. Kreegipuu A., Blom N., and Brunak S. PhosphoBase, a database of phosphorylation sites: release 2.0. Nucleic Acids Res 1999; 27: 237-239.

25. Derynck R., Akhurst R.J., and Balmain A. TGF-β signaling in tumor suppression and cancer progression. Nat Genet 2001; 29: 117-129.

26. Susalka S.J., Nikulina K., Salata M.W., Vaughan P.S., King S.M., Vaughan K.T., and Pfister K.K. The roadblock light chain binds a novel region of the cytoplasmic dynein intermediate chain. J Biol Chem 2002; 277: 32939-32946.

27. Nagano F., Orita S., Sasaki T., Naito A., Sakaguchi G., Maeda M., Watanabe T., Kominami E., Uchiyama Y., and Takai Y. Interaction of Doc2 with tctex-1, a light chain of cytoplasmic dynein. J Biol Chem 1998; 273: 30065-30068.

28. Campbell K.S., Cooper S., Dessing M., Yates S., and Buder A. Interaction of p59fyn kinase with the dynein light chain, Tctex-1, and colocalization during cytokinesis. J Immunol 1998; 161: 1728-1737.

29. Tai A.W., Chuang J.Z., Bode C., Wolfrum U., and Sung C.H. Rhodopsin's carboxy-terminal cytoplasmic tail acts as a membrane receptor for cytoplasmic dynein by binding to the dynein light chain Tctex-1. Cell 1999; 97: 877-887.

30. Jaffrey S.R. and Snyder S.H. PIN: an associated protein inhibitor of neuronal nitric oxide synthase. Science. 1996; 274: 774-777.

31. Fan J.S., Zhang Q., Li M., Tochio H., Yamazaki T., Shimizu M., and Zhang M. Protein inhibitor of neuronal nitric-oxide synthase, PIN, binds to a 17-amino acid residue fragment of the enzyme. J Biol Chem 1998; 273: 33472-33481.

32. Puthalakath H., Huang D.C., O'Reilly L.A., King S.M., and Strasser A. The proapoptotic activity of the Bcl-2 family member Bim is regulated by interaction with the dynein motor complex. Mol Cell 1999; 3: 287-296.

33. Schnorrer F., Bohmann K., and Nusslein-Volhard C. The molecular motor dynein is involved in targeting swallow and bicoid RNA to the anterior pole of Drosophila oocytes. Nat Cell Biol 2000; 2:185-190.

34. Rodriguez-Crespo I., Yelamos B., Roncal F., Albar J.P., Ortiz de Montellano P.R., and Gavilanes F. Identification of novel cellular proteins that bind to the LC8 dynein light chain using a pepscan technique. FEBS Lett 2001; 503: 135-141.

35. Lo K.W., Naisbitt S., Fan J.S., Sheng M., and Zhang M. The 8-kDa dynein light chain binds to its targets via a conserved (K/R)XTQT motif. J Biol Chem 2001; 276: 14059-14066.

36. Martinez-Moreno M., Navarro-Lerida I., Roncal F., Albar J.P., Alonso C., Gavilanes F., and Rodriguez-Crespo I. Recognition of novel viral sequences that associate with the dynein light chain LC8 identified through a pepscan technique. FEBS Lett 2003; 544: 262-267.

37. Massague J. and Wotton D. Transcriptional control by the TGF-β/Smad signaling system. EMBO J 2000; 19: 1745-1754.

38. Landis S.H., Murray T., Bolden S., and Wingo P.A. Cancer statistics, 2001. CA Cancer J Clin 2001; 51: 15-36.

39. Yamada S.D., Baldwin R.L., and Karlan B.Y. Ovarian carcinoma cell cultures are resistant to TGF-β1-mediated growth inhibition despite expression of functional receptors. Gynecol Oncol 1999; 75: 72-77.

40. Hu W., Wu W., Nash M.A., Freedman R.S., Kavanagh J.J., and Verschraegen C.F. Anomalies of the TGF-β postreceptor signaling pathway in ovarian cancer cell lines. Anticancer Res 2000; 20: 729-733.

41. Lync M.A., Nakashima R., Song H.J., Degroff V.L., Wang D., Enomoto T., and Weghorst C.M. Mutational analysis of the transforming growth factor β receptor type II gene in human ovarian carcinoma. Cancer Res 1998; 58: 4227-4232.

42. Chen T., Triplett J., Dehner B., Hurst B., Colligan B., Pemberton J., Graff J.R., and Carter J.H. Transforming growth factor-β receptor type I gene is frequently mutated in ovarian carcinomas. Cancer Res 2001; 61: 4679-4682.

43. Cadillo M.R., Yap E., and Castagna G. Molecular genetic analysis of TGF-β1 in ovarian neoplasia. J Exp Clin Cancer Res 1997; 16: 49-56.

44. Wang D., Kanuma T., Muzunuma H., Takama F., Ibuki Y., Wake N., Mogi A., Shitara Y., and Takenoshita S. Analysis of specific gene mutations in the transforming growth factor-β signal transduction pathway in human ovarian cancer. Cancer Res 2000; 60: 4507-4512.

45. Zeinoun Z., Teugels E., De Bleser P.J., Neyns B., Geerts A., and De Greve J. Insufficient TGF-β1 production inactivates the autocrine growth suppressive circuit in human ovarian cancer cell lines. Anticancer Res 1999; 19: 413-420.

46. Jozan S., Guerrin M., Mazars P., Dutaur M., Monsarrat B., Cheutin F., Bugat R., Martel P., and Valette A. Transforming growth factor β1 (TGF-β1) inhibits growth of a human ovarian carcinoma cell line (OVCCR1) and is expressed in human ovarian tumors. Int J Cancer 1992; 52: 766-770.

47. Bowman A.B., Patel-King R.S., Benashski S.E., McCaffery J.M., Goldstein L.S., and King S.M. Drosophila roadblock and Chlamydomonas LC7: a conserved family of dynein-associated proteins involved in axonal transport, flagellar motility, and mitosis. J Cell Biol 1999; 146:165-180.